Drinking Water Engineering

Drinking Water Engineering

Editor: Danice Coy

NY RESEARCH
P R E S S

New York

Published by NY Research Press
118-35 Queens Blvd., Suite 400,
Forest Hills, NY 11375, USA
www.nyresearchpress.com

Drinking Water Engineering
Edited by Danice Coy

© 2017 NY Research Press

International Standard Book Number: 978-1-63238-532-1 (Hardback)

Cataloging-in-publication Data

Drinking water engineering / edited by Danice Coy.
 p. cm.
Includes bibliographical references and index.
ISBN 978-1-63238-532-1
1. Drinking water--Purification. 2. Water-supply engineering. 3. Water--Distribution. I. Coy, Danice.
TD430 .D75 2017
628.162--dc23

Printed in the United States of America.

Contents

Preface

Drinking water is necessary for all forms of life. It also plays a crucial role in food preservation. Drinking water engineering aims at devising new techniques to ensure the quality of water and making it safe for consumption by living organisms. The various sub-fields of drinking water engineering along with technological progress that have future implications are glanced at in this book. It aims to provide a general view of the different areas of drinking water engineering, and its applications. While understanding the long-term perspectives of the topics, the book makes an effort in highlighting their impact as a modern tool for the growth of the discipline. Scientists and students actively engaged in this field will find this text full of crucial and unexplored concepts.

After months of intensive research and writing, this book is the end result of all who devoted their time and efforts in the initiation and progress of this book. It will surely be a source of reference in enhancing the required knowledge of the new developments in the area. During the course of developing this book, certain measures such as accuracy, authenticity and research focused analytical studies were given preference in order to produce a comprehensive book in the area of study.

This book would not have been possible without the efforts of the authors and the publisher. I extend my sincere thanks to them. Secondly, I express my gratitude to my family and well-wishers. And most importantly, I thank my students for constantly expressing their willingness and curiosity in enhancing their knowledge in the field, which encourages me to take up further research projects for the advancement of the area.

Editor

Impact of decreasing water demand on bank filtration in Saxony, Germany

T. Grischek, D. Schoenheinz, C. Syhre, and K. Saupe

University of Applied Sciences, Faculty of Civil Engineering/Architecture, Division of Water Sciences,
Friedrich-List-Platz 1, 01069 Dresden, Germany

Abstract. Bank filtration has been of great importance to the drinking water supply in Germany for many decades. The water quality of pumped raw water from bank filtration sites depends to a high degree on the water quality of the infiltrating surface water and the landside groundwater, the mixed portion of both, as well as the flow and transport conditions in the aquifer. Following the improvement of river water quality and a drastic decrease in water demand during the last 20 years in Germany, the influence of landside groundwater quality has become more important for the raw water quality of waterworks relying on bank filtration. The hydrogeologic analysis of three bank filtration sites in Saxony and the management of abstraction rates and well operation in response to fluctuating water demand are discussed.

1 Bank filtration in Germany

In Germany, only 3% of the annually available water resources are needed for the public water supply. In 2007, the main sources were groundwater and spring water (70%), followed by surface water and artificial infiltrate (22%), and bank filtrate (8%) (FSA, 2009).

In place of direct surface water abstraction, bank filtration with subsequent natural aquifer treatment has been used for water supply purposes since the 1870s. Two of the oldest exploited bank filtration sites are the Düsseldorf-Flehe Waterworks on the River Rhine and the Dresden-Saloppe Waterworks on the River Elbe. Both have been providing daily drinking water for several hundred thousand people since 1870 and 1875, respectively.

The water abstraction relies mostly on vertical wells, except for the waterworks along the River Rhine where horizontal wells are also installed. Typical characteristics for bank filtration sites in Germany are given in Table 1.

Abstracted raw water is a mixture of bank filtrate and landside groundwater. Thus, raw water quality does not only depend on river water quality but also on landside groundwater quality. Due to the intensive agricultural use of the landside catchment, high portions of landside groundwater may result

in increased concentrations of nitrate, sulphate, hardness and pesticides.

Poor quality of river water, in combination with high loads of suspended matter, may also result in river bed clogging. While the clogging layer comprises the biologically most active layer, it also strongly reduces water infiltration if the hydraulic conductivity decreases.

2 Bank filtration in Saxony

2.1 Water demand and bank filtrate abstraction

The Federal State of Saxony is located in the Southeast of Germany, bordered by the Czech Republic to the south and Poland to the east (Fig. 1). Bank filtrate as a raw water source is used for both drinking water supply and process water. Bank filtration sites are concentrated in the lower lands at the rivers Mulde and Elbe and their tributaries, but also at the Lausitzer Neisse (Fig. 1).

In one study, 19 bank filtration sites for the production of drinking water and 13 bank filtration sites for the production of process water were investigated. The bank filtration characteristics found in Saxony are typical for Germany, though no horizontal wells are operated. The abstraction rates of the waterworks producing drinking water range from 4000 m³/a (Koltzschen Waterworks on the River Zwickauer Mulde) to 21 650 000 m³/a (Torgau Waterworks on the River Elbe). Waterworks supplying process water abstract between

Figure 1. Bank filtration sites in Saxony.

Table 1. Characteristics of bank filtration sites in Germany (Kühn and Müller, 2005).

Condition	Typical range
aquifer thickness	4 to 70 m
hydraulic conductivity	0.0001 to 0.05 m/s
distance bank – well	20 to 860 m
well fields length along river	1 to 2 km
travel times	3 days to 0.5 years

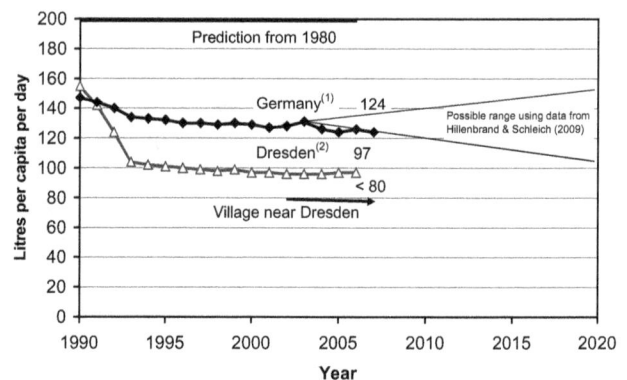

Figure 2. Development of water demand per capita per day in Germany and Saxony. (1) BDEW (2009); (2) City of Dresden (2009).

$18\,250\,\mathrm{m^3/a}$ to $2\,453\,000\,\mathrm{m^3/a}$. The portion of bank filtrate accounts for 15 to 96% of the pumped raw water. For many waterworks used for drinking water production, travel times of bank filtrate have been determined by using groundwater flow modelling, measurements of temperature and electrical conductivity and investigations using chloride and persistent trace organics.

Before 1990, the planning and construction of waterworks and well galleries was based on a predicted water demand of up to 200 litres per capita per day. Since 1990 many waterworks, especially in East Germany, have been facing drastic reductions in water consumption. Mean water abstraction rates for public water supply decreased due to political changes, the "water price shock" after the re-unification of Germany, demographic changes, and changes in con-

sumption patterns (Fig. 2). While in Saxony in 1991 the production of raw water from bank filtrate totalled about $40 \times 10^6\,\mathrm{m^3/a}$, since 2001 the abstraction has been only about $20 \times 10^6\,\mathrm{m^3/a}$. In some regions, water use decreased by more than 50% within 10 years. Due to the expected demographic development, a further decline in water consumption and thus water production is expected.

At many bank filtration sites, reduced water abstraction results in a lower portion of bank filtrate in the abstracted raw water. Thus, the quality of the landside groundwater becomes more important for the subsequent water treatment.

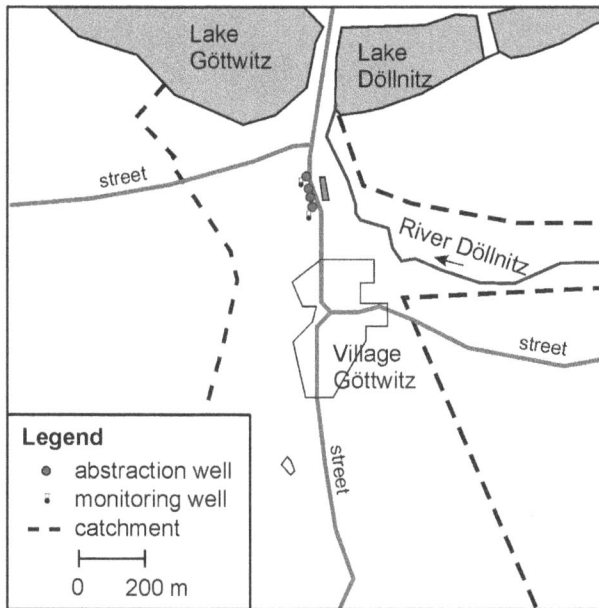

Figure 3. Scheme of the bank filtration site in Göttwitz.

Table 2. Geohydraulic conditions for the Göttwitz site.

Condition	Average value
aquifer thickness	10 m
hydraulic conductivity	2×10^{-4} m/s
distance between lake bank and abstraction wells	200 m
number of wells	4
maximum total abstraction rate	1700 m³/d

To mitigate this effect, a reduction in the number of wells seems reasonable. However, since peak demands still have to be met, this option is not everywhere practicable.

In the following section, the impact of a reduced mean water abstraction on bank filtrate portions and travel times will be presented for three waterworks in Saxony and a hypothetical RBF site typical of Saxony.

2.2 Bank filtration site in Göttwitz

The bank filtration site in Göttwitz is located in the lowlands of northwest Saxony (Fig. 1) close to the Döllnitz stream in the east and Lake Göttwitz in the north (Fig. 3). In accordance with the general discussion, water abstraction decreased between the early 1990s and 2008 from about 1700 m³/d to about 1000 m³/d due to a lower water demand per capita per day (water savings) and reduction of leakage in the drinking water distribution systems. A sandy aquifer of low thickness (Table 2) is overlain by a mighty clay layer, limiting infiltration from the Döllnitz stream.

Table 3. Geohydraulic conditions for the Görlitz-Weinhübel site.

Condition	Average value
aquifer thickness	10 m
hydraulic conductivity	1×10^{-3} m/s
distance between river bank and abstraction wells	50 to 150 m
number of wells	32
maximum total abstraction rate	11 248 m³/d

For an abstraction rate of 1700 m³/d, the corresponding bank filtration portion was determined to 13% based on groundwater flow modelling using Processing Modflow (Schwanke, 2008). The bank filtrate originates both from the Lake Göttwitz (12%) and the Döllnitz stream (1%). The low percentage of bank filtrate in the pumped raw water is caused by the morphology of the confining layer of the upper aquifer and management of lake water levels. Lake Göttwitz has a higher water level than Lake Döllnitz, which periodically is emptied.

A reduced water abstraction of about 1000 m³/d results in a lower bank filtration portion of about 4%. The travel time between the lake and the northern well was calculated to be longer than 4 months for an abstraction rate of 1700 m³/d and longer than 20 months for 1000 m³/d. The travel time for the infiltrating stream water (1–2% of abstracted water) has been calculated to be 2–3 months for both scenarios. If the water demand decreases further, the portion of bank filtrate will likewise be reduced, and only groundwater will be abstracted. This may cause problems in the future because nitrate and sulphate concentrations in the groundwater in the nearby agricultural area are high, whereas the nitrate concentration of bank filtrate is very low. Operation of only one well located nearest to the lake bank instead of all four wells would not have a noticeable effect on the bank filtrate portion if the total abstraction is low.

2.3 Görlitz-Weinhübel Waterworks

The Görlitz-Weinhübel Waterworks is located in East Saxony, at the German-Polish border, in the floodplain of the River Lausitzer Neisse (Fig. 1). The waterworks supplies drinking water from bank filtrate and artificially recharged groundwater. A lake and several artificial infiltration basins are fed by river water to augment the quantity of available water in the abstraction wells (Fig. 4). This technique was mainly practised during the 1980s, when water demand and consumption were at a relatively high level. However, recent investigations into the infiltration capacity of the lake proved that the bottom of the lake is almost completely clogged.

The main hydrogeological and technical information for the site is given in Table 3.

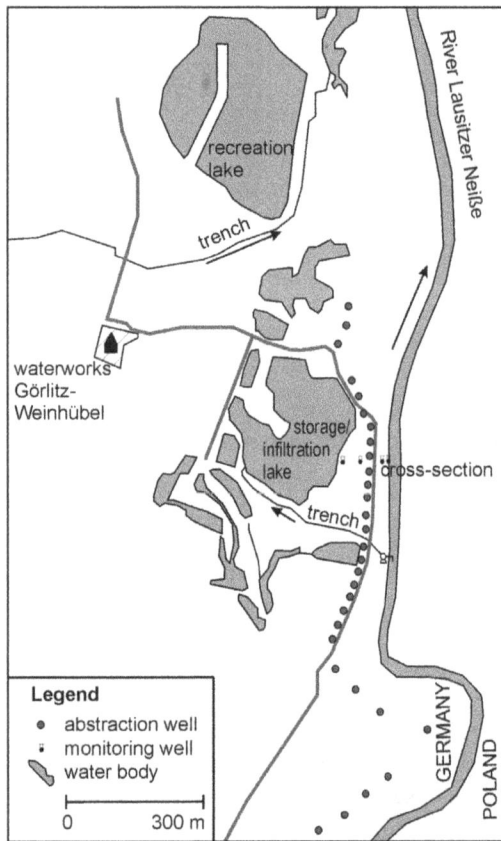

Figure 4. Scheme of the bank filtration site in Görlitz-Weinhübel.

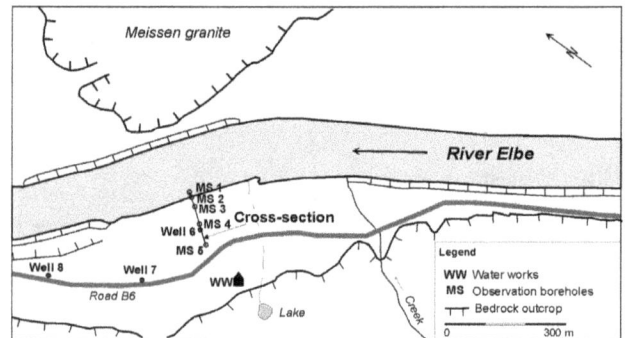

Figure 5. Scheme of the bank filtration site in Meissen-Siebeneichen.

Table 4. Geohydraulic conditions for the Meissen-Siebeneichen site.

Condition	Average value
aquifer thickness	18 m
hydraulic conductivity	1.2 to 1.4×10^{-3} m/s
distance between river bank and abstraction wells	100 m
number of wells	3
max. total abstraction rate	3800 m^3/d

The shortest travel occurs between the river and the well north of the cross-section (Fig. 4). Based on an analysis of seasonal temperature data for river water and groundwater as well as hydraulic gradients the travel time of bank filtrate was determined to be 8 to 24 days depending on river stage. Lower abstraction from the siphon pipe system results in a longer travel time. Originally, the impact of landside water quality was low, because river water would infiltrate in the lakes and basins. However, due to significant clogging of the lake bed, the portion of landside groundwater abstracted is expected to increase. This would result in higher manganese (greater than 1 mg/L) and ammonia concentrations (greater than 0.5 mg/L) in the pumped raw water.

2.4 Meissen-Siebeneichen Waterworks

The river bank filtration site in Meissen-Siebeneichen is situated on the River Elbe, in a transverse valley between the Spaar Mountains and the Meissen granite bedrock (Fig. 5). The valley is filled with Pleistocene deposits to a depth of 5–20 m that comprise interfingered glaciofluviatile sediments ranging from fine sand to medium sand and gravel. The deposits are overlain by Holocene meadow loam (2–7 m

thick). Along the monitored cross-section, the thickness of the aquifer is about 18 m (Table 4).

The mean discharge of the River Elbe at Meissen is about 300 m^3/s. Groundwater recharge in the 200–300 m wide strip between the river and the bedrock is very low due to the meadow loam cover. Groundwater flow from faults in the bedrock and from the other side of the river beneath the river bed towards the production wells was assumed to be negligible. Thus, the only source of the pumped raw water was thought to be river water via bank filtration.

Groundwater samples were taken from the cross-section at a quarterly interval in 1993–1995 and on 12 May 2001 (Fig. 6). Groundwater was obtained using either a mobile submersible pump lowered into 120 mm diameter observation wells or from in situ membrane pumps used to sample from directly below the bed of the River Elbe. In the field, well head measurements included temperature, pH, alkalinity and O_2. In the laboratory, analyses included major ions, iron and manganese. Analyses were carried out according to German guidelines and DIN methods. Prior to analysis, water samples were filtered through a 0.45 µm cellulose-acetate filter. The concentrations of anions Cl^-, NO_3^-, and SO_4^{2-} were determined using ion-chromatography.

The results of field investigations between 1993 and 1995 showed very high nitrate (up to 130 mg/L), chloride and sulphate concentrations at the bottom of the aquifer between the production well and the River Elbe (Fig. 7a).

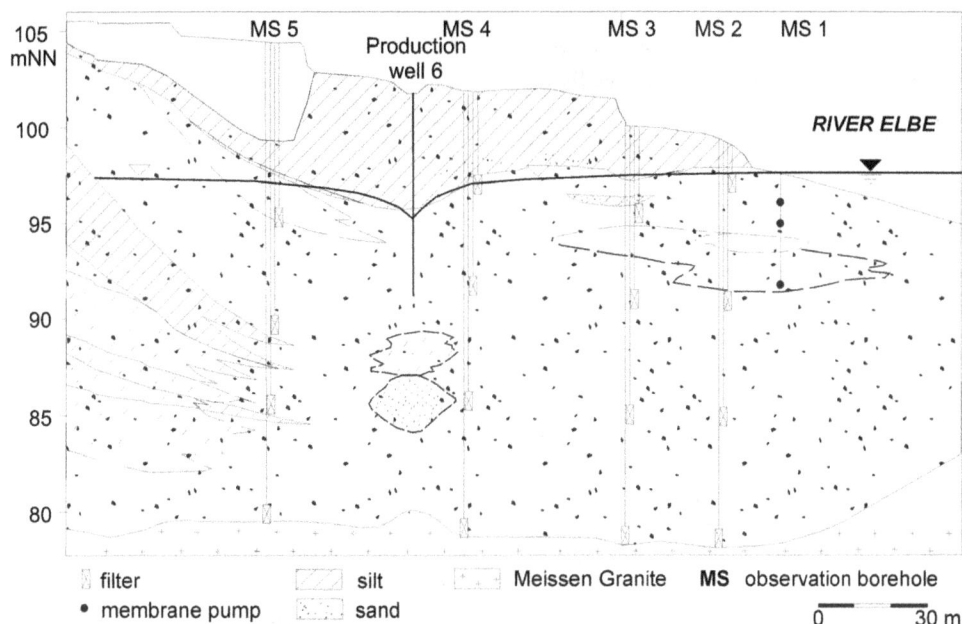

Figure 6. Location of observation boreholes along the cross-section.

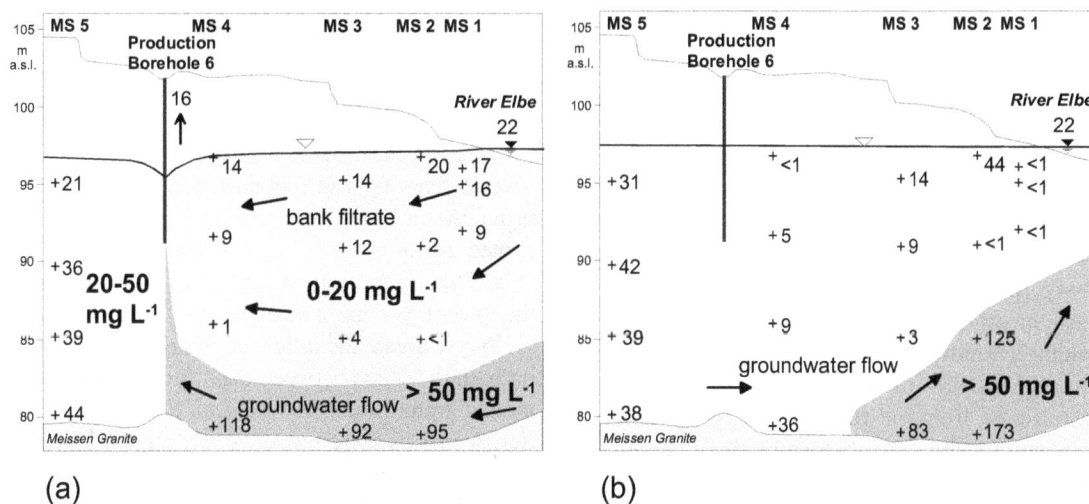

Figure 7. Nitrate concentrations (mg/L) at observation points along the cross-section. (**a**) mean values 1993–1995, (**b**) sampling on 12 May 2001.

Whereas chloride and sulphate concentrations were also high landward of the production well, nitrate concentration was only about 40 mg/L. Only along the right-hand side of the river (looking downstream), the groundwater is highly polluted by nitrate (up to 170 mg/L) due to fertilizer applications at vineyards and greenhouses. It was thus realised that the groundwater flows through the lower layer of the aquifer from the right side of the river beneath the river bed towards the production well. In the upper and middle layer of the aquifer, nitrate concentrations were lower than in river water due to denitrification processes along the flow path of bank filtrate.

The indication of groundwater flow beneath the river bed was further supported by results of the determination of dissolved organic carbon (DOC) and ethylenediaminetetraacetate (EDTA). In the lower layer of the aquifer, low DOC and EDTA concentrations were found indicating mixing of groundwater and a small proportion of bank filtrate (Grischek et al., 1994). Low DOC values in this layer also suggested a high proportion of weakly or non-biodegradable organic matter resulting in a carbon limitation of the denitrification rate.

During the exploitation of full production capacity, the waterworks abstracted in total $0.044\,\mathrm{m^3/s}$ from the wells 6, 7, and 8. Since 1996, water abstraction from these wells has been limited to only a few days per year. A groundwater sampling campaign in 2001 showed that as a consequence the bank filtrate in the aquifer was slowly replaced by groundwater. However, there were still high nitrate concentrations in the lower layer of the aquifer and near the river, indicating a very slow process of replacement and the effects of occasional pumping (Fig. 7b). High sulphate and chloride concentrations, found at sampling points next to river, indicated an up-coning of the groundwater from the lower layer and exfiltration into the river.

The geohydraulic assumptions based on hydrochemical investigations were verified by groundwater flow modelling using Processing Modflow for Windows (Chiang and Kinzelbach, 2001), and transport modelling with MT3D using the finite difference method (Zhen, 1990). The model area extended over $1790 \times 800\,\mathrm{m^2}$ and four layers of 5 m thickness each. The groundwater recharge from the River Elbe is controlled by the piezometric heads in the aquifer, the river water levels along the river course and the hydraulic conductance of the river bed. The horizontal conductivity value was set evenly to $1.4 \times 10^{-3}\,\mathrm{m/s}$. The corresponding vertical hydraulic conductivities were chosen to be an order of magnitude smaller than the horizontal conductivities due to anisotropy (the ratio between horizontal and vertical hydraulic conductivity). This allowed for model validation using results from hydrochemical investigations in the four different layers of the cross-section.

The choice of model boundaries was based on the bedrock geomorphology, piezometric contour maps and results from hydrochemical investigations. Neumann type boundaries were used along groundwater flow lines taken from piezometric contour maps ($Q = 0$), for groundwater flow into the model calculated from the related recharge areas and recharge rates, and for the three abstraction wells. Groundwater recharge from precipitation was set to $2.2\,\mathrm{L/s/km^2}$. The river was not included as a fixed head boundary but as a Cauchy boundary with a clogging layer in the river bed of 0.1 m thickness and a hydraulic conductivity of $1 \times 10^{-5}\,\mathrm{m/s}$.

Water head measurements from two sampling campaigns were used for model calibration. Hydraulic conductivities and proportions of pumped water from the production wells were altered slightly to achieve a better fit of measured and computed piezometric heads. A value of 0.25 for the effective porosity is considered reasonable for the alluvial sediments. The longitudinal dispersivity was estimated from the scale length of the transport phenomenon. Values of 10 m for the longitudinal dispersivity and of 1 m for both the horizontal and the vertical transverse dispersivities were defined for the present transport model.

The results from groundwater flow and transport modelling reinforced the theory of groundwater flow beneath the river. Figure 8 shows examples for flow path lines of particles

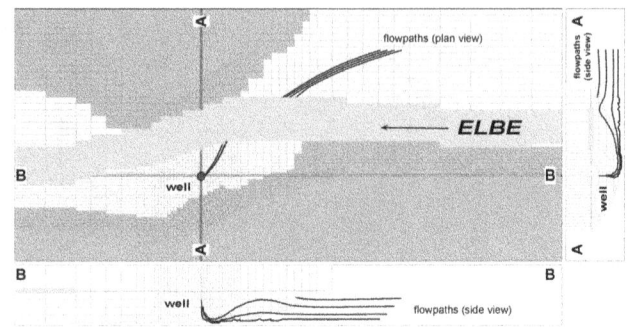

Figure 8. Flow path lines of groundwater from the opposite side of the river.

starting at the right side of the River Elbe in each model layer. Cross-sections A-A and B-B show that only particles starting in the upper layer of the aquifer are flowing into the river. All other particles turn away before they reach the river but flow towards the production well. Thus, high nitrate concentrations in the lower layer of the aquifer between the river and the production well can be explained by groundwater flow from the opposite side of the river.

Due to the small surface and subsurface catchment, the bank filtrate portion is very high. As a function of the clogging intensity in the river bed, the bank filtrate portion varies between 86 and 96%.

Using the groundwater transport model, a time scale of several years for full replenishment of the bank filtrate and groundwater from the opposite side of the river was confirmed by non-steady state calculations. Thus, in 2001 (Fig. 7b), no full replenishment of the nitrate-rich water in the lower layer could be expected.

To investigate the influence of decreasing well abstraction on both the ratios between bank filtrate, landside groundwater and groundwater flow beneath the river bed, four abstraction scenarios were modelled with the MT3D model. Scenario 1 is relevant to the original average abstraction rate of $0.0176\,\mathrm{m^3/s}$ from the well number 6 and contemporarily an abstraction of $0.0142\,\mathrm{m^3/s}$ from both well number 7 and well number 8. Scenarios 2 to 4 consider only water abstraction from well number 6 with quantities of $0.0176\,\mathrm{m^3/s}$, $0.0088\,\mathrm{m^3/s}$, and $0.0044\,\mathrm{m^3/s}$, respectively.

The closure of wells number 7 and 8 results in 4% less bank filtrate, thus there is only a little effect on the mixing conditions at well 6. However, the comparison of all scenarios (Table 5) shows clearly that decreasing water abstraction reduces the bank filtration portion from 85% (scenario 2) to 73% (scenario 3) and to 63% (scenario 4). Consequently, both the portion of landside groundwater and of groundwater flow beneath the river bed increased relative to the total abstraction (Table 5). The total numbers and the relative values for the changes in mixing of various components are shown in Fig. 9.

Table 5. Mixing conditions at well 6 as function of well abstraction, Meissen-Siebeneichen Waterworks.

Scenario		1	2	3	4
total well abstraction	m³/s	0.0176	0.0176	0.0088	0.0044
bank filtrate portion	m³/s	0.0157	0.015	0.0064	0.0028
	%	89	85	73	63
Landside portion	m³/s	0.0005	0.0009	0.0009	0.0006
	%	3	5	10	14
groundwater flow beneath river	m³/s	0.0014	0.0018	0.0015	0.001
	%	8	10	17	22
travel time	days	30–160	45–210	70–420	210–950

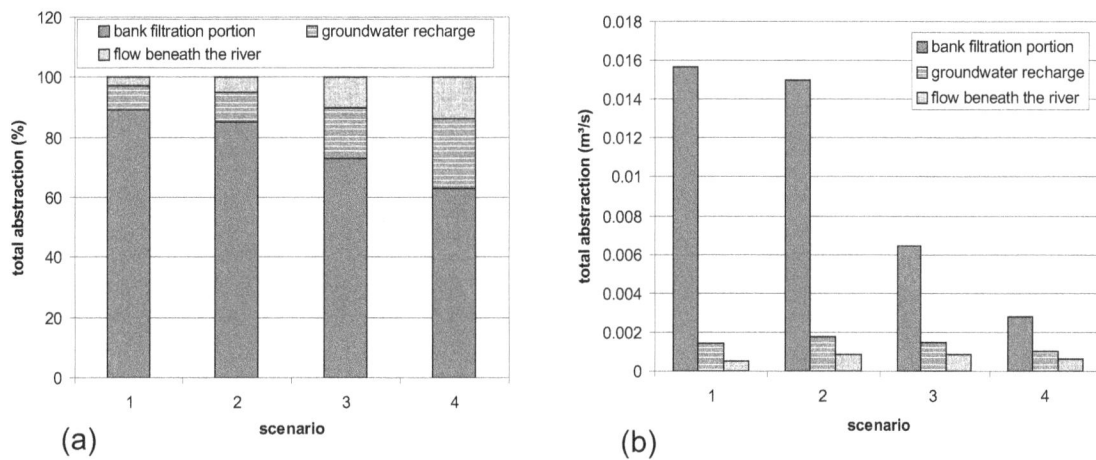

Figure 9. Total abstraction rates and portions of raw water components. (**a**) relative portions (**b**) absolute portions.

Naturally, the travel times between the river and the abstraction well become significantly longer with decreasing abstraction rates. For scenario 1, the bank filtrate travels as a function of the flow paths 30 to 160 days, and for scenario 4 it takes 210 to 950 days (Table 5). These longer travel times as a consequence of less water abstraction mean longer contact times and thus higher removal rates for dissolved organic carbon (Schoenheinz et al., 2004).

2.5 Generalisation of results based on a hypothetical groundwater flow and transport model

A hypothetical model covering a recharge area of $2000 \times 1500 \, \text{m}^2$ and representing an aquifer with two layers was established. The model size and the parameters were chosen from typical conditions for RBF sites in Saxony. To limit the conclusions to the effect of abstraction rates on mixing ratio of landside groundwater and bank filtrate only, full communication between river and aquifer was postulated and no groundwater flow beneath the river bed was considered.

Geohydraulic conditions are documented in Table 6. The hydraulic conductivity was set to 5×10^{-4} m/s, the hydraulic

Table 6. Geohydraulic conditions for the hypothetical RBF site.

Condition	Average value
aquifer thickness	50 m
hydraulic conductivity	5×10^{-4} m/s
distance between river bank and abstraction wells	200 m
number of wells	4
max. total abstraction rate	13 824 m³/d

gradient towards the river was set to 0.16%. Boundary conditions are shown in Fig. 10.

Four abstraction wells are located parallel to the river at a distance of 200 m from the bank. The distance between each of the 4 wells was set to 100 m. The influence of both reduced abstraction rates and the number of abstraction wells on the mixing behaviour between bank filtrate and landside groundwater are of interest. Therefore, 3 scenarios with 4 wells, 2 wells and 1 well were defined (Table 7).

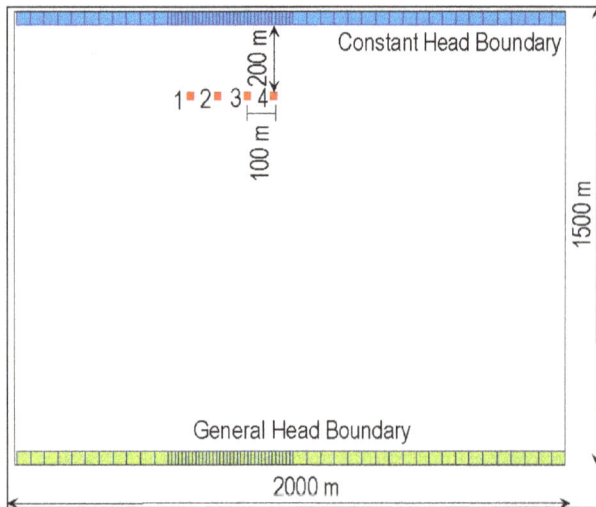

Figure 10. Set-up of hypothetical RBF model.

Assuming a strong decrease in water demand, a subsequent decrease in total water abstraction was modelled. When four wells are in operation, a decrease in total abstraction from 0.16 m³/s to 0.08 m³/s and 0.04 m³/s resulted in a reduction of the portion of bank filtrate from 51% to 29% and 6%, respectively (Table 7). If the same amount of water were abstracted from only two wells spaced 300 m apart, the portion of bank filtrate would be slightly lower. If only one well is operated with a high abstraction rate, the portion of bank filtrate would be slightly higher.

To concentrate on selected operation of a lower number of wells could be an advantage in receiving a certain portion of bank filtrate. However, the effect is strongly dependent on the specific geohydraulic site conditions and especially on the distance between the wells and the river bank. Waterworks operating a large number of wells could select wells located nearest to the river bank or within a meander for continuous operation (Grischek et al., 2009, 2003).

3   Managing bank filtration sites with respect to a fluctuating water demand

The management of waterworks depends on various limiting factors. Satisfying the fluctuating water demand is of prime importance, although this is often opposed to the preferred continuous operation of wells to ensure stable flow conditions and removal rates. Another important factor is the mixing of river bank filtrate and groundwater to obtain an optimum quality with regard to raw water treatment. At most sites, the main aims of water quality management include achieving the maximum attenuation of organic compounds during aquifer passage and low concentrations of DOC, dissolved iron and nitrate in raw water. At all sites with long flow paths, mixing ratios of bank filtrate and groundwater

Table 7. Impact of well operation on the portion of bank filtrate.

Scenario	Total abstraction m³/s	Number of wells	Abstraction from each well m³/s	Portion of bank filtrate %
A	0.16	4	0.04	51
		2	0.08	50
		1	0.16	52
B	0.08	4	0.02	29
		2	0.04	28.5
		1	0.08	32
C	0.04	4	0.01	6
		2	0.02	4
		1	0.04	10

were found to be of primary importance to the concentration of DOC, nitrate, sulphate, dissolved iron and manganese in the abstracted raw water.

Figure 11 and Table 8 give an overview of options for management measures at bank filtration sites.

(I) In the short term, optimisation of production well operation is the most promising and probably the most economic way to handle higher portions of landside groundwater and changes in raw water quality due to changing mixing ratios. This entails controlling the flow path length and travel time as well as the mixing by the selection of the most suitable wells from a well field. For that, a detailed investigation of groundwater flow conditions and portions of bank filtrate in the raw water is very important to decisions concerning the most effective water quality management measures.

(II) Knowing the hydrogeologic conditions of an existing bank filtration site, the most convincing way to tackle raw water quality problems related to unfavourable flow or mixing conditions between river and groundwater might be to replace wells. However, during a period of decreasing water demand and lower income of the waterworks, investments necessary to construct new wells – with locations adapted to achieve the principal aims in water pre-treatment – are limited and rare.

(III) Mostly, the frequently occurring seasonal floods in Saxon rivers erode the river bed and limit the clogging. Furthermore, better surface water quality and lower abstraction rates tend to reduce river bed clogging. Consequently, technical measures in the river bed of the River Elbe are not required. For bank filtration sites on rivers with high or increasing river bed clogging due to the organic load or the low water velocity, either frequent river bed cleaning (Martin-Alonso, 2003) or the set-up of recharge basins fed by river water are possible solutions. The latter one was practised in Görlitz and Dresden where an increase in water abstraction was required before the 1990s but was not obtainable by bank filtration only.

Table 8. Management actions in catchments of bank filtration sites.

	Management action	Actors	Evaluation
I	optimisation of well operation	waterworks	short-term, high efficiency
II	new well construction (location, depth)	waterworks	expensive
III	technical measures in the riverbed	waterworks	continuous measure, expensive
IV	changes in land use in the catchment	politics, farmers, waterworks	long-term measure

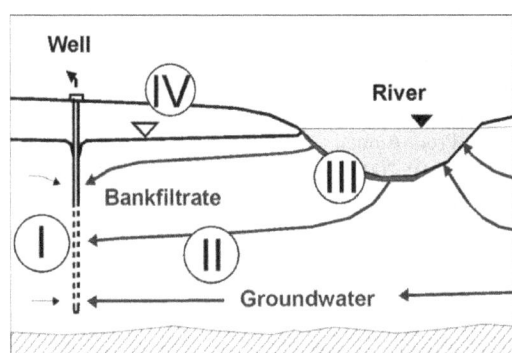

Figure 11. Options for managing river bank filtration in quantity and quality.

(IV) Collaboration with farmers in the catchment area, agreements on compensation payments to limit the application of fertilizers (especially N) and pesticides or buying the land are appropriate measures on a long term basis to reduce nitrate, sulphate and pesticide concentrations in landside groundwater. Such measures have been practised with different intensities and varying degrees of success. They may not be successful in balancing the higher nitrate and sulphate concentration of pumped raw water if the portion of landside water is increasing due to lower water abstraction.

4 Summary

Bank filtration sites in Germany have been designed decades ago to replace direct abstraction of polluted surface water. Whereas from the 1970s to the 1980s, river water quality and removal of organic pollutants was the major concern in management of bank filtration sites, the situation has changed since 1990. The ongoing decrease in water demand in many regions in Germany, especially in Saxony, leads to lower portions of bank filtrate in the abstracted raw water. Landside groundwater quality becomes more important for the subsequent treatment for drinking water production. At some sites nitrate concentrations greater than 80 mg/L and sulphate concentrations greater than 300 mg/L in landside groundwater demand specific mixing ratios between bank filtrate and groundwater in the aquifer to prevent otherwise neces-

sary treatment to remove nitrate and/or sulphate. The identification and determination of flow path indicators require not only the knowledge of hydrogeological conditions but also an integrated consideration of land and water use in the catchment of the bank filtration sites.

As a function of the river bed conditions, the land use on the opposite side of the river is also important because groundwater flow beneath the river towards the wells is not a priori negligible. At sites with high organic loads in the river water and less erosive conditions, clogging may enhance groundwater flow beneath the river bed. Of course, with a decrease in water abstraction the clogging induced by the operation of wells becomes less important due to lower infiltration rates and thus lower input of particles per square meter of river bed. However, even for rivers without any clogging of the bed, the anisotropy of the aquifer can result in groundwater flow beneath the river. Under such conditions, not only the catchment protection on the bank where the well field is located but also the land use and thus the groundwater quality on the far bank might play an important role.

Generally, it can be concluded that lower abstraction rates decrease the bank filtration portion and increase the influence of landside groundwater on the raw water quality. Thus, optimisation of well operation to cover the lower mean demand and the remaining peak demand becomes a key issue in management of bank filtration sites.

Under other hydrogeological or other socio-political conditions as discussed herein, the flow conditions such as mixing behaviour, groundwater flow beneath the river and travel times should be investigated by groundwater flow modelling prior to any planning, construction or optimisation of bank filtration sites.

Acknowledgements. The authors are grateful to the Ministry for Science and Art, Saxony, for financial support (grant no. 475316002512007), and to Mr. Deike (waterworks Görlitz), Mrs. Böhm (OEWA GmbH) and Mr. Fischer (DREWAG GmbH) for professional support.

References

BDEW: German Federal Association of Energy and Water Resources Management, www.bdew.de, last access: 1 July 2009 (in German).

City of Dresden: Drinking Water Abstraction, www.dresden.de, last access: 1 July 2009.

Chiang, W.-H. and Kinzelbach, W.: 3D-Groundwater Modeling with PMWIN, Springer, Berlin, Heidelberg, New York, 2001.

FSA: Public water supply by water source, Federal Statistical Authority Germany, www.destatsi.de, last access: 1 July 2009 (in German).

Grischek, T. and Ray, C.: Bank filtration as managed surface – groundwater interaction, Int. J. Water, 5(2), 125–139, 2009.

Grischek, T., Schoenheinz, D., and Ray, C.: Siting and design issues for riverbank filtration schemes, in: Riverbank Filtration: Improving Source-Water Quality, edited by: Ray, C., Melin, G., and Linsky, R., Kluwer Academic Publ., 291–302, 2003.

Grischek, T., Schoenheinz, D., and Nestler, W.: Unexpected groundwater flow beneath the River Elbe at the bank filtration site Meissen, Germany, in: Water resources and environment research, edited by: Schmitz, G. H., ICWRER 2002, 1, 116–120, 2002.

Grischek, T., Nestler, W., Dehnert, J., and Neitzel, P.: Groundwater/river interaction in the Elbe River basin in Saxony, in: Ground Water Ecology, edited by: Stanford, J. A. and Vallett, H. M., American Water Resources Association, Herndon, 309–318, 1994.

Hillenbrand, T. and Schleich, J.: Determinants of water demand in Germany, energie wasser-praxis, 6, 38–42, 2009 (in German).

Kühn, W. and Müller, U. (Eds.): Export oriented research and development in the field of water supply and treatment, part drinking water, DVGW, Germany, 2005 (in German).

Martin-Alonso, J.: Combined use of surface water and groundwater for drinking-water production in the Barcelona metropolitan area, Proc. 2nd Int. Riverbank Filtration Conf., 16–19.09.2003, Cincinnati, 217–222, 2003.

Schoenheinz, D., Grischek, T., and Worch, E.: Investigations on temperature effects on the DOC degradation during river bank filtration, Proc. Annual Conference of the Water Chemistry Society, 17.–19.05.2004, Bad Saarow, 150–154, 2004 (in German).

Schwanke, M.: Groundwater flow modelling and investigations into the development of the water balance at waterworks Göttwitz, Master thesis, University of Applied Sciences Dresden, Faculty of Civil Engineering/Architecture, 2008 (in German).

Zhen, C.: MT3D, a modular three-dimensional transport model, S.S. Papadopulos and Associates, Inc., Rockville, MD, 1990.

2

Development of water use scenarios as a tool for adaptation to climate change

R. Jacinto, M. J. Cruz, and F. D. Santos

CCIAM, SIM, Faculty of Sciences, University of Lisbon, C1, Sala 1.4.39,
Campo Grande, 1749-016 Lisbon, Portugal

Correspondence to: R. Jacinto (jacinto.rita@gmail.com)

Abstract. The project ADAPTACLIMA, promoted by EPAL, the largest Portuguese Water Supply Utility, aims to provide the company with an adaptation strategy in the medium and long term to reduce the vulnerability of its activities to climate change. We used the four scenarios (A1, A2, B1, B2) adopted in the Special Report Emissions Scenarios (SRES) of the IPCC (Intergovernmental Panel on Climate Change) to produce local scenarios of water use. Available population SRES for Portugal were downscaled to the study area using a linear approach. Local land use scenarios were produced using the following steps: (1) characterization of the present land use for each municipality of the study area using Corine Land Cover and adaptation of the CLC classes to those used in the SRES; (2) identification of recent tendencies in land use change for the study area; (3) identification of SRES tendencies for land use change in Europe; and (4) production of local scenarios of land use. Water use scenarios were derived considering both population and land use scenarios as well as scenarios of change in other parameters (technological developments, increases in efficiency, climate changes, or political and behavioural changes).

The A2 scenario forecasts an increase in population (+16 %) in the study area while the other scenarios show a reduction in the resident population (−6 to 8 %). All scenarios, but especially A1, show a reduction in agricultural area and an increase in urban area. Regardless of the scenario, water use will progressively be reduced until 2100. These reductions are mainly due to increased water use efficiency and the reduction of irrigated land. The results accord with several projects modelling water use at regional and global level.

1 Introduction

Climate change has already had effects on the natural environment and on socio-economic activities, and much bigger changes are projected for this century (IPCC, 2007; EEA, 2008). In Europe, especially in the south, water availability and water stress are already a concern (Henrichs et al., 2002; EEA, 2008). Climate change will further affect water availability and water consumption (EEA, 2008). Thus, there is a growing scientific and policy interest for water resources analyses on the global and regional scale (Märker et al., 2003; Menzel et al., 2007). Climate and socioeconomic drivers such as population and economic growth, land use and technological evolution, may have a great impact on water resources and stress (Menzel et al., 2007; Shen et al., 2008).

At the European level, several studies have looked into water resources availability and stress in the future (Henrichs et al., 2002; Mäerker et al., 2003; Menzel et al., 2007). However, the results differ between authors. Henrichs et al. (2002) projected a reduction in total water withdrawals in Western Europe and attributed that reduction to technological evolution and to households and industrial efficiency. Menzel et al. (2007) concluded that in Southern Europe the water stress and water withdrawals will increase due to irrigation demand and climate change.

In Portugal, there are published projections for the future population, such as the National Statistics Institute

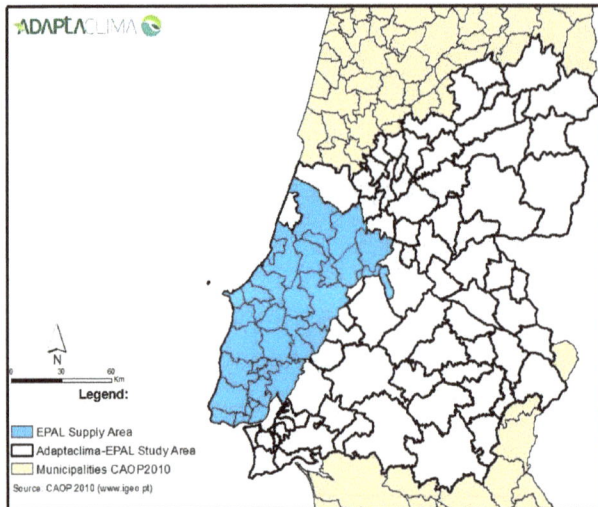

Figure 1. Study Area. Source: Adaptaclima-EPAL project report (2011).

projections made for the country up to the middle of the 21st century (INE, 2008) and the study "Climate change: Mitigation Strategies in Portugal" (MISP, 2007) with projections of the SRES scenarios downscaled for Portugal up to the end of the 21st century.

The project ADAPTACLIMA-EPAL aims to provide socioeconomic scenarios for population growth, land use and water use for an area which includes the Portuguese basin of the Tagus River (Fig. 1) with all the municipalities that are supplied directly or indirectly by EPAL, and the west aquifers of continental territory of Portugal. In total, 106 municipalities were included in the study area, of which, 34 are supplied by EPAL. Projections of future water withdrawals will contribute to EPAL, as well as to other stakeholders and policy makers. Other tasks of the project will include modelling future climate and water availability. All the generated data will be analysed and used to provide information to put in place a robust adaptation strategy to reduce vulnerability to future changes. In this paper future water stress for EPAL is evaluated by presenting socioeconomic scenarios for population growth, land use and water use for the project area.

2 Materials and methods

Local scenarios[1] of water use until the end of the century were explored, not only for the EPAL system, but also for other sectors, namely agriculture or industry. We used the four IPCC SRES (A1, A2, B1 and B2) (IPCC, 2000) to produce climate and hydrological scenarios for the project. The IPCC SRES sets up changes in population, economic growth,

---

[1] The terminology local scenario was adopted for our results while for IPCC scenarios the term SRES was used. At the national scale, there are also INE's population projections.

technology, energy consumption, and greenhouse gas emissions (IPCC, 2007; Shen et al., 2008). Several studies have produced water consumption scenarios for Europe for the 21st century based on SRES (e.g. Seckler et al., 1998; Märker et al., 2003; Flörke et al., 2005; Alcamo et al., 2007; Bates et al., 2008; Shen et al., 2008; Kok et al., 2009).

Linear regionalization (Fig. 2) of the SRES for the study area was applied, because it is the most transparent and simple method available; it assumes that the evolution on the study area will be similar to the evolution on the biggest area (i.e. Portugal) (van Vuuren et al., 2007, 2010).

- Consistency with local observed data (for the historic or base period);

- Consistency with the original scenario (SRES scenarios);

- Transparency;

- Plausibility of the results.

The main steps of the methodology are summarised in Fig. 3. The evolution of water use depends on several factors such as population and land uses (e.g. Boland, 1998; Houghton-Carr et al., 2008). In addition, other parameters were taken into account, namely developments in technology, as well as changes in climate and consumer behaviour.

Downscaling methods transform worldwide, continental or country level scenarios on useful and applicable scenarios at more detailed scales of analysis. The downscaling for water use was done following three steps which are outlined below.

2.1 Present water use by activity sector

Data of water use per capita, by sector and municipality, was collected from the following sources: National Water Plan (INAG, 2001), Instituto Nacional de Estatistica (INE), EPAL (internal data for the period 2000–2010) and National Inventory of Water Supply Systems and Waste water (INSAAR, 2007).

2.2 SRES scenarios – coefficients of change in demand factors

The future water consumption was determined by demand factors. These factors are the following ones:

a. *Population* – The CIESIN (Center for International Earth Science Information Network), has developed a database containing regionalized SRES for population and GDP at the country scale (CIESIN, 2002). Using the population on the base year, from the National Statistics Institute (INE) estimated for 2005 (INE, 2006), local population scenarios for each municipality were built for the study area by applying the CIESIN growth rates

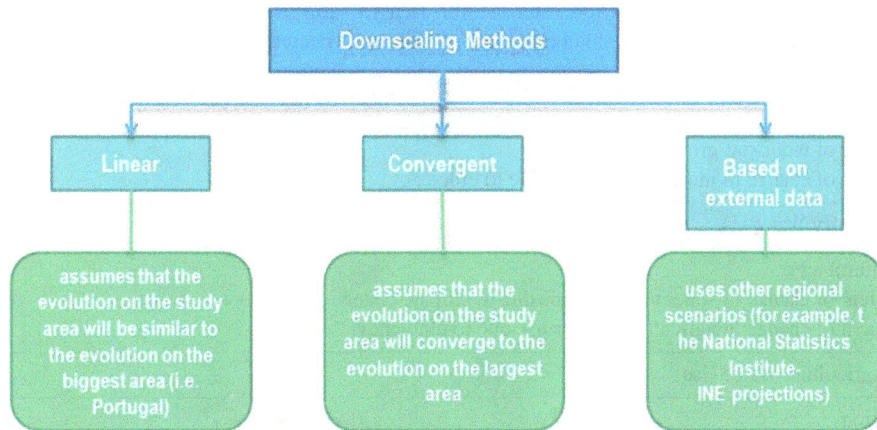

Figure 2. Downscaling methods in Socioeconomic Scenarios. Source: adapted from van Vuuren et al. (2010).

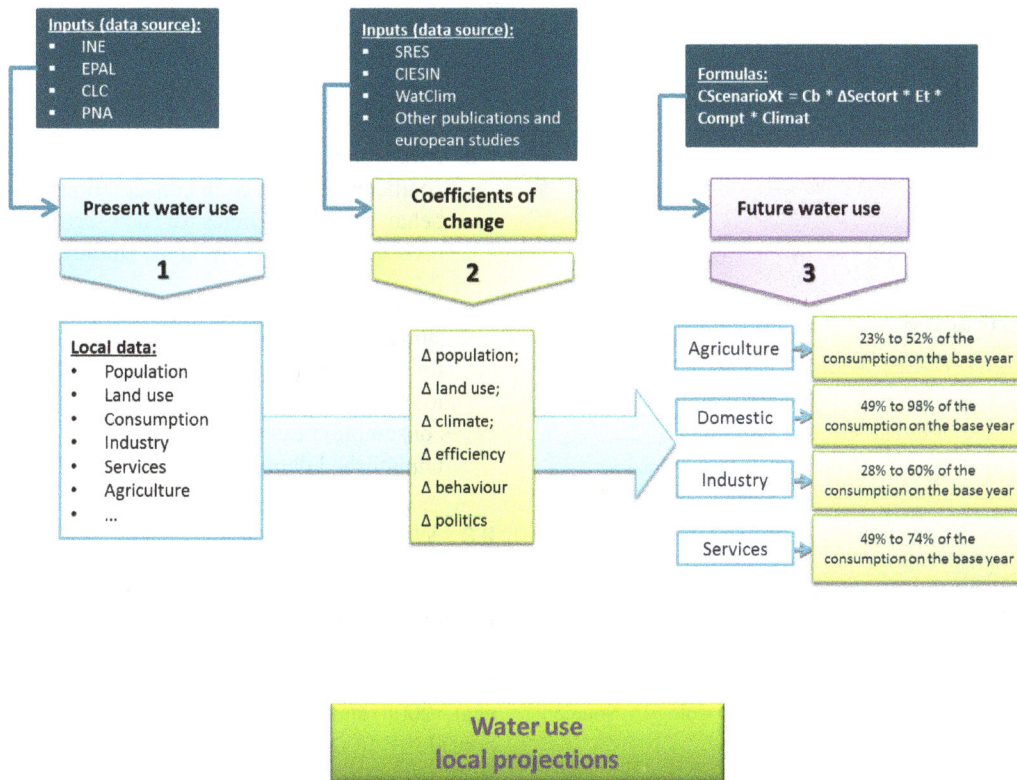

Figure 3. Resumed methodology for water withdrawals projections. Source: Adaptaclima-EPAL project report (2011).

downscaled for Portugal. Given the size of the study area and homogeneous characteristics in relation to the larger area, e.g. Portugal, the linear method of downscaling (O'Neill et al., 2001; Graffin et al., 2004) was considered the most appropriate;

b. *Land use* – Corine Land Cover land use classes were reclassified by grouping classes taking into consideration the classes used in SRES (Fig. 3). The sectors of water use considered in the National Water Plan (INAG,

2001) were taken into account. Local land use scenarios change (Fig. 4) in percentages for each municipality of the study area were produced by applying linear downscaling on the trends of the SRES scenarios for Europe and considering the characteristics of each county and the trends observed in recent years (IPCC, 2007; Schröter et al., 2005; Verburg et al., 2006; Rounsevell et al., 2006). Since Lisbon municipality is already almost only urban, the urban expansion for Lisbon municipality was redistributed to neighbouring municipalities. This

Table 1. Main trends in the various forcing factors for water use for each scenario. Legend: (+) indicates an increase, (++) indicates a substantial increase, (≈) indicates no change or very small changes, (–) indicates a decrease; (– –) indicates a substantial decrease.

	A1	A2	B1	B2
Coefficient of industrial growth	+	≈	≈	≈
Coefficient of change in industry water use due to efficiency	– –	–	– –	–
Total Industry Water Use	–	–	–	–
Agricultural area	– –	– –		–
Coefficient of change in water use for crops due to climate change	++	++	+	+
Coefficient of change in agricultural water use due to efficiency	–	–	–	–
Total Agriculture Water Use	– –	– –	–	–
Population	–	+	–	–
Coefficient of change in behaviour	≈	≈	–	–
Coefficient of change in consumption due to climate change	+	+	+	+
Coefficient of change in domestic water consumption due to efficiency	–	–	–	–
Total Domestic Water Use	–	≈	– –	– –

Source: Adaptaclima-EPAL project report (2011).

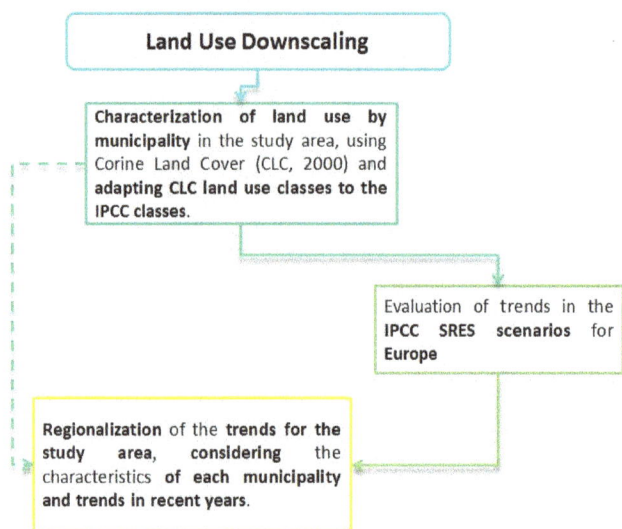

Figure 4. Land Use Downscaling Methodology. Source: Adaptaclima-EPAL project report (2011).

redistribution to the neighbouring municipalities was done by analysing recent tendencies (where the closest areas are more likely to grow) and applying the same tendencies for the local scenarios.

c. *Water use efficiency* – we calculated coefficients of change in consumption efficiency for each sector of water use (Fig. 3) and for each scenario considering available studies on scenarios of water consumption in Europe for the next century (Sect. 3) (e.g. Seckler et al., 1998; Märker et al., 2003; Flörke et al., 2005; Alcamo et al., 2007; Bates et al., 2008; Shen et al., 2008; Kok et al., 2009);

d. *Behaviour and water pricing policies* – we calculated a coefficient of change in consumption due to changes in behaviours and policies for each sector of water use and each scenario considering the studies referred under (c);

e. *Climate* – we calculated a coefficient of change in consumption due to climate change for each scenario and for each sector considering the available climate scenarios for the study area and correlations obtained between consumption rates and climate (SIAM II, 2006; EPAL, unpublished data).

2.3 Water use projections

None of the studies which produced scenarios for water use in Europe for the 21st century could be directly applied to this region for three different reasons. Firstly, these studies use different assumptions and scenarios. For example, Shen et al. (2008) and Märker et al. (2003) do not consider changes considered in the SRES-like reductions of agricultural area in all scenarios, evaluating only the changes in consumption as a function of changes in climate and behaviour. Secondly, they present scenarios for different time periods. For example, Flörke et al. (2005) presents scenarios only for the 2000–2030 period. And the third reason is that some of these studies do not consider local characteristics and different dynamics between the different European countries.

Analysis of those studies was done to assess the general trends presented but with the intention of applying them to the SRES scenarios and the real situation of the study area. Regardless of the study, general trends could be identified (see also Table 1):

1. Changes in land use, agricultural and industrial production are major forces of changes in water consumption (Mark et al., 2003; Houghton-Carr et al., 2008). All SRES scenarios indicate a reduction of the agricultural area (Table 1, Sect. 2);

2. The effects of global warming increase the need for both domestic and agriculture water consumption (Bates et al., 2008). However, these increases should be small, around 5 % in 50 yr (Bates et al., 2008). Scenarios A1 and A2 have the highest temperature increases (and larger extensions of dry periods) and thus the largest increases in water consumption needs (Table 1);

3. Technologies and behaviours might have effects on water consumption. It could drop in all sectors due to technological developments, such as increases in recycled water, improvements in irrigation efficiency and reduction of losses (Mark et al., 2003; Flörke et al., 2005; Kok et al., 2009). Technological improvements will be higher in scenarios A1 and A2. These trends are forced by worsening water shortages. According to Kok et al. (2009), it is likely that the increasing water scarcity in the southern countries of Europe will lead national governments to implement more efficient water rates, helping to reduce demand. The EU will impose more savings and water recycling, as well as sustainable urban drainage systems on a large scale for cities that are in water stress (especially in scenarios A2 and B1). The same authors reported the possibility of using desalination in coastal areas around the Mediterranean to provide water to some metropolitan areas. Water markets provide higher levels of funding for the government to invest in water saving devices at different scales (family, neighbourhood, city, etc.) (Kok et al., 2009). In B1 and B2 scenarios, there will be a tendency to improve water sustainability, reducing waste and reducing consumption per capita;

4. The evolution of population will also have influence in water consumption. This will decrease in all scenarios except A2.

Scenarios were produced for the whole study area (106 municipalities) for water use in the domestic, agriculture, industry and services sectors. Scenarios were also produced for the sub-area supplied by EPAL (34 municipalities) and for the municipalities for which we had data for water use by sector (Lisbon and Batalha). For each sector the following formula was used. Values for each parameter in the formula were obtained considering available publications.

$$CScenarioX_t = C_b \cdot \Delta Sector_t \cdot E_t \cdot Comp_t \cdot Clima_t \qquad (1)$$

where $CScenarioX_t$ is the consumption in the scenario X and year t, C_b is the consumption of water in the base year, $\Delta Sector_t$ is the net change in the demand sector between the

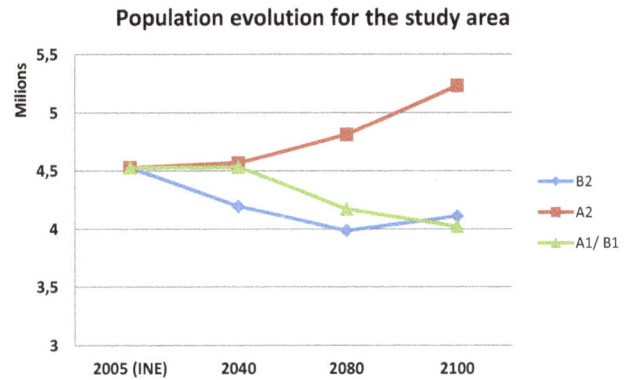

Figure 5. Population local scenarios. Source: Adaptaclima-EPAL project report (2011).

base year and year t (e.g. change in population, agricultural area or industrial activity), E_t is the coefficient of efficiency gains between the base year and year t, $Comp_t$ is the coefficient of change in behaviour between the base year and year t, and $Clima_t$ is the coefficient of change in consumption due to changes in climate between the base year and year t.

3 Results and discussion

In Table 1 we compiled the main trends in the forcing factors (land use, population, climate, efficiency, behaviour, politics) which affect water use. Only one scenario (A2) indicates an increase in the resident population of the study area of around 16 % by the end of the century (Fig. 5). The remaining scenarios indicate a reduction in the population from 6 to 8 %. All scenarios, but especially A1, indicate a reduction in agricultural area and an increase in urban area (Fig. 6).

The main trends of the SRES for land use changes in Europe have been identified in several publications (e.g. Schröter et al., 2005; Verburg et al., 2006; Rounsevell et al., 2006; IPCC, 2007) and are summarised in Table 3. All scenarios, but especially A1, show a significant reduction in agricultural area and an increase in urban area.

Comparing the results for our local scenarios in the Fig. 6 and the SRES European tendencies from different authors (Schröter et al., 2005; Verburg et al., 2006; Rounsevell et al., 2006; IPCC, 2007) from Table 2, can be concluded that our local scenarios coincide with the urban and agriculture land use, but not in grassland. Our results show that this last type of land use will diminish while the European SRES tendencies for grassland will be maintained on the same level.

Regardless of the scenario, water use will progressively be reduced until 2100 (Table 4). These reductions are mainly due to increased water use efficiency and a reduction in irrigated land. Even if a warmer and drier future climate will lead to a rise in water demand for irrigation and for domestic use, none of the scenarios showed an increase in water use. Even when we consider a large increase in population

Table 2. Percentage of Growth on the INE (2008) projections and scenarios of CIESIN (2002).

	CIESIN B2	CIESIN A2	CIESIN A1 e B1	INE central	INE low	INE high	INE without migrations
% total grow 2005–2060	−8.33 %	3.76 %	−0.86 %	−1.94 %	−15.70 %	13.47 %	−23.32 %
total grow 2005–2100	−6.07 %	19.61 %	−8.09 %				

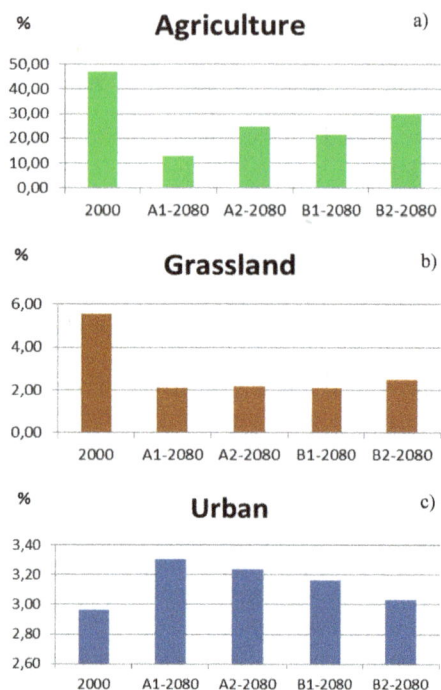

Figure 6. Local scenarios for land use change for: **(a)** agriculture, **(b)** grassland and **(c)** urban. Source: Adaptaclima-EPAL project report (2011).

Table 3. Main trends in land use changes in the SRES for Europe.

	A2	A2	B1	B2
Agriculture	− −	− −	− −	−
Urban	+ + +	+ +	+ +	+
Industry	+	+	+	≈
Biofuel	+	+	+ +	+
Forests	+ +	+	+	+ + +
Grassland	≈	≈	≈	≈
Protected Areas	+	≈	+ +	+ +

Table 4. Water consumption by sector in the project study area Adaptaclima-EPAL in the base year (2000) and scenarios (percent of base year values) for 2080.

	Base year (m^3)	A1	A2	B1	B2
Agriculture[1]	1 443 548 957 m^3	22.7 %	46.6 %	37.5 %	52.2 %
Industry[2]	2 194 069 m^3	60.2 %	59.4 %	28 %	58.3 %
Domestic[2]	237 702 531 m^3	87.6 %	98.3 %	48.7 %	49.8 %
Services[2]	6 466 211 m^3	65.7 %	63.9 %	48.7 %	74.1 %

[1] Baseline data: National Water Plan 2001, Corine Land Cover 2000; [2] Baseline data: INSAAR (2007), Corine Land Cover (2000).

(scenario A2), overall water consumption will be reduced. Our results thus seem to indicate that behavioural and technological factors will be the main factors in shaping future water use in the study area.

The projections of future water withdrawals presented in this paper were based on a range of assumptions regarding agricultural, industrial, and domestic water use. That is why we used four different scenarios, which indicate different possible futures in terms of socio-economic developments. All scenarios projected a reduced water use in all the analysed sectors, which increases confidence in the results. Furthermore, our results agree with several projects that model water use at a regional or global level (e.g. Seckler et al., 1998; Menzel et al., 2007). Henrichs et al. (2002), have modelled water use up to 2070 by considering a simple scenario where present water use trends were prolonged into the future and have obtained a decreased in water withdrawals in Western Europe, mainly resulting from gains in the efficiency of water use. The experiences of many coun-

tries show that water use intensity decreases after reaching a saturation amount, mainly due to industrial restructuring and water use efficiency improvements in production processes (Shen et al., 2008). Thus, although projected changes in water withdrawals strongly depend on the assumptions regarding socio-economic factors such as economic and industrial growth (Henrichs et al., 2002), which are largely uncertain, it seems that most scenarios indicate the same long-term trends in reduction of water use.

However, these results need to be considered carefully, also taking into account scenarios of climate and hydrological change. In Portugal, water use is highly correlated to temperature, and the warm season is also the dry season, with higher consumption rates. Therefore there is already some predisposition to water stress, especially in the south of the continental territory and during the th summer months. During the 20th century there has been a clear trend towards drier conditions, with decreases in rainfall and moisture availability in most Mediterranean regions; severe drought episodes (from both meteorological and hydrological contexts) have become more frequent and persistent, namely in the Iberia area (Sousa et al., 2011). In Portugal the total annual precipitation has been considerably reduced during the last decades as it rained for less number of days (SIAM II, 2006). Global climate models project a decrease in precipitation, increases in temperature and evapotranspiration, and an enlarged dry season for the study area (SIAM II, 2006). Our results support that socioeconomic changes may reduce water consumption, and therefore the risk of increased water stress due to climate change may be lower than previously expected.

4 Conclusions

The results of this study show that all different scenarios indicate reductions in water use in the study area. One of the main results is that behavioural and technological factors are determinant in shaping future water use in the study area. In Portugal, water use inefficiency corresponds to about 41 % of the total withdrawals as referred to in the National Water Plan (2001). Thus, it is not surprising that there can be enormous gains in water efficiency, especially in the agricultural and industrial sectors.

The scenarios are a useful framework for thinking about the future and a fundamental step for EPAL to prepare a long-term adaptation strategy to climate change which is meant to contribute to the reduction of vulnerability of different future societies. EPAL's successful adaptation to climate change is vital for economic activity and the local population. The tasks that look at the impacts on the resources (quantity and quality of water) will also need to be taken into account in the process, so that the adaptation occurs in a sustainable way. To build a consistent strategy for climate change adaptation, both climate and socioeconomic scenarios are important, while socioeconomic scenarios can decrease the probability of decision failures.

Acknowledgements. This study was supported by EPAL. We are thankful to EPAL's Climate Change working group, who has supplied valuable information. We are also thankful to Nuno Grosso, Mário Pulquério, David Avelar and Tiago Lourenço for their contributions in this project.

References

Aguiar, R. and Santos, F. D.: MISP, Energia e Emissões de Gases com Efeito de Estufa: Um Exercício de Prospectiva para Portugal até 2070. Projecto MISP: Mitigation Strategies in Portugal, Fundação Calouste Gulbenkian and Instituto D. Luis, Lisboa, 48 pp., 2007.

Alcamo, J., Flörke, M., and Märker, M.: Future long-term changes in global water resources driven by socio-economic and climatic changes, Hydrolog. Sci. J., 52, 247–275, doi:10.1623/hysj.52.2.247, 2007.

Bates, B. C., Kundzewicz, Z. W., Wu, S., and Palutikof, J. P. (Eds.): Climate Change and Water. Technical Paper of the Intergovernmental Panel on Climate Change, IPCC Secretariat, Geneva, 210 pp., http://www.ipcc.ch/pdf/technical-papers/ccw/chapter3.pdf (last access: December 2010), 2008.

Boland, J. J.: Water Supply and Climate Uncertainty, Water supply planning practice in the U.S. Universities Council on Water Resources: 112, http://ucowr.org/files/Achieved_Journal_Issues/V112_A10Water%20Supply%20and%20Climate%20Unvertainty.pdf (last access: October 2011), 1998.

CIESIN (Center for International Earth Science Information Network): Country-level Population and Downscaled Projections, http://www.ciesin.columbia.edu/datasets/downscaled (last access: 2 December 2010), 2002.

CLC (Corine Land cover): Carta de Ocupação do solo revista, http://www.igeo.pt/e-IGEO/egeo_downloads.htm (last access: January 2011), 2000.

EEA, JRC and WHO: Impacts of Europe's changing climate – 2008 indicator-based assessment, European Environment Agency, Copenhagen, 2008.

Flörke, M., Isoard, Stéphane, I., and Alcamo, J.: Outlook on Water Use in Europe in 2030, Environmental Info 2005 (Brno) Masaryk University Brno, ISBN: 80-210-3780-6, 57–62, 2005.

Graffin, S. R., Rosenzweig, C. R., Xing, X., and Yetman, G.: Downscaling and Geo-spatial Gridding of Socio-Economic Projections from the IPCC Special Report on Emissions Scenarios (SRES), CIESIN, Center for Climate Systems Research, Columbia University, 2004.

Henrichs, T., Lehner, B., and Alcamo, J.: An Integrated Analysis of Changes in Water Stress in Europe, Center for Environmental Systems Research, University Kassel, Germany, Integrated Assessment, 3, 15–29, 2002.

Houghton-Carr, H. A., Fry, M. J., and Farquharson, F. A. K.: Driving forces of future water availability and water use within Europe, BHS 10th National Hydrology Symposium, Exeter, 2008.

INAG: Plano Nacional da Água, Instituto Nacional da Água, Ministério do Ambiente, do Ordenamento do Território e do Desenvolvimento Regional, http://www.inag.pt/index.php?option=com_content&view=article&id=9&Itemid=69 (last access: December 2010), 2001.

INE: Anuário Estatístico 2005, Instituto Nacional de Estatística, available at: http://www.ine.pt/xportal/xmain?xpid=INE&xpgid=ine_destaques&DESTAQUESdest_boui=74644&DESTAQUESmodo=2; (last access: December 2010), 2006a.

INE: Estimativas Provisórias de População Residente Portugal 2005, NUTS II, NUTS III e Municípios, ISSN: 1645-8389, http://www.ine.pt (last access: December 2010), 2006b.

INE: Projecções de População Residente em Portugal 2008–2060 – Metodologia, Instituto Nacional de Estatística – Departamento de Estatísticas Demográficas e Sociais (http://www.ine.pt/xportal/xmain?xpid=INE&xpgid=ine_publicacoes&PUBLICACOESpub_boui=6188154&PUBLICACOESmodo=2), 2008.

IPCC: Emissions Scenarios. A Special Report of Working Group III of the Intergovernmental Panel on Climate Change, Cambridge University Press, Cambridge, 2000.

IPCC: IPCC Fourth Assessment Report (AR4), edited by: Watson, R. T. and the Core Writing Team, IPCC, Geneva, Switzerland, 184 pp., 2007.

INSAAR: Inventário Nacional de Sistemas de Abastecimento de Águas e Águas Residuais, available at: http://insaar.inag. pt/index.php?id=21&year=2007 (last access: December 2010), 2007.

Kok, K., Vliet, M., Bärlund, I., Sendzimir, J., and Dubel, A.: First ("first-order") draft of pan-European storylines – results from the second pan-European stakeholder workshop, SCENES Deliverable 2.6, Wageningen Universitiy, Wageningen, http://www.environment.fi/download.asp?contentid= 113374&lan=en (last access: April 2011), 2009.

Märker, M., Flörke, M., Vassolo, S., and Alcamo, J.: Preliminary assessment of IPCC-SRES scenarios on future water resources using the WaterGAP 2 model, Center of Environmental Systems Research, University of Kassel, http://www.mssanz.org.au/ MODSIM03/Volume_01/A07/03_Maerker.pdf (last access: April 2011), 2003.

Menzel, L., Flörke, M., Matovelle, A., and Alcamo, J.: Impacts of socio-economic development and climate change on water resources and water stress, Proc. 1st International Conference on Adaptative and Integrative Water Management (CAIWA), 2007.

O'Neill, B., Balk, D., Brickman, M., and Ezra, M.: A guide to global population projections, Demogr. Res., 56, 203–288, doi:10.4054/DemRes.2001.4.8, 2001.

Rounsevell, M. D. A., Reginster, I., Araújo, M. B., Carter, T. R., Dendoncker, N., Ewert, F., House, J. I., Kankaanpää, S., Leemans, R., Metzger, M. J., Schmit, C., Smith, P., and Tuck, G.: A coherent set of future land use change scenarios for Europe, Agr. Ecosyst. Environ., 114, 57–68, doi:10.1016/j.agee.2005.11.027, 2006.

Schröter, D., Cramer, W., Leemans, R., Prentice, I. C., Araújo, M. B., Arnell, N. W., Bondeau, A., Bugmann, H., Carter, T. R., Gracia, C. A., de la Vega-Leinert, A. C., Erhard, M., Ewert, F., Glendining, M., House, J. I., Kankaanpää, S., Klein, R. J. T., Lavorel, S., Lindner, M., Metzger, M. J., Meyer, J., Mitchell, N. W., Reginster, T. D. I., Rounsevell, M., Sabaté, S., Sitch, S., Smith, B., Smith, J., Smith, P., Sykes, M. T., Thonicke, K., Thuiller, W., Tuck, G., Zaehle, S., and Zierl, B.: Ecosystem Service Supply and Vulnerability to Global Change in Europe, http://www.sciencemag.org/content/suppl/ 2005/11/21/1115233.DC1/Schroter.SOM.REV.pdf (last access: January 2011), 2005.

Seckler, D., Molden, U. A. D., Silva, R., and Barker, R.: World water demand and supply, 1990 to 2025: Scenarios and issues. Research Report 19. Colombo, Sri Lanka, International Water Management Institute, ISBN: 92-9090354-6, http://www.iwmi.cgiar.org/Publications/ IWMI_Research_Reports/PDF/PUB019/REPORT19.PDF (last access: January 2011), 1998.

SIAM II: Alterações Climáticas em Portugal – Cenários, Impactos e Medidas de Adaptação, Projecto SIAM, edited by: Fase II, F. D. S. and Miranda, P., Editorial Gradiva, Lisbon, http://siam.fc. ul.pt/siamII_pdf/SIAMII.pdf, 2006.

Shen, Y., Oki, T., Utsumi, N., Kanae, S., and Nahasaki, N.: Projection of future world water resources under SRES scenarios: water withdrawal, Hydrolog. Sci. J., 53, 11–33, doi:10.1623/hysj.53.1.11, 2008.

Sousa, P. M., Trigo, R. M., Aizpurua, P., Nieto, R., Gimeno, L., and Garcia-Herrera, R.: Trends and extremes of drought indices throughout the 20th century in the Mediterranean, Nat. Hazards Earth Syst. Sci., 11, 33–51, doi:10.5194/nhess-11-33-2011, 2011.

van Vuuren, D. P., Lucas, P. L., and Hilderink, H. B. M.: Downscaling drivers of global environmental change – Enabling use of global SRES scenarios at the national and grid levels, Report 550025001, Netherlands Environmental Assessment Agency, http://www.rivm.nl/bibliotheek/rapporten/ 550025001.pdf (last access: January 2011), 2007.

van Vuuren, D. P., Smith, S. J., Riahi, Keuwan: Downscaling socioeconomic and emissions scenarios for global environmental change research: a review, John Wiley & Sons, Ltd, 1, 393–404, doi:10.1002/wcc.50, 2010.

Verburg, P. H., Schulp, C. J. E., Witte, N., and Veldkamp, A.: Downscaling of land use change scenarios to assess the dynamics of European landscapes, Agr. Ecosyst. Environ., 114, 39–56, doi:10.1016/j.agee.2005.11.024, 2006.

3

A new model for the simplification of particle counting data

M. F. Fadal[1]**, J. Haarhoff**[1]**, and S. Marais**[2]

[1]Department of Civil Engineering Science, University of Johannesburg, South Africa
[2]Process Technology Department, Rand Water, South Africa

Correspondence to: J. Haarhoff (jhaarhoff@uj.ac.za)

Abstract. This paper proposes a three-parameter mathematical model to describe the particle size distribution in a water sample. The proposed model offers some conceptual advantages over two other models reported on previously, and also provides a better fit to the particle counting data obtained from 321 water samples taken over three years at a large South African drinking water supplier. Using the data from raw water samples taken from a moderately turbid, large surface impoundment, as well as samples from the same water after treatment, typical ranges of the model parameters are presented for both raw and treated water. Once calibrated, the model allows the calculation and comparison of total particle number and volumes over any randomly selected size interval of interest.

1 Introduction

The power of particle counters to provide a detailed description of the numbers and sizes of particles in a suspension is often not fully exploited, although its potential had been realised some decades ago in fields as diverse as phycology and water treatment (Sheldon, 1979; Lewis and Manz, 1991). The counters produce a count and the size limits for numerous channels, and the full meaning of the analysis is often obscured by a sheer weight of numbers. A method is required to compact the multitude of numbers from every count to as few as possible parameters to offer a reliable description of the particle size distribution. A second useful application of a generalised description is to allow the comparison of particle counts made by different particle counters with their own unique channel size settings. Such models have been proposed and used in the past. It is the objective of this paper to firstly propose the use of a new, improved model, aiming to overcome some of the weaknesses of earlier proposals. Secondly, the model will be applied to a large data set of particle counts collected before and after treatment at a large South African drinking water supplier.

2 Theoretical development

2.1 The power law

The commonly used power law is simply a straight line defining the normalised particle counts N (y-axis) in terms of the geometric mean size d of each counting channel (x-axis) on a log-log plane. The power law has some very real conceptual weaknesses – at small particle sizes, the particle number tends to infinity; at large sizes, the particle volume tends to infinity (Wilczak et al., 1992). The model and its calibration equations are, for n channels:

$$N = A \cdot d^{\beta}$$

$$\begin{bmatrix} \ln A \\ \beta \end{bmatrix} = \begin{bmatrix} n & \sum \ln d \\ \sum \ln d & \sum (\ln d)^2 \end{bmatrix}^{-1} \cdot \begin{bmatrix} \sum \ln N \\ \sum \{(\ln N)(\ln d)\} \end{bmatrix} \quad (1)$$

2.2 The variable-β model

To rectify the weaknesses of the power law, a variable-β model was proposed in conceptual form with no calibration data (Lawler, 1997). On a log-log plane, the variable-β model plots as an inverted parabola, centred about an axis at

$d = 1\,\mu m$. The variable-β model and its calibration equations are, for n channels:

$$N = A \cdot d^{\beta \ln d}$$

$$\begin{bmatrix} \ln A \\ \beta \end{bmatrix} = \begin{bmatrix} n & \sum (\ln d)^2 \\ \sum (\ln d)^2 & \sum (\ln d)^4 \end{bmatrix}^{-1} \cdot \begin{bmatrix} \sum \ln N \\ \sum \{(\ln N)(\ln d)^2\} \end{bmatrix} \quad (2)$$

The variable-β model was calibrated and compared to the power law in an exhaustive study which used the particle counts from 1432 water samples, ranging from raw surface water to treated drinking water, including samples from the intermediate treatment steps (Ceronio and Haarhoff, 2005). It was conclusively demonstrated that the variable-β model provided a better fit than the power law.

2.3 A proposed refinement to the variable-β model

Despite the improved fit provided by the variable-β model, it was pointed out that the variable-β model has an important limitation (Ceronio and Haarhoff, 2005). Regardless of the values of A and β, the maximum N would always be found at a size of $d = 1\,\mu m$, regardless of the nature of the suspension. To remove this limitation, a further conceptual improvement was offered, without any further development or validation. The suggested three-parameter model is called the Ceronio model in this paper and plots as an inverted parabola on a log-log plane, without any constraints on the position of the vertical axis. The Ceronio model and its calibration equations are, for n channels:

$$N = A \cdot d^{\beta \ln d + C}$$

$$\begin{bmatrix} \ln A \\ \beta \\ C \end{bmatrix} = \begin{bmatrix} n & \sum (\ln d)^2 & \sum \ln d \\ \sum (\ln d)^2 & \sum (\ln d)^4 & \sum (\ln d)^3 \\ \sum \ln d & \sum (\ln d)^3 & \sum (\ln d)^2 \end{bmatrix}^{-1} \cdot \begin{bmatrix} \sum \ln N \\ \sum \{(\ln N)(\ln d)^2\} \\ \sum \{(\ln N)(\ln d)\} \end{bmatrix} \quad (3)$$

It is noted in passing that the first matrix on the right-hand side of the calibration equations (for all the models above) is a function of the channel settings of the particle counter only, without being affected by the counts. The onerous inversion of the matrix has therefore only to be performed once for every instrument setting. Figure 1 illustrates the three models for a randomly selected particle count.

Once the models are calibrated, they can be used to rapidly obtain any desired property of the suspension. To obtain the total number of particles in any random size interval between d_1 and d_2, using the Ceronio model for illustration:

$$\#_{d_1, d_2} = A \int_{d_1}^{d_2} d^{\beta \cdot \ln d + C} \cdot dd \quad (4)$$

The corresponding total particle volume (assuming the particles to be spheres) in any random size interval between d_1 and d_2 is calculated with:

$$V_{d_1, d_2} = \frac{\pi A}{6} \int_{d_1}^{d_2} d^{\beta \cdot \ln d + C + 3} \cdot dd \quad (5)$$

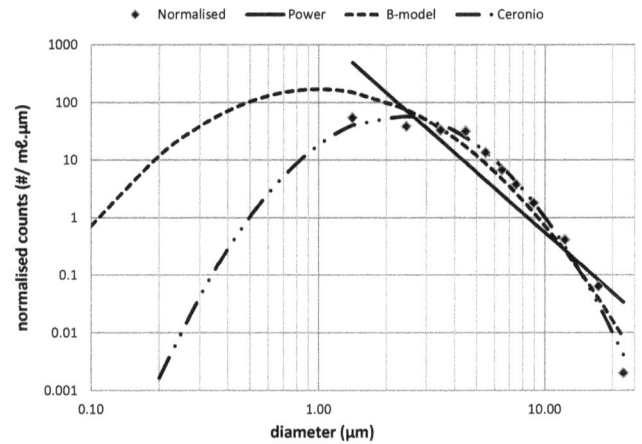

Figure 1. The power law, the variable-β model and the Ceronio model, fitted to the same particle count.

The power of the Ceronio model lies predominantly in its ability to model suspensions where the maximum normalised counts deviate from $d = 1\,\mu m$. The diameter where the normalised count reaches a maximum is provided by:

$$d_{\max} = e^{-C/2\beta} \quad (6)$$

3 Model application

3.1 Particle counting data collection

Rand Water is a drinking water supplier supplying about 3.7 million m^3/day to a population of roughly 11 million people in the Gauteng Province, as well as parts of the Mpumalanga, Free State and North West provinces of South Africa. Its primary water source is Vaal Dam, an impoundment of 2536 million m^3. Raw water is conveyed from the Vaal Dam by open channel and pipes to two treatment plants – Zuikerbosch (ZB) and Vereeniging (VG). At Zuikerbosch, the bulk of the raw water is first retained in a large balancing tank before it proceeds to treatment; at Vereeniging there is no balancing tank and the water is treated directly upon arrival.

Particle counting is performed on samples taken directly from the sampling taps on the incoming and outgoing pipelines. This paper utilises the counts for four sampling positions, namely the two raw water sampling points (ZB Raw and VG Raw) and two selected points on the treated water (pipelines B10 and A20, henceforth labelled ZB Final and VG Final), yielding a total of 321 samples. The data used was collected at roughly fortnightly intervals from August 2006 to September 2009, covering slightly more than three years. The samples were transported in glass containers to the laboratory and counted forthwith with the same particle counter.

Table 1. Percentage of samples that recorded zero counts in the channels indicated.

Channel	d_{min}	d_{max}	ZB Raw	ZB Final	VG Raw	VG Final
9	10	15	0 %	0 %	0 %	0 %
10	15	20	0 %	0 %	0 %	0 %
11	20	25	6 %	14 %	16 %	9 %
12	25	30	37 %	44 %	46 %	38 %
13	30	40	52 %	46 %	64 %	46 %
14	40	50	87 %	71 %	86 %	78 %
15	50	100	92 %	88 %	93 %	86 %

3.2 Data screening

The particle counter used was a PAMAS 3116 FM with 16 channels, covering the range from 1 μm upwards. The channels used are separated at 1; 2; 3; 4; 5; 6; 7; 8; 10; 15; 20; 25; 30; 40; 50, and 100 μm. From these channel boundaries, the geometric mean of each channel was calculated to obtain the "d" required for the calibration matrices provided earlier. From the differential counts in each channel, the normalised counts were calculated by dividing them by the width of each channel, to obtain the "N" in the calibration matrices. The d- and N-values were used for further analysis.

For both the raw and treated water samples, there were very few counts in the higher size ranges. A necessary data screening step was to eliminate those larger channels which returned zero values, therefore not contributing to a meaningful fit of the data. The results are shown in Table 1. The channels in the lower half of Table 1 were eliminated from further consideration, based on the large percentage of samples having zero particle counts. There were therefore $15-4 = 11$ data points available for each calibration. After accounting for the three parameters in the Ceronio model, this left $11-3 = 8$ degrees of freedom, which is considered adequate for the purpose of reliable model calibration. The few zero counts in channel 11 were replaced by values of "1" to prevent the calibration procedure from trying to take the logarithm of zero.

3.3 Comparison of the Ceronio and variable-β models

Both the Ceronio and variable-β models were calibrated for each of the 321 samples. The goodness of fit for each sample was determined from the sum of squares SS, i.e. the sum of the squared differences between the logarithm of the actual count and logarithm of the modelled count. Figures 2 and 3 show the cumulative distributions for the SS for the Zuikerbosch and Vereeniging samples respectively. They clearly show the improvement in fit brought about by the Ceronio model. The sums of squares were reduced by 30 to 40 % in all cases, the improvement thus being about the same for both treatment plants, and for both raw and treated water.

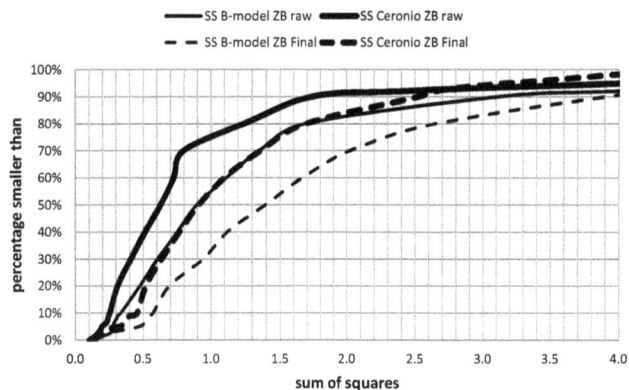

Figure 2. Cumulative sum of square for the ZB treatment plant.

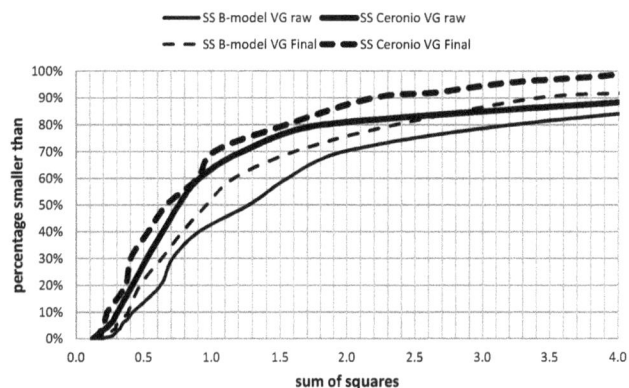

Figure 3. Cumulative sum of square for the VG treatment plant.

4 Discussion of the Ceronio model

Some of the samples included in Figs. 2 and 3 were not modelled very accurately, as evidenced by their large sum of squares. For this section, where general guidelines for the Ceronio model constants are discussed, the data had to be further screened to include only those samples which could be modelled within a sum of squares of 2. This filtering step removed 56 (17 %) samples from the data set, about evenly spread amongst the four sampling positions, which left 265 samples (VG Raw: $n = 65$; VG Final: $n = 67$; ZB Raw: $n = 69$; ZB Final: $n = 64$). The parameter values of these samples were used to determine the cumulative distributions for the four sampling points discussed below.

Parameter A determines the height of the size distribution, as shown in Fig. 4. As expected for a surface water impoundment subject to sharp seasonal turbidity variations, this parameter covers a broad range. The value of A corresponds directly to the normalised count at $d = 1$ μm, similar to its interpretation for the variable-β model. From Fig. 5, the raw water samples had A-values about two orders of magnitude higher than the final treated water samples.

Parameter β, as shown in Fig. 6, determines the curvature of the size distribution. The interesting observation from

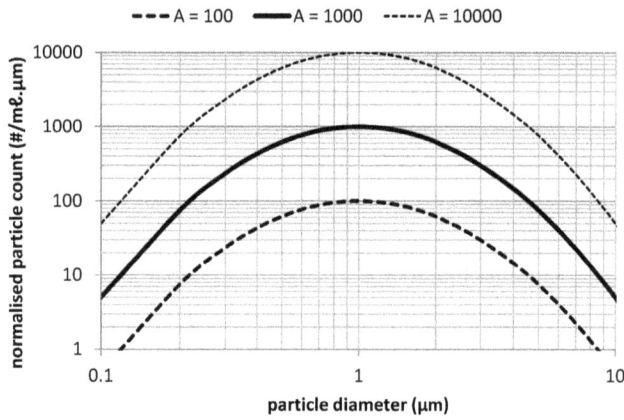

Figure 4. The effect of model parameter A on the particle size distribution, with $\beta = -1$ and $C = 0$.

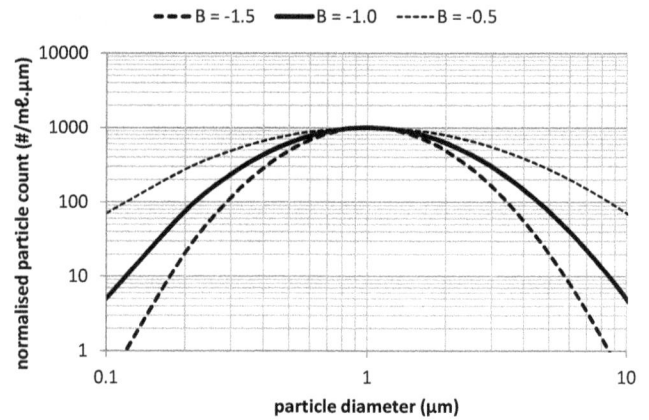

Figure 6. The effect of model parameter β on the particle size distribution, with $A = 1000$ and $C = 0$.

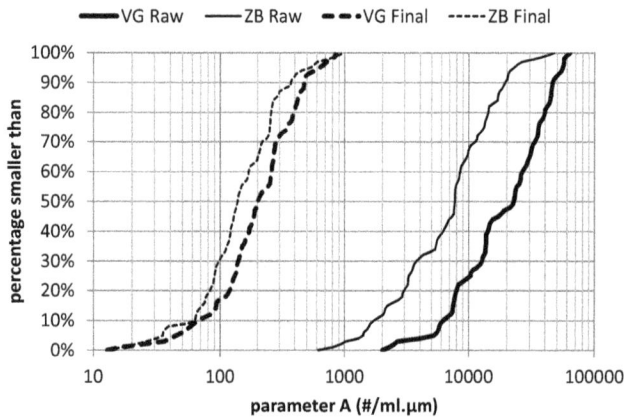

Figure 5. The cumulative distribution of model parameter A.

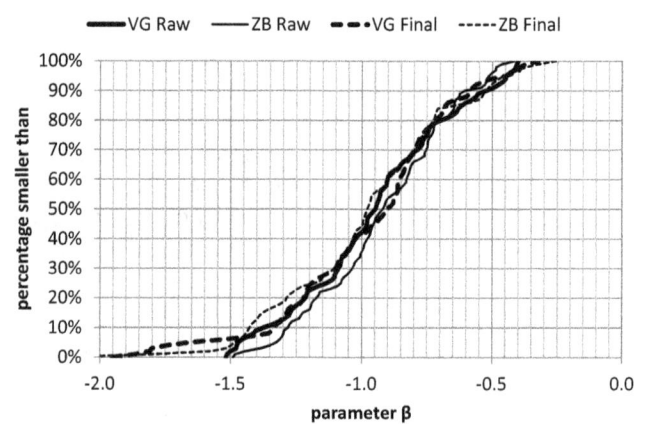

Figure 7. The cumulative distribution of model parameter β.

Fig. 7 is that the cumulative distributions for the raw and treated waters are not different. Using the 10th and 90th percentiles as guidelines, the range of β was from -1.4 to -0.5.

Parameter C moves the size distribution from left to right, as shown in Fig. 8. Figure 9 indicates that the range of C is between -1.5 to $+1.5$. The C-values of the raw water samples lie consistently to the left of the C-value of the treated water samples, indicating that the raw waters had relatively more small particles than the treated water samples. The median value of d_{max} for the raw water samples is at $0.8 \, \mu m$ (10th percentile $0.4 \, \mu m$, 90th percentile $1.6 \, \mu m$). Although the median value of d_{max} for the treated water samples is at $1.0 \, \mu m$ (10th percentile $0.3 \, \mu m$; 90th percentile $1.7 \, \mu m$), its variability emphasises the weakness of the variable-β model, which forces d_{max} to be $1 \, \mu m$ in all cases.

It is pointed out that the d_{max} values discussed in the previous paragraph are values predicted from the Ceronio model, and are not directly supported by the particle counting data. The smallest particles that could be counted, due to the technological limitations of the particle counter, are in the interval between 1 and $2 \, \mu m$, which are characterised by a mean

particle diameter of $1.4 \, \mu m$. This means that the values for d_{max} are mostly just smaller than the smallest particles that could actually be counted, with no data points to validate the shape and position of the apex of the predicted particle count. This is an unfortunate limitation, which of course applies equally to the validation of the variable-β and Ceronio models. This implies that, until this validation can be done by including smaller particles, not too much weight should be lent to the exact value of d_{max}. For purposes involving particle sizes above $2 \, \mu m$, the general conclusions regarding the Ceronio constants are firmly supported by the data sets used.

5   Summary and conclusions

- Every sample processed by a modern electronic particle counter, produces an extensive list of numbers specific to the particular instrument setting. Such numbers are not easy to interpret, and also do not allow comparison

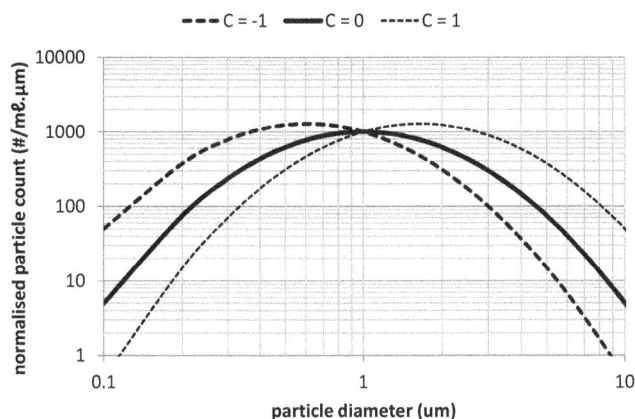

Figure 8. The effect of model parameter C on the particle size distribution, with $A = 1000$ and $\beta = -1$.

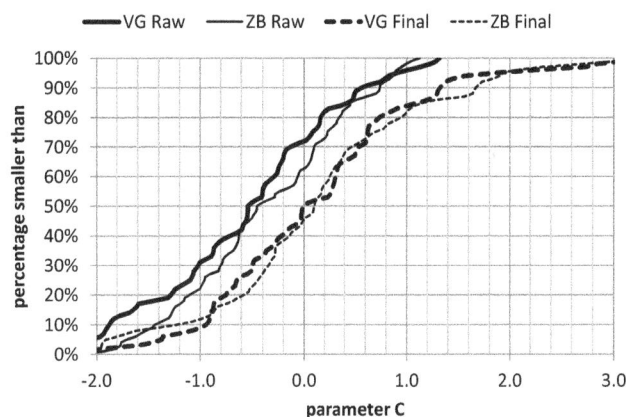

Figure 9. The cumulative distribution of model parameter C.

amongst particle counting results done with different instrument settings.

- Mathematical models present an opportunity to compress the particle counting data into two or three model parameters, which are much more amenable to analysis and interpretation. Moreover, the models are calibrated using all the channels with non-zero counts, thus providing a more robust description than could be obtained from using single counts from individual channels.

- The power law, commonly used, has two serious shortcomings which prohibit its use at both ends of the size spectrum. As the size gets smaller, the number of particles approaches infinity; as the size gets larger, the volume of particles approaches infinity.

- The variable-β model removes the shortcomings of the power law, but has its own limitation. Regardless of the sample, the maximum normalised particle count is always found at a size of 1 μm.

- This paper developed an earlier proposal to remove the limitation of the variable-β model, here called the Ceronio model. By introducing a third model parameter, the Ceronio model can describe a distribution with the maximum normalised count at any size.

- Using 321 samples of both raw and treated water from a large South African drinking water supplier, the Ceronio model consistently provided a better fit than the variable-β model, reducing the sum-of-squares by between 30 % and 40 %.

- A systematic analysis of the three parameters describing the Ceronio model provided useful insights. Parameter A is a parameter closely related to the number of particles in a sample; parameter B provides a measure of the ratio of smaller to larger particles; and parameter C

predominantly influences at which size the maximum normalised particle count is found.

- Potentially, the most important predictions of the Ceronio model deals with those small particles between 0.1 μm and 2 μm. This is the size range where the maximum normalised particles counts are found, and also where the fundamental transport mechanisms of particles in water are at their weakest. Unfortunately, the particle counter used could not measure sufficient data points in this range to provide solid experimental verification of the Ceronio model, indicating an important research need.

- For larger particles, the Ceronio model provides a robust tool for the computation and comparison of particle numbers and volumes at any desired particle size interval.

Acknowledgements. The authors express their gratitude to Rand Water for making available the particle counting data and for assistance in the preparation of this paper.

References

Ceronio, A. D. and Haarhoff, J.: An improvement on the power law for the description of particle size distribution in potable water treatment, Water Res., 39, 305–313, 2005.

Lawler, D. F.: Particle size distributions in treatment processes: theory and practice. Proceedings of the Fourth International Conference: The role of particle characteristics in separation processes, IAWQ-IWSA Joint Specialist Group on Particle Separation, Jerusalem, 29–30 October 1996, 1997.

Lewis, C. M. and Manz, D. H.: Light-scatter particle counting: Improving filtered-water quality, J. Environ. Eng., 117, 15 pp., 1991.

Sheldon, R. W.: Measurement of phytoplankton growth by particle counting, Limnol. Oceanogr., 24, 760–767, 1979.

Wilczak, A., Howe, E. W., Aleta, E. M., and Lee, R. G.: How pre-oxidation affects particle removal during clarification and filtration, Journal of the American Water Works Association, 84, 85–94, 1992.

4

Modelling water quality in drinking water distribution networks from real-time direction data

S. Nazarovs, S. Dejus, and T. Juhna

Riga Technical University, Riga, Latvia

Correspondence to: S. Nazarovs (nazarow@yahoo.com)

Abstract. Modelling of contamination spread and location of a contamination source in a water distribution network is an important task. There are several simulation tools developed, however the significant part of them is based on hydraulic models that need node demands as input data that sometimes may result in false negative results and put users at risk. The paper considers applicability of a real-time flow direction data based model for contaminant transport in a distribution network of a city and evaluates the optimal number of flow direction sensors. Simulation data suggest that the model is applicable for the distribution network of the city of Riga and that the optimal number of sensors in this case is around 200.

1 Introduction

Development of comprehensive tools for simulation of contamination transport and location of the contamination source in a water distribution network is a subject of scientific and engineering interest. In case of contamination outbreak contamination transport modelling tools enable more accurate determination of the area that is affected by the outbreak and requires cleaning. Modelling tools can also be used for locating the source of contamination for further elimination of the source and improvement of the network security that is particularly important in case of deliberate contamination of a water distribution network. Reports of deliberate contamination of networks in the past are summarized by Gleik (2006). Ostfeld (2005) presented a review of water quality modelling methods.

Contamination transport simulation tools research started in the early 80's (Ostfeld, 2005) and aims in three main directions:

– Development of governing equations for transport of contaminating agents, interactions with walls and reactions.

– Development of methods for solution of the equations.

– Development of practical network models capable of predicting which parts of network can be affected once the contamination is spotted at some point of the network.

The development of comprehensive mathematical models was focused on various types of contaminants; for modelling chlorine decay (Rossman et al., 1994; Clark et al., 1995; Ozdemir and Ger, 1999; Al-Omari and Chaudhry, 2001; Ozdemir and Ucak, 2002) and trihalomethanes formation (Clark, 1998; Elshorbagy et al., 2000; Li and Zhao, 2005) in water distribution networks. To model bacteria spread and regrowth in water distribution systems, Digiano and Zhang (2004) developed a mechanistic model. A mechanistic model with higher distinction between attached and bulk bacteria was proposed by Munavalli and Mohan Kumar (2005).

Mechanistic models for bacterial or chemical contamination have a significant drawback that impedes practical application of the models in the wake of a contamination incident. The models are based on equations of hydraulics, in other words a hydraulic model is used to calculate flows and pressure distribution in the system. Precise modelling of hydraulics requires accurate information about water demand in each node of a network. Therefore a water distribution network model is usually calibrated using flow and pressure data from the network as well as estimates of demand loads, based on population density and typical consumption patterns. Although a calibrated model describes network behavior well during normal operation, it may be inapplicable at small timescale like hours and minutes after contamination incident. Water demand in different parts of the network may fluctuate significantly at such a small timescale. In case contamination is detected in the system, fluctuations of demand patterns will be caused by issue of public health notices and directives in the first place. Changed colour and odour of contaminated water may prevent citizens from using water and contribute to demand reduction too.

The effect of above said demand fluctuations was demonstrated by Davidson et al. (2005) by simulating a contamination scenario close to one that occurred in Glasgow, Scotland in December 1997 where diesel fuel got into the water distribution network. Simulation results indicate that in the scenario involving issue of health notices and reduction of demand in a part of the network by 30 %, flow reversal in some pipes takes place that is not taken into account by the model with static demand.

Therefore flow patterns may change significantly due to reduction of demand after issue of public health notices. Public health notices, in particular an advice to boil water before drinking, are to be published in newspapers in case bacterial contamination is detected in the samples. If one of the water treatment plants gets contaminated, citizens in the area, the plant supplies water to, will be warned and advised not to drink tap water. If this happens, water consumption in the area may drop and flow direction in some pipes will be reversed allowing contaminated water to enter the areas normally supplied by other water treatment plants. Pipes where flow reversal is possible due to reduced consumption because of day-night cycle or health notices are the most sensitive parts of the network that may affect evaluation of contamination spread. Most of these pipes are located around the borders of segments supplied by different water treatment plants. Sensors may be installed on these pipes to monitor flow direction and see in which direction the contamination will spread.

Precise calculations of flow magnitude and direction in pipes after contamination incident may be virtually impossible. Errors in flow magnitude calculations will result in errors during simulation of contamination spread leading to false negative or false positive results at given segment of a water distribution network. The effect of false negative results involves some contaminated areas being declared clean thus putting general population at risk. False positive results may increase decontamination costs and cause unnecessary disruption in operation of water distribution network.

Therefore there is a need of more practical simulation tools based, to some extent, on real-time data rather than on estimated demands that would reduce probability of false negative and false positive results.

A practical approach to seeking a contamination source was offered by Besner at al. (2005). The idea is to gather statistical data on distribution system operation such as valve closures, repairs; consumer complains and water quality measurements. The statistical data can be used to track back the events that occurred in the system prior to a contamination incident, and, hopefully, indicate the contamination source. Nagib and Adel (2009) proposed to apply the predictive control algorithm to a water distribution system. The algorithm uses real-time data including parameters of active elements (e.g. pump speed, valve status) for modelling operation of the network. Kang and Lansey (2011) describe method of using real-time data (flow rates and pressure heads) for estimation of demands and pipe roughness coefficient.

An interesting approach to development of a practical tool for contamination tracking and source seeking has been presented by Davidson et al. (2005). The method eliminates need for demand estimations as real-time data on flow directions in pipes are used instead. The authors suggest that the size and location of the affected area in case of contamination incident depends primarily on flow directions in pipes rather than flow magnitudes.

The method proposed by Davidson et al. (2005) is based on the assumption that contamination can travel downstream only. Therefore, if flow direction sensors are installed in the network and real-time data on flow directions in pipes are available, it is possible to find all downstream pipes and junctions of the point where contamination has been detected. Given the contamination can travel downstream only, the downstream pipes are the only ones that may be contaminated. This does not necessarily mean that all downstream nodes and pipes will be contaminated, but the possibility of contamination cannot be ruled out for them. So real-time flow direction data makes it possible to tell how the contamination may be spreading from the node where it has been detected and also rule out parts of the water distribution network that are not located downstream of the contaminated node and therefore are not affected. In other words the proposed technique provides the worst-case scenario of possible contamination spread.

The flow direction data make it possible to find all upstream nodes and pipes too, so the data can provide suggestions where the contamination source may be located.

The data used for contamination spread simulation are the real-time data from several flow direction sensors as well as information on location of closed valves and check valves.

However the technique offered in the paper by Davidson et al. (2005) has a drawback. With too few sensors installed on the network the results may be inaccurate containing a large false positive error. By increasing the number of sensors one can improve accuracy of the model; however the costs of sensor installation and operation will also increase. Therefore a tradeoff should be found between installation costs of sensors and modelling accuracy.

The evaluation of the modelling error can be done with help of a hydraulic model of a water distribution network in

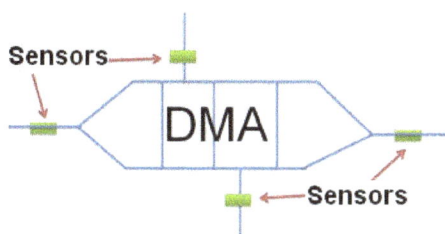

Figure 1. DMA concept.

the following way. Given the location of the contamination source is known, a water demand pattern can be fixed and simulations can be run on the network model so the affected part of the network is found for this pattern and the total contaminated pipe length is calculated. Once the affected area of the network has been determined, it is possible to select several pipes and assume that they have flow direction sensors. Then the simulations are run again taking into account that the flow direction in pipes without sensors is uncertain and if such a pipe is in contact with a contaminated node, the pipe is considered contaminated too. The total contaminated pipe length is calculated and the result is compared with the result of the simulation with the fixed demand pattern thus providing estimation of the error.

Choosing the optimal location of the sensors is crucial. One of the possible approaches is to install the sensors in such a way that the network is split into district metering areas (DMA) as shown in Fig. 1. The term "DMA" in this paper represents a part of water distribution system that is connected to the rest of the network through pipes with flow direction sensors. The more sensors are installed in the network, the smaller DMA areas may be. Each DMA has flow direction sensors on every pipe connecting it with other DMAs. There are no flow sensors inside the DMA.

Davidson et al. (2005) consider applicability of their proposed approach for a subdivision of a municipal water distribution network rather than for a whole city. The objective of this paper is to test applicability of this approach to a whole water distribution network of the city of Riga (Latvia) by running several contamination scenarios and find the optimal number of sensors for this network in order to obtain sufficient accuracy and reasonable installation and operation costs.

2 Methods

This paper considers several scenarios a bacterial contamination being introduced into the Riga distribution network. It is assumed that the bacterial contamination travels downstream with the flow and can also attach to pipe walls.

A model of the Riga water distribution network and demand pattern was used for simulations (Rubulis et al., 2012). The Riga network provides drinking water to approximately 700 000 inhabitants. The distribution network is supplied by three groundwater supply stations and by water treatment plant "Daugava" that takes water from the river Daugava. Water quality control and contamination detection is secured by a constant monitoring of water quality parameters in particular turbidity, pH and chlorine concentration at the exit of water treatment plants. Many contaminating agents are known to alter properties of water such as turbidity and pH (Inamori and Fujimoto, 2004).

Besides that bacterial contamination is monitored by taking water samples on a daily basis from various points of the network.

The Riga water distribution system model used in the paper includes 919 pipes and 574 junctions. The total length of the pipes in the model is 538 km. Pipes with diameter smaller than 200 mm are not included into the model. Three contamination scenarios with three different contamination sources were considered. Flow direction was found for each pipe in the model during the simulation, as if a flow direction sensor was installed on each of 919 pipes. All downstream pipes and nodes of the contamination source were found using the obtained flow direction data and total length of contaminated pipes was calculated by summing lengths of the downstream pipes. The result has been recorded and later used to normalize the data. Then some pipes were selected and marked as the ones having flow direction sensors. The pipes were selected in such a way that the network was split into DMAs. The placement of sensors was optimized. Several sites in the network where the sensors did not serve their purpose in an effective way because of very close location to each other were removed. To minimize the number of sensors installed, several optimization methods were used. First, for sites in the network where a link splits into 2 links with the same flow direction and sensors, the two sensors were removed and a new sensor was installed on the first link prior the splitting point (Fig. 2a). On sites where two consequent sensors were installed, the one that was closer to a treatment plant, was removed (Fig. 2b). It goes without saying that careful consideration should be made before each optimization step. After the sensor placement optimization step simulations were run again. Flow direction data from the first run were used for pipes marked with flow direction sensors. The pipes without sensors (the ones inside DMA) were considered contaminated if at least one pipe supplying the DMA was contaminated.

Flow direction sensors typically have a simple relay output. As soon as flow direction is changed or the flow is stopped, a signal may be sent to the data acquisition station via wireless communication device such as a GPRS modem. The time lapse between the moment when the flow direction is changed and the moment when the data are available to the user is mainly limited by data transfer rate through GPRS that may be relatively low (56 kilobit per second). However as amount of data sent from one sensor is just a few bytes (sensor address and status) time lapse between change in flow

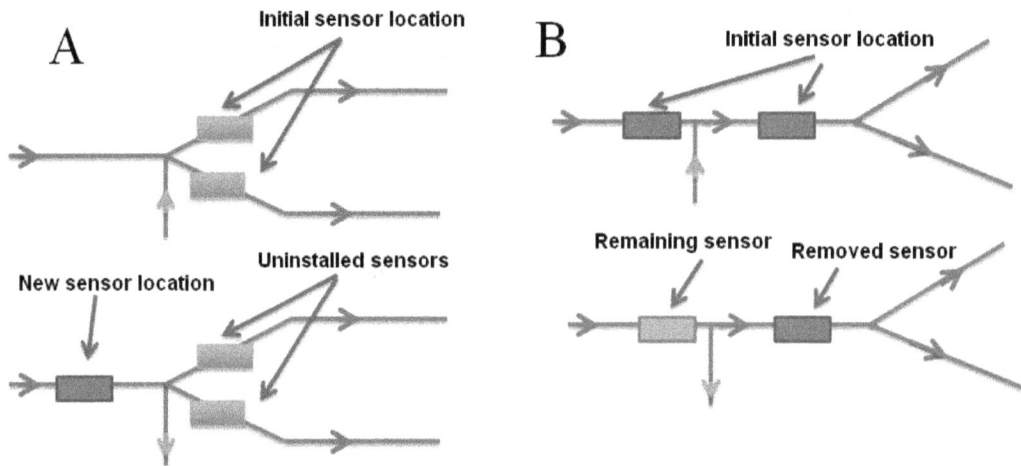

Figure 2. Methods of optimization of number of sensors.

Figure 3. Contamination sources for different scenarios.

direction and data availability in the evaluation tool is in the range of several seconds.

Flow direction sensors are typically activated when flow is about $0.01\,\mathrm{m\,s^{-1}}$ or more. Pipes with smaller flows will be considered as "zero flow" pipes.

Effectiveness of several sensor installation patterns was evaluated by running simulations of 3 different contamination scenarios. Scenario 1: water treatment plant "Daugava" that produces about 50 % of drinking water in Riga city gets contaminated. Scenario 2: one of water reservoirs of Riga city water distribution network gets contaminated. Scenario 3: random node of the water distribution network gets contaminated. Locations of contamination sources in the hydraulic model are shown in Fig. 3.

Simulations were made for several number of sensors. Case studies with 57, 67, 160, 185 and 207 sensors were considered. Total contaminated pipe length was calculated for each number of sensors for every scenario.

3 Results and discussion

The data presented in this section represent total contaminated pipe length obtained from simulations of three different contamination scenarios for various numbers of flow direction sensors. The data are normalized for each scenario by taking the total contaminated pipe length for 919 sensors as unity. Averaged data points are obtained by calculating an average of total contaminated pipe lengths of three scenarios

Table 1. Simulation results.

Number of sensors installed	Scenario 1				Scenario 2				Scenario 3				Average normalized contamination
	Calculated contamination			Normalized Length	Calculated contamination			Normalized Length	Calculated contamination			Normalized Length	
	Pipes	Length (km)	Nodes		Pipes	Length (km)	Nodes		Pipes	Length (km)	Nodes		
919	558	256.09	348	1.0	66	30.50	48	1.0	234	99.38	157	1.0	1.0
207	609	284.70	378	1.1	66	30.50	48	1.0	354	172.76	238	1.7	1.3
185	621	305.64	387	1.2	66	30.50	48	1.0	401	198.67	260	2.0	1.4
160	648	317.04	401	1.2	82	35.76	57	1.2	420	209.10	271	2.1	1.5
67	618	305.43	384	1.2	128	54.89	91	1.8	587	340.75	374	3.4	2.1
57	630	309.88	391	1.2	140	60.70	97	2.0	592	342.35	377	3.4	2.2

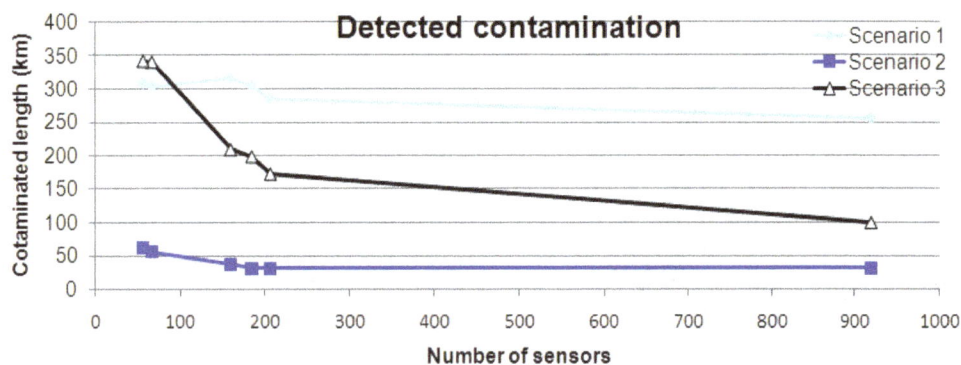

Figure 4. Simulated contamination length for various numbers of sensors.

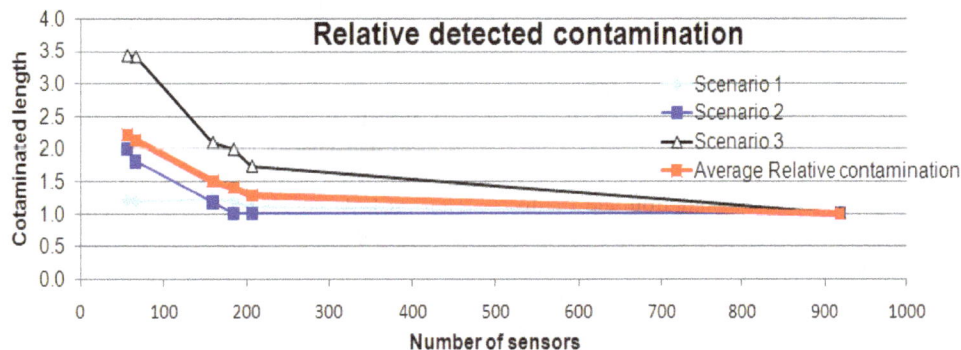

Figure 5. Normalized contamination length for various numbers of sensors.

for every number of sensors. The averaged curve is also normalized by taking average contaminated pipe length for 919 pipes.

The relationship between the number of sensors and total contaminated length for each scenario is presented in Fig. 4 and Table 1. Normalized data and average data for all three scenarios are presented in Fig. 5 and Table 1.

The simulation results suggest that optimal number of sensors for Riga network is around 200. Further increase in sensor number has little effect on simulated contamination length. For Riga model, installation of 207 sensors allows splitting the network into 25 DMAs. As mentioned be-

fore, sites where flows, supplied by different treatment plants, meet, are of special interest (Fig. 6). There are 67 such sites in Riga model.

The suggested flow direction sensor placement pattern for Riga network is shown in Fig. 7. Pipes with sensors installed are designated with triangles and marked in blue.

The obtained results demonstrate that number of sensors can be significantly reduced without major decrease of simulation results accuracy. According to Fig. 5, total simulated contaminated length on average increases only by about 20–30 % if number of sensors is reduced from 919 to 207 thus

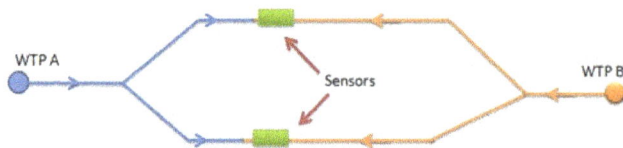

Figure 6. "Collision" sites for water from different treatment plants.

Figure 7. Suggested DMA distribution at Riga water distribution network. Sensor installation sites shown in blue.

more than four times cutting costs of installation and maintenance.

It should be also mentioned that flow direction sensors are relatively cheap and do not need sophisticated signal conditioning circuits. These properties of flow direction sensors allow reduction of costs compared to volumetric flow sensors.

4 Conclusions

Advances in communications technologies such as data transfer through GPRS stations and increasing calculation power of computers made it feasible to collect with practically meaningful frequency flow data from a water distribution network. Therefore it is possible to use real-time data from the network to track possible contamination spread as well as locate the source of contamination. Results presented in the paper suggest that flow direction-based method described by Davidson et al. (2005) may be applicable to the water distribution network of a city of a size of Riga (around 700 000 citizens). The network in this case can be split to

about 25 DMA areas that require around 200 flow direction sensors. Splitting the network into distribution management areas allows reducing number of sensors. Usage of flow direction sensors instead of volumetric or velocity sensors helps to reduce installation and maintenance costs. Number of sensors can also be reduced by optimizing location of sensors.

Acknowledgements. This work has been undertaken as a part of the research project SECUREAU (Nr. 217976) which is supported by the European Union within the 7th Framework Programme. There hereby follows a disclaimer stating that the authors are solely responsible for the work. It does not represent the opinion of the Community and the Community is not responsible for any use that might be made of data appearing herein.

References

Al-Omari, A. S. and Chaudhry, M. H.: Unsteady-state inverse chlorine modeling in pipe networks, J. Hydraul. Eng.-ASCE, 127, 669–677, 2001.

Besner, M. C., Gauthier, V., Trépanier, M., Leclair, M., and Prévost, M.: Interactive analyser for understanding water quality problems in distribution systems, Urban Water J., 2, 93–105, 2005.

Clark, R. M.: Chlorine demand and TTHM formation kinetics: a second-order model, J. Environ. Eng., 124, 16–24, 1998.

Clark, R. M., Rossman, L. A., and Wymer, L. J.: Modeling distribution system water quality: regulatory implications, J. Water Res. Pl.-ASCE, 121, 423–428, 1995.

Davidson, J., Bouchart, F., Cavill, S., and Jowitt, P.: Real-Time Connectivity Modeling of Water Distribution Networks to Predict Contamination Spread, J. Comput. Civil Eng., 19, 377–386, 2005.

Digiano, F. A. and Zhang, W.: Uncertainty analysis in a mechanistic model of bacterial regrowth in distribution systems, Environ. Sci. Technol., 38, 5925–5931, 2004.

Elshorbagy, W. E., Abu-Qdais, H., and Elsheamy, M. K.: Simulation of THM species in water distribution systems, Water Res., 34, 3431–3439, 2000.

Gleik, P. H.: Water and terrorism, Water Policy, 8, 481–503, 2006.

Inamori, Y. and Fujimoto N.: Physical/Mechanical Contamination of Water, Water Quality and Standards, Vol. 2, in: Encyclopedia of Life Support Systems (EOLSS), Developed under the Auspices of the UNESCO, Eolss Publishers, Oxford, UK, http://www.eolss.net, 2004.

Kang, D. and Lansey, K.: Demand and roughness estimation in water distribution systems, J. Water Res. Pl.-ASCE, 137, 20–30, 2011.

Li, X. and Zhao, H.: Development of a model for predicting trihalomethanes propagation in water distribution systems, Chemosphere, 62, 1028–1032, 2005.

Munavalli, G. R. and Mohan Kumar, M. S.: Water quality parameter estimation in a distribution system under dynamic state, Water Res., 39, 4287–4298, 2005.

Nagib, G. N. M. and Adel, A.: Water Supply Network System Control Based on Model Predictive Control, Proceedings of the

International MultiConference of Engineers and Computer Scientists (2), IMECS 2009, 18–20 March, Hong Kong, 2009.

Ostfeld, A.: A review of modeling water quality in distribution systems, Urban Water J., 2, 107–114, 2005.

Ozdemir, O. N. and Ger, A. M.: Unsteady 2-D chlorine transport in water supply pipes, Water Res., 33, 3637–3645, 1999.

Ozdemir, O. N. and Ucak, A.: Simulation of chlorine decay in drinking-water distribution systems, J. Environ. Eng., 128, 31–39, 2002.

Rossman, L. A., Clark, R. M., and Grayman, W. M.: Modeling chlorine residuals in drinking- water distribution systems, J. Environ. Eng., 120, 803–820, 1994.

Rubulis, J., Dejus, S., and Meksa, R.: Online measurement usage for predicting water age from tracer tests to validate a hydraulic model, Water Distribution Systems Analysis 2010 – Proceedings of the 12th International Conference, WSDA 2010, 1488–1497, 2012.

5

Numerical and experimental investigation of leaks in viscoelastic pressurized pipe flow

S. Meniconi, B. Brunone, M. Ferrante, and C. Massari

Dipartimento di Ingegneria Civile ed Ambientale, Università degli Studi di Perugia, Perugia, Italy

Correspondence to: S. Meniconi (silvia.meniconi@unipg.it) and B. Brunone (brunone@unipg.it)

Abstract. This paper extends the analysis concerning the importance in numerical models of unsteady friction and viscoelasticity to transients in plastic pipes with an external flow due to a leak. In fact recently such a benchmarking analysis has been executed for the cases of a constant diameter pipe (Duan et al., 2010), a pipe with a partially closed in-line valve (Meniconi et al., 2012a), and a pipe with cross-section changes in series (Meniconi et al., 2012b). Tests are based on laboratory experiments carried out at the Water Engineering Laboratory (WEL) of the University of Perugia, Italy, and the use of different numerical models. The results show that it is crucial to take into account the viscoelasticity to simulate the main characteristics of the examined transients.

1 Introduction

If the transients in a pressurized constant diameter pipe supplied by a constant head reservoir – hereafter referred to as single pipe – due to the instantaneous closure of a valve placed at the downstream end section, are simulated by means of the Allievi-Joukowsky analytical model (Fig. 1a, continuous line), no decay of the pressure peaks takes place after the end of the manoeuvre. In fact, in this case friction effects are neglected and an elastic behaviour of the pipe material is assumed. In Fig. 1, the time-history of the dimensionless piezometric head – hereafter referred to as dimensionless pressure signal – at the end section of the pipe, $h = (H - H_0)/\Delta H_{AJ}$ is shown, where $H =$ piezometric head, $\Delta H_{AJ} = a_i Q_0/(Ag)$ is the Allievi-Joukowsky overpressure, $a_i =$ instantaneous elastic pressure wave speed, $Q =$ discharge, $A =$ pipe area, $g =$ acceleration gravity, $\theta = t/\tau$ is the dimensionless time with $t =$ time evaluated from the beginning of manoeuvre, $\tau = 2L/a$ is the characteristic time of the pipe, $L =$ pipe length, $a =$ mean pressure wave speed, and the subscript 0 refers quantities to the initial conditions.

When friction forces in an elastic pipe are evaluated by means of the uniform flow formulas, i.e. within the so called steady-state approach (Fig. 1a, dashed line almost undistinguishable from the continuous line), no valuable differ-

ence occurs with respect to the Allievi-Joukowsky model in terms of decay and rounding of pressure peaks. Thus nor the Allievi-Joukowsky model nor the one based on the steady-state approach, simulate properly the strong decay and rounding of pressure peaks of experimental data both in elastic (Fig. 1b) and viscoelastic (Fig. 1c) pipes. This result has motivated the intense research activity in the field of unsteady friction (Adamkowski and Lewandowski, 2006; Bergant et al., 2001; Brunone et al., 1991, 1995; Brunone and Berni, 2010; Ghidaoui et al., 2005; Pezzinga, 2000, 2009; Storli and Nielsen, 2011; Zielke, 1968) and viscoelasticity (Covas et al., 2004, 2005; Ferrante et al., 2011; Franke and Seyler, 1983; Ghilardi and Paoletti, 1986; Meniconi et al., 2012a, b; Soares et al., 2008) modelling during transients in pressurized pipes in the last two decades. In fact the damping and rounding of pressure peaks in a single pipe are ascribed to the effect of unsteady friction in elastic pipes and to both unsteady friction and viscoelasticity in plastic ones. Only recently, Duan et al. (2010) have shown quantitatively that in plastic pipes the role of unsteady friction is relevant only in the first phases of the transients. In other words, the viscoelastic effect becomes more and more dominant with respect to unsteady friction, as time progresses.

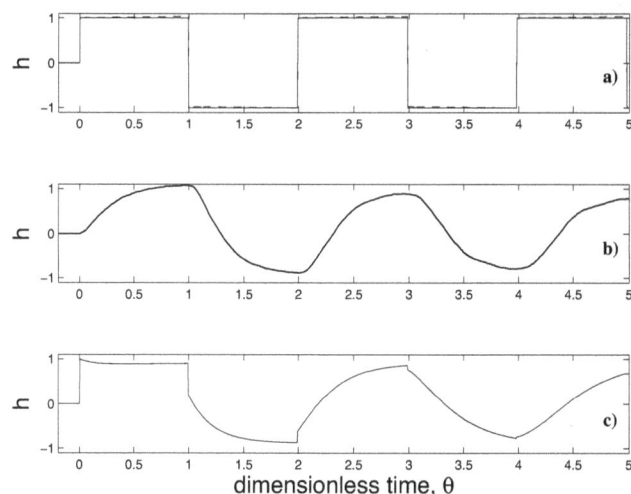

Figure 1. Dimensionless pressure signal in a single pipe due to the instantaneous closure of the end valve: (**a**) continuous line: Allievi-Joukowsky model (frictionless and elastic pipe) and dashed line: steady-state approach (steady-state friction and elastic pipe), (**b**) typical experimental data in elastic pipes, and (**c**) typical experimental data in viscoelastic pipes.

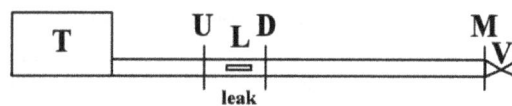

Figure 2. Sketch of the experimental setup (T = supply tank, U = section upstream of the leak, L = leak device, D = section downstream of the leak, M = section immediately upstream of the manoeuvre valve, V = manoeuvre valve).

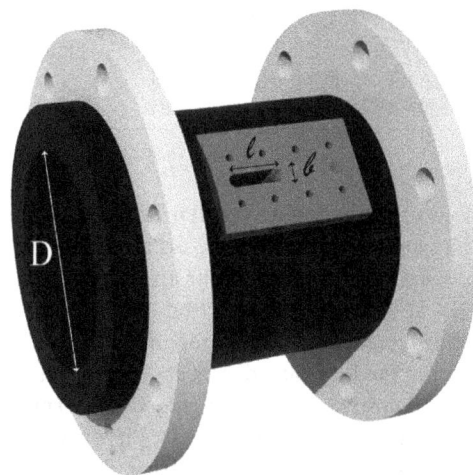

Figure 3. Device used to simulate leaks.

The interest in the simulation of transients in pressurized pipes is not due only to the importance of evaluating properly the extreme values of the pressure. In fact in the last two decades, starting from the pioneering paper by Liggett and Chen (1994), transient test-based techniques for leak detection and sizing have been more and more used because of their reliability and cheapness (Colombo et al., 2009). Within such techniques several distinctions can be made about the role played in the diagnosis procedure by the equations governing transients. Precisely, to locate and size leaks, the momentum and continuity equations can be solved and the numerical results are compared with the experimental data; alternatively, only the properties of the pressure waves are utilized. In the former approach, i.e., the Inverse Transient Analysis (Liggett and Chen, 1994), the governing equations are integrated in the time domain (e.g., Covas and Ramos, 2010) or converted into the frequency domain after having linearized the friction term and the nonlinear boundary conditions (e.g., the ones at the manoeuvre valve and at the leak) to reduce the needed amount of computer time (Covas et al., 2005; Ferrante and Brunone, 2003a; Lee et al., 2005a, b, 2006; Mpesha et al., 2001; Wang et al., 2002). In the latter approach, the characteristics of the possible leaks are inferred directly from transient data – in most cases pressure traces – by measuring the arrival time and the entity of pressure waves at the measurement sections, particularly those reflected by the leaks (Brunone, 1999; Brunone and Ferrante, 2001; Covas and Ramos, 2010; Ferrante and Brunone, 2003b; Ferrante et al., 2009a; Jönsson, 2001; Jönsson and Larson, 1992). A detailed literature review is presented in Colombo et al. (2009).

In the present paper, attention is focused on the numerical simulation of transients in a viscoelastic single pipe with external flow, i.e., a leak – hereafter referred to as damaged pipe.

Precisely, the interaction between a pressure wave and different leaks in a plastic pipe is examined during the first phases of the transients. In the first part, laboratory experiments are discussed by pointing out the effect of a leak; in the second part the results of different 1-D numerical models are compared.

2 Experimental setup

Experimental tests have been carried out at the Water Engineering Laboratory (WEL) of the University of Perugia, Italy. The experimental setup (Fig. 2) comprises a high density polyethylene (HDPE) pipe with $L = 166.28$ m, internal diameter $D = 93.3$ mm, nominal diameter DN110, and wall thickness $e = 8.1$ mm. This pipe connects the upstream tank to the downstream manoeuvre valve – ball valve DN50 – that discharges in the air.

To investigate the effect of a leak discharging into the atmosphere on the pressure signal, a new device (Fig. 3) with an orifice at its wall is installed at a distance $L' = 105.44$ m from the manoeuvre valve. With respect to the previous laboratory arrangement (Brunone and Ferrante, 2001; Ferrante and Brunone, 2003b; Ferrante el al., 2009a, b) the new

Table 1. Geometrical characteristics of leaks used in tests.

no.	steel plate with the hole	hole area, A_L (mm^2)
1		52.52
2		116.64

Figure 4. Damaged pipe with leak no. 1 (Table 1): pressure signal at section M for different values of the discharge at the end valve, Q_0.

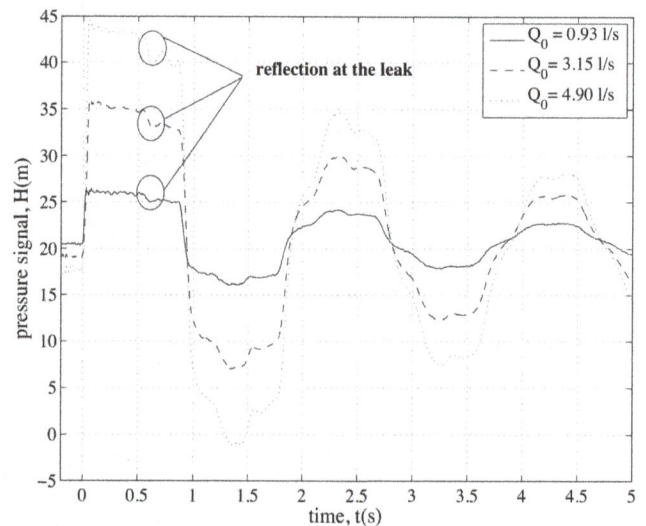

Figure 5. Damaged pipe with leak no. 2 (Table 1): pressure signal at section M for different values of the discharge at the end valve, Q_0.

device allows to simulate rectangular leaks of different size by changing the steel plate with the orifice (Table 1).

Pressure signal is measured with a frequency acquisition of 1024 Hz at four sections: section M, immediately upstream of the manoeuvre valve, sections D and U, placed downstream of the leak (at a distance $s_D = 97.50$ m from the end valve) and upstream of the leak ($s_U = 138.50$ m), respectively, and at the supply tank (Fig. 1). Piezoresistive transducers with a full scale of 3.5, 7 or 10 bar, depending on the pressure maximum value during the transient test, are used. The steady-state discharge at the end valve, Q_0, is measured by means of a magnetic flow meter.

3 Experimental pressure signals

Figures 4 and 5 show pressure signals, H, measured at section M in the damaged pipe with the leak no. 1 and no. 2, respectively. During tests, different values of Q_0, and then of ΔH_{AJ}, have been considered.

According to literature (e.g., Brunone and Ferrante, 2001; Covas and Ramos, 2010), for a given leak, the larger ΔH_{AJ}, the larger the pressure wave reflected by the leak.

In Fig. 6 the pressure signals in the single pipe and damaged pipes (with the leaks no. 1 and no. 2) for about the same Q_0 (~ 3 L s^{-1}), are compared. It can be noted the larger damping occurring in the damaged pipe, as well as that the larger the leak, the larger such a damping (Collins et al., 2012; Colombo et al., 2009).

4 1-D numerical models

According to literature (Covas et al., 2004, 2005; Franke and Seyler, 1983; Ghilardi and Paoletti, 1986; Meniconi et al., 2012a, b; Soares et al., 2008), the complete 1-D numerical model to simulate transients in pressurized viscoelastic pipes is based on the continuity equation:

$$\frac{\partial H}{\partial t} + \frac{(a_i)^2}{gA}\frac{\partial Q}{\partial s} + \frac{2(a_i)^2}{g}\frac{d\epsilon_r}{dt} = 0, \qquad (1)$$

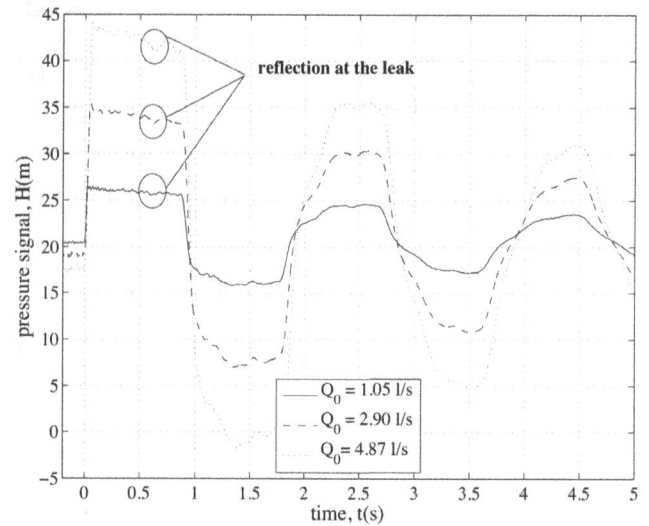

and the momentum equation:

$$\frac{\partial H}{\partial s} + \frac{Q}{gA^2}\frac{\partial Q}{\partial s} + \frac{1}{gA}\frac{\partial Q}{\partial t} + J = 0, \qquad (2)$$

with s = axial co-ordinate, J = total friction term, and ϵ_r = retarded strain. More details on the model and the calibration procedure are reported in Meniconi et al. (2011, 2012a, b). It is worthy of noting that J is regarded as the sum of two components. That is,

$$J = J_s + J_u, \qquad (3)$$

where J_s is the quasi steady-state value based on the instantaneous mean flow velocity, and J_u, the additional term

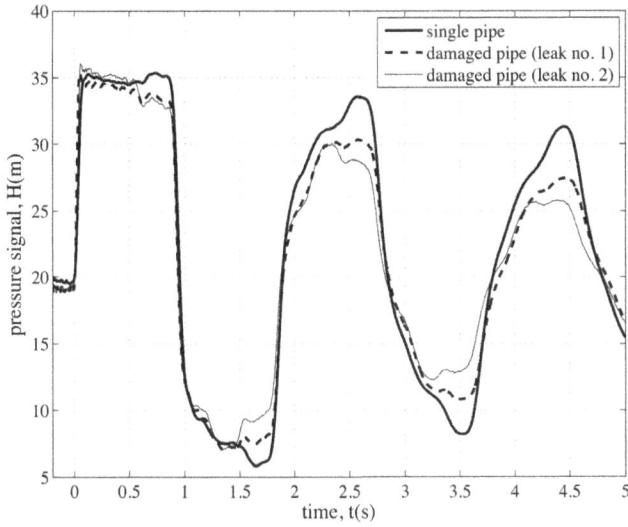

Figure 6. Pressure signal at section M for approximately the same value of the discharge at the end valve, Q_0 ($\cong 3\,\mathrm{L\,s^{-1}}$) in the case of: single pipe, damaged pipe with leak no. 1 and 2, respectively.

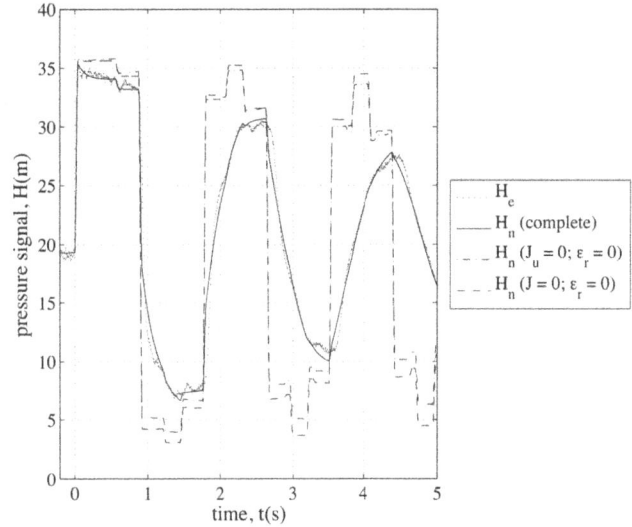

Figure 7. Damaged pipe with leak no. 1 ($Q_0 = 2.90\,\mathrm{L\,s^{-1}}$): experimental pressure signal at section M vs. numerical model simulations.

due to unsteadiness (Vardy and Brown, 2003, 2004; Ghidaoui et al., 2005), is evaluated within an instantaneous acceleration-based model by means of the following relationship (Brunone et al., 1991, 1995; Bergant et al., 2001; Pezzinga, 2000):

$$J_\mathrm{u} = \frac{k_\mathrm{d}}{2gA}\left(\frac{\partial Q}{\partial t} + a^\mathrm{i}\mathrm{sign}(Q)\left|\frac{\partial Q}{\partial s}\right|\right), \tag{4}$$

in which k_d = decay coefficient and sign $(\underline{Q})$ = (+1 for $Q \geq 0$ or −1 for $Q < 0$).

Such a complete model can be simplified by neglecting:

1. the unsteady friction ($J_\mathrm{u} = 0$);

2. the unsteady friction and the viscoelasticity ($J_\mathrm{u} = 0$ and $\varepsilon_\mathrm{r} = 0$);

3. the friction term and the viscoelasticity ($J = 0$ and $\varepsilon_\mathrm{r} = 0$).

The simplified model no. 1 takes into account the predominant effect of viscoelasticity (Duan et al., 2010; Meniconi et al., 2012a, b). The simplified model no. 2 coincides with the steady-state approach for transients in elastic pipes; whereas the last one derives from the classical Allievi-Joukowski theory.

In all models the boundary condition at the leak is given by the Torricelli equation:

$$Q_\mathrm{L} = C_\mathrm{L}A_\mathrm{L}\sqrt{2g(H_\mathrm{L} - z_\mathrm{L})}, \tag{5}$$

where C = discharge coefficient, and z = elevation, with the subscript L referring quantities to the leak.

5 Numerical experiments for a damaged pipe

Figures 7 and 8 show the comparison between experimental pressure signal, H_e, and the numerical simulations, H_n, given by the models described in Sect. 4 within the Method of Characteristics (Wylie and Streeter, 1993).

As for the single pipe (Fig. 1), when the simplified models in which the viscoelasticity is neglected are used – i.e., models no. 2 and no. 3 – the main characteristics of the experimental traces are not captured: nor the damping, nor the rounding. With regard to the pressure waves reflected by the leak, only the one occurring during the first characteristic time of the pipe is quite well simulated. On the contrary, the results given by the simplified model no. 1 (not reported in Figs. 7 and 8) are almost indistinguishable from those of the complete model. This confirms the predominance in plastic pipes of viscoelasticity with respect to unsteady friction also for a damaged pipe. A quantitative measure of such behaviours is given in Fig. 9, where, with regard to the first 5 s of the transient, the determination coefficient R^2 – denoting the strength of the association between H_e and H_n – is reported. For leak no. 1, R^2 assumes the same maximum value (= 0.984) for the complete and simplified model no. 1 ($J_\mathrm{u} = 0$). This means that the 98.84 % of the total variation in H_e can be explained by the linear relationship between H_e and H_n. The other 1.17 % of the total variation in H_e remains unexplained by the numerical models. The quality of the numerical simulation deteriorates when simplified models no. 2 ($J_\mathrm{u} = 0$ and $\varepsilon_\mathrm{r} = 0$) and no. 3 ($J = 0$ and $\varepsilon_\mathrm{r} = 0$) are used. Particularly, for the simplified model no. 2, R^2 breaks down to 0.498, whereas it becomes 0.475 for the simplified model no. 3. R^2 exhibits almost the same behaviour for the damaged pipe with leak no. 2.

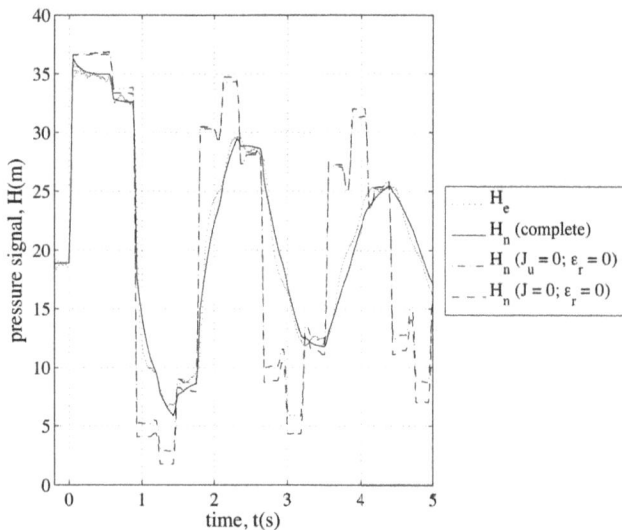

Figure 8. Damaged pipe with leak no. 2 ($Q_0 = 3.15 \, \mathrm{L \, s^{-1}}$): experimental pressure signal at section M vs. numerical model simulations.

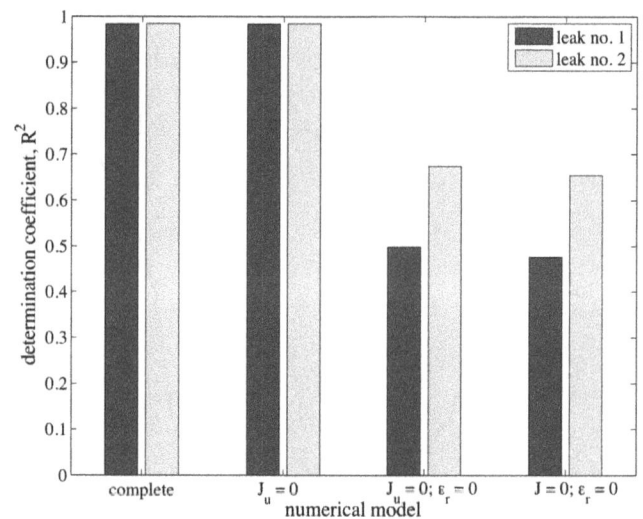

Figure 9. Determination coefficient of the different numerical models in the case of a damaged pipe, with the leak no. 1 (dark grey) and no. 2 (light gray).

6 Conclusions

This paper can be included in the research activity focused on the analysis of the role that unsteady friction and viscoelasticity play in the numerical simulation of transients in plastic pipes. In fact, up to few years ago there was no clear delineation between when and when not include such effects in transient flow numerical models not even for the simplest case of the constant diameter pipe (single pipe). In recent contributions (Duan et al., 2010; Meniconi et al., 2012a, b) such a benchmarking analysis has been executed for the cases of a single pipe, a pipe with a partially closed in-line valve, and a pipe with cross-section changes in series. The important case of a pipe with an external flow due to a leak deserved less attention.

In the first part of this paper the results of tests executed at the Water Engineering Laboratory (WEL) of the University of Perugia, Italy, with a more flexible device to simulate leaks, are discussed. Precisely, it is pointed out the effect of the leak on the pressure signal both in terms of the reflected pressure wave and the damping of the extreme values of the pressure.

In the second part, different numerical models are presented: the complete model, that takes into account both unsteady friction and viscoelasticity, and three simplified models in which the unsteady friction (model no. 1), both the unsteady friction and viscoelasticity (model no. 2), the friction term and the viscoelasticity (model no. 3) are neglected, respectively.

By comparing experimental and numerical results, it is demonstrated the crucial role played by the viscoelasticity with respect to unsteady friction also in the considered case of a plastic pipe with a leak. From a quantitative point of

view, such a behaviour of numerical models is pointed out by considering the value of the determination coefficient R^2, which denotes the strength of the association between the numerical and experimental pressure traces during a specified period of time.

Acknowledgements. This research has been supported by Fondazione Cassa Risparmio Perugia under the Project "Leaks and blockages detection techniques for reducing energy and natural resources wastage". The support of A. Cirimbilli, C. Del Principe and A. Adorisio in the laboratory activity is highly appreciated.

References

Adamkowski A. and Lewandowski, M.: Experimental examination of unsteady friction models for transient pipe flow simulation, J. Fluid. Eng., 128, 1351–1363, 2006.

Bergant, A., Simpson, A., and Vitkovsky, J.: Developments in unsteady pipe flow friction modelling, J. Hydraul. Res., 39, 249–257, 2001.

Brunone, B.: A transient test-based technique for leak detection in outfall pipes, J. Water Res. Pl., 125, 302–306, 1999.

Brunone, B. and Berni, A.: Wall shear stress in transient turbulent pipe flow by local velocity measurement, J. Hydraul. Eng., 136, 716–726, 2010.

Brunone, B. and Ferrante, M.: Detecting leaks in pressurised pipes by means of transients, J. Hydraul. Res., 39, 539–547, 2001.

Brunone, B., Golia, U. M., and Greco, M.: Some remarks on the momentum equation for fast transients, in: Proc. Int. Meeting on "Hydraulic Transients and Water Column Separation", edited by: Cabrera, E. and Fanelli, M., 201–209, 1991.

Brunone, B., Golia, U. M., and Greco, M.: Effects of two-dimensionality on pipe transients modeling, J. Hydraul. Eng., 121, 906–912, 1995.

Collins, R. P., Boxall, J. B., Karney, B. W., Brunone, B., and Meniconi, S.: How severe can transients be after a sudden depressurization?, J. Am. Water Works Ass., 104, E243–E251, 2012.

Colombo, A. F., Lee, P., and Karney, B. W.: A selective literature review of transient-based leak detection methods, J. Hydro-Environment Res., 2, 212–227, 2009.

Covas, D. and Ramos, H.: Case studies of leak detection and location in water pipe systems by inverse transient analysis, J. Water Res. Pl., 136, 248–257, 2010.

Covas, D., Stoianov, I., Mano, J., Ramos, H., Graham, N., and Maksimovic, C.: The dynamic effect of pipe-wall viscoelasticity in hydraulic transients. Part I – Experimental analysis and creep characterization, J. Hydraul. Res., 42, 516–530, 2004.

Covas, D., Stoianov, I., Mano, J., Ramos, H., Graham, N., and Maksimovic, C.: The dynamic effect of pipe-wall viscoelasticity in hydraulic transients. Part II – Model development, calibration and verification, J. Hydraul. Res., 43, 56–70, 2005.

Duan, H., Ghidaoui, M., Lee, P. J., and Tung, Y. K.: Unsteady friction and visco-elasticity in pipe fluid transients, J. Hydraul. Res., 48, 354–362, 2010.

Ferrante, M. and Brunone, B.: Pipe system diagnosis and leak detection by unsteady-state tests. 1 Harmonic analysis, Adv. Water Res., 26, 95–105, 2003a.

Ferrante, M. and Brunone, B.: Pipe system diagnosis and leak detection by unsteady-state tests. 2 Wavelet analysis, Adv. Water Res., 26, 107–116, 2003b.

Ferrante, M., Brunone, B., and Meniconi, S.: Leak detection in branched pipe systems coupling wavelet analysis and a lagrangian model, J. Water Supply Res. T., 58, 95–106, 2009a.

Ferrante, M., Brunone, B., and Meniconi, S.: Leak-edge detection, J. Hydraul. Res., 47, 233–241, 2009b.

Ferrante, M., Massari, C., Brunone, B., and Meniconi, S.: Experimental evidence of hysteresis in the head-discharge relationship for a leak in a polyethylene pipe, J. Hydraul. Eng., 137, 775–781, 2011.

Franke, P. and Seyler, F.: Computation of unsteady pipe flow with respect to visco-elastic material properties, J. Hydraul. Res., 21, 345–353, 1983.

Ghidaoui, M. S., Zhao, M., McInnis, D. A., and Axworthy, D. H.: A review of water hammer theory and practice, Appl. Mech. Rev., 58, 49–76, 2005.

Ghilardi, P. and Paoletti, A.: Additional visco-elastic pipes as pressure surges suppressors, in: Proc. 5th Int. Conf. on "Pressure Surges", 113–121, 1986.

Jönsson, L.: Experimental studies of leak detection using hydraulic transients, in: Proc. 29th IAHR Congress, 559–565, 2001.

Jönsson, L. and Larson, M.: Leak detection through hydraulic transient analysis, in: Proc. Int. Conf. on "Pipeline Systems", edited by: Coulbeck, B. and Evans, E., Kluwer Academic Publishers, 273–286, 1992.

Lee, P. J., Vitkovsky, J. P., Simpson, A. R., Lambert, M. F., and Liggett, J.: Frequency domain analysis for detecting pipelines leaks, J. Hydraul. Eng., 131, 596–604, 2005a.

Lee, P. J., Vitkovsky, J. P., Lambert, M. F., Simpson, A. R., and Liggett, J.: Leak location using the pattern of the frequency response diagram in pipelines: a numerical study, J. Sound Vib., 284, 1051–1073, 2005b.

Lee, P. J., Lambert, M., Simpson, A., and Vitkovsky, J. P.: Experimental verification of the frequency response method of leak detection, J. Hydraul. Res., 44, 451–468, 2006.

Liggett, J. A. and Chen, L.-C.: Inverse transient analysis in pipe networks, J. Hydraul. Eng., 120, 934–955, 1994.

Meniconi, S., Brunone, B., Ferrante, M., and Massari, C.: Long period analysis of transient pressure signals for in-line valve checkin, in: Proc. 11th Int. Conf. on "Computing and Control for the Water Industry – CCWI2011 Urban Water in Management: Challanges and Opportunities", Exeter, edited by: Savic, D. A., Kapelan, Z., and Butler, D., 3, 787–792, 2011.

Meniconi, S., Brunone, B., Ferrante, M., and Massari, C.: Transient hydrodynamics of in-line valves in viscoelastic pressurized pipes: long-period analysis, Exp. Fluids, 53, 265–275, 2012a.

Meniconi, S., Brunone, B., and Ferrante, M.: Water hammer pressure waves at cross-section changes in series in viscoelastic pipes, J. Fluid. Struct., 33, 44–58, doi:10.1016/j.jfluidstructs.2012.05.007, 2012b.

Mpesha, W., Gassman, S. L., and Chaudhry, M. H.: Leak detection in pipes by frequency response method, J. Hydraul. Eng., 127, 134–147, 2001.

Pezzinga, G.: Evaluation of unsteady flow resistances by quasi-2D or 1D models, J. Hydraul. Eng., 126, 778–785, 2000.

Pezzinga, G.: Local balance unsteady friction model, J. Hydraul. Eng., 135, 45–56, 2009.

Soares, A. K., Covas, D., and Reis, L. F. R.: Analysis of PVC pipe-wall viscoelasticity during water hammer, J. Hydraul. Eng., 134, 1389–1395, 2008.

Storli, P. T. and Nielsen, T. K.: Transient friction in pressurized pipes. II: Two-coefficient instantaneous acceleration-based model, J. Hydraul. Eng., 137, 679–695, 2011.

Vardy, A. and Brown, J.: Transient turbulent friction in smooth pipe flows, J. Sound Vib., 259, 1011–1036, 2003.

Vardy, A. and Brown, J.: Transient turbulent friction in fully rough pipe flows, J. Sound Vib., 270, 233–257, 2004.

Wang, X. J., Lambert, M. F., Simpson, A. R., Ligett, J. A., and Vitkovsky, J. P.: Leak detection in pipelines using the damping of fluid transients, J. Hydraul. Eng., 128, 697–711, 2002.

Wylie, E. B. and Streeter, V. L.: Fluid transients in systems, Prentice Hall, Englewood Cliffs, N.J., 1993.

Zielke, W.: Frequency dependent friction in transient pipe flow, J. Basic Eng., 90, 109–115, 1968.

6

Effects of ozonation and temperature on the biodegradation of natural organic matter in biological granular activated carbon filters

L. T. J. van der Aa[1,2], L. C. Rietveld[2], and J. C. van Dijk[2]

[1]Waternet, P.O. Box 94370, 1090 GJ Amsterdam, The Netherlands
[2]Delft University of Technology, P.O. Box 5048, 2600 GA Delft, The Netherlands

Abstract. Four pilot (biological) granular activated carbon ((B)GAC) filters were operated to quantify the effects of ozonation and water temperature on the biodegradation of natural organic matter (NOM) in (B)GAC filters. The removal of dissolved organic carbon (DOC), assimilable organic carbon (AOC) and oxygen and the production of carbon dioxide were taken as indicators for NOM biodegradation. Ozonation stimulated DOC and AOC removal in the BGAC filters, but had no significant effect on oxygen consumption or carbon dioxide production. The temperature had no significant effect on DOC and AOC removal, while it had a positive effect on oxygen consumption and carbon dioxide production. Multivariate linear regression was used to quantify these relationships. In summer, the ratio between oxygen consumption and DOC removal was approximately 2 times the theoretical maximum of $2.6\,\mathrm{g\,O_2\,g\,C^{-1}}$ and the ratio between carbon dioxide production and DOC removal was approximately 1.5 times the theoretical maximum of $3.7\,\mathrm{g\,CO_2\,g\,C^{-1}}$. The production and loss of biomass, the degassing of (B)GAC filters, the decrease in the NOM reduction degree and the temperature effects on NOM adsorption could only partly explain these excesses and the non-correlation between DOC and AOC removal and oxygen consumption and carbon dioxide production. It was demonstrated that bioregeneration of NOM could explain the excesses and the non-correlation. Therefore, it was likely that bioregeneration of NOM did occur in the (B)GAC pilot filters.

1 Introduction

The main reasons to use activated carbon (AC) in drinking water treatment are for the removal of organic micro-pollutants, for the removal of precursors for disinfection byproducts and for the removal of organic compounds causing color, taste and odor. Granular activated carbon (GAC) filtration is often preceded by a pre-oxidation step such as ozonation. Pre-oxidation is used for disinfection, oxidation of micro-pollutants and the improvement of color, taste and odor (Kruithof and Maschelein, 1999). Another important effect of pre-oxidation is that it increases the biodegradability of natural organic matter (NOM) (Carlson and Amy, 1997; Hammes et al., 2006; Rietveld et al., 2008; Siddiqui et al., 1997; van der Kooij et al., 1989; Yavich et al., 2004).

This enhances the growth of biomass in subsequent treatment steps and results in biological granular activated carbon (BGAC) filtration, also known as biological activated carbon filtration (Graveland, 1994; Jekel, 1979; Sontheimer et al., 1988). In BGAC filters, adsorption and biodegradation of NOM occur simultaneously. Biomass grows on the external surface of the AC grains and in the macro-pores. Micro-pores are too small for bacteria to enter (Labouyrie et al., 1997; Walker and Weatherley, 1998). Adsorption of NOM seriously hinders the adsorption of organic micro-pollutants due to pore blocking, competition and pre-loading (Carter et al., 1992; Carter and Weber, 1994; Knappe et al., 1997). When some of the NOM is biodegraded, adsorption of organic micro-pollutants is less hindered. As a result BGAC filtration is found to be more effective than GAC filtration for the removal of NOM (Graveland, 1994; Sontheimer et al., 1988), atrazine (Orlandini, 1999), micro-cystines (Wang et al., 2007) and other organic micro-pollutants (van der Aa, 2011). If the biodegradable NOM, produced by the pre-oxidation, is not sufficiently removed in subsequent treatment steps, this will have a negative impact on the biostability of the treated water. In the Netherlands, drinking water is often distributed without any residual disinfectant. Therefore, much attention is given to biofilm formation and regrowth in the distribution system (Escobar et al., 2001; Servais et al., 2004; van der Kooij, 1992). In BGAC filters, biomass has a large influence on the removal of biodegradable NOM and also, therefore, on biostability.

NOM consists of a mixture of compounds (Leenheer and Croué, 2003). Heterotrophic bacteria take up some of these organic substances to build up cell material (assimilation) and to oxidize it to generate energy for growth and maintenance (dissimilation). During the biodegradation of NOM, heterotrophic bacteria consume oxygen and produce carbon dioxide and water (van der Kooij et al., 1982b). Aerobic biodegradation can be assessed by measuring oxygen consumption (Urfer and Huck, 2001) and carbon dioxide production. Some of the NOM compounds are biodegraded easily, while others are hardly biodegraded at all. In natural waters used for drinking water production, NOM is usually not easily biodegradable because bacteria already have consumed the easily biodegradable part. Pre-oxidation increases the biodegradability of NOM, indicated by increased concentrations of biodegradable dissolved organic carbon (BDOC), easily assimilable organic carbon (AOC) and organic acids, such as oxalate, formiate and acetate (Carlson and Amy, 1997; Hammes et al., 2006; Siddiqui et al., 1997; van der Kooij et al., 1989; Yavich et al., 2004). AOC is typically less than 10% of the BDOC in a water sample (Hammes et al., 2006; van der Kooij et al., 1982a, 1989). It is assumed that in drinking water the carbon source is usually the limiting factor for biodegradation by heterotrophic bacteria. However, in some cases also phosphate limitation was reported (Juhna and Rubulis, 2004).

Biodegradation in (B)GAC filters increases with increased water temperature and with empty bed contact time (EBCT). Exhausted (B)GAC filters typically remove 10% to 35% of the NOM from the influent (Graham, 1999; Sontheimer et al., 1988). However, extreme removal ratios were reported from practically no dissolved organic carbon (DOC) removal (Juhna and Rubulis, 2004) to 65% removal (Jekel, 1979).

Nowadays, many drinking water treatment plants (DWTPs) have installed (B)GAC filters. Because influent water quality and operations vary, the need for dynamic models for optimization of operational conditions has increased (van der Helm et al., 2008; Bosklopper et al., 2004). Because biodegradation of NOM is a key process in BGAC filtration, it should be incorporated in such models. The objective of this study was to quantify the effects of ozonation and water temperature on the biodegradation of NOM in (B)GAC filters. Because reported biodegradation in (B)GAC filters varies a lot, this study was conducted specifically for the Weesperkarspel pilot plant. One GAC and three BGAC pilot filters were operated to treat surface water. After approximately half a year of operation, the (B)GAC filters were assumed not to adsorb NOM any more. The removal of DOC, AOC and oxygen and the production of carbon dioxide were taken as indicators for NOM biodegradation. An additional objective was to determine the correlation between these indicators.

2 Materials and methods

2.1 Weesperkarspel pilot plant: surface water

At Loenen – Weesperkarspel DWTP of Waternet, the water cycle company of Amsterdam and surrounding areas, the raw water originates from seepage water from the Bethune polder, with a high concentration of humic substances (Baghoth et al., 2008). The DOC concentration varies between 8 and $10 \, \mathrm{g \, C \, m^{-3}}$ and the water temperature between 3 and 21 °C. This water is pre-treated in the full-scale plant with coagulation, sedimentation, a reservoir with an average residence time of 100 days and rapid sand filtration. The pilot plant consisted of three ozone bubble columns plus contact chambers, followed by four identical gravity filters operated in down-flow mode. The influent waters of filters W-GAC8, W-BGAC5, W-BGAC6 and W-BGAC7 received net ozone doses of 0, 0.5, 1.5 and $2.5 \, \mathrm{g \, O_3 \, m^{-3}}$, respectively. The filter medium in the (B)GAC filters was Norit GAC830, with a density of $450 \times 10^3 \, \mathrm{g \, AC \, m^{-3}}$. The flow rates were kept constant at $0.88 \times 10^{-3} \, \mathrm{m \, s^{-1}}$ and the EBCTs were 40 min. Figure 1 and Table 1 provide an overview of the pilot plant and its characteristics. During operation, the (B)GAC filters were backwashed with air and water every 4 to 20 days, depending on filter bed resistance. The columns were monitored twice per month on influent, effluent and over-the-bed height. The measured parameters were DOC, extinction of ultraviolet light at a wavelength of 254 nm, AOC, pH, oxygen and temperature.

2.2 Analytical methods

Ozone was analyzed according to the diethyl-p-phenylenediamine method (Gilbert, 1981). AOC was measured in duplicate, applying the simultaneous incubation of strains P17 and NOX (van der Kooij et al., 1982b). For measurements of adenosine tri-phosphate (ATP) on AC, the method described by Magic-Knezev and van der Kooij (2004) was used. Other water quality parameters were determined according to standard procedures (Standard Methods, 1998). Carbon dioxide was not analyzed, but calculated from the carbonic equilibrium using pH, conductivity, temperature and bicarbonate concentration (van Schagen et al., 2008).

Table 1. Characteristics and influent water qualities pilot plant Weesperkarpsel.

Column	W-GAC8	W-BGAC5	W-BGAC6	W-BGAC7
Column diameter [m]	0.25	0.25	0.25	0.25
Bed height [m]	2.10	2.10	2.10	2.10
Carbon type [–]	GAC830	GAC830	GAC830	GAC830
Operation time [s]	586·24·3600	600·24·3600	559·24·3600	586·24·3600
Flow [$m^3 s^{-1}$]	4.3×10^{-5}	4.3×10^{-5}	4.3×10^{-5}	4.3×10^{-5}
Surface load [$m s^{-1}$]	8.7×10^{-4}	8.7×10^{-4}	8.7×10^{-4}	8.8×10^{-4}
EBCT [s]	40·60	40·60	40·60	40·60
Net ozone dose [$g O_3 m^{-3}$]	–	0.5	1.5	2.5
Temp [°C]	12.8 (3.0–20.6)	12.2 (3.6–20.6)	11.6 (3.8–20.6)	11.6 (3.8–20.6)
pH [–]	7.7 (7.3–7.9)	7.7 (7.3–7.8)	7.7 (7.1–7.8)	7.7 (7.6–7.8)
DOC [$g C m^{-3}$]	6.0 (5.0–6.9)	6.0 (5.2–7.6)	5.9 (4.9–7.5)	5.8 (4.9–6.7)
UV254 [m^{-1}]	14.4 (11.7–17.8)	13.5 (11.6–16.9)	10.8 (5.4–14.4)	8.5 (6.4–10.4)
Oxygen [$g O_2 m^{-3}$]	10.0 (7.9–12.9)	11.0 (8.3–14.2)	11.1 (7.4–13.7)	11.6 (9.5–14.3)
Carbon dioxide [$g CO_2 m^{-3}$]	8.1 (4.6–10.7)	7.7 (0.1–15.7)	8.6 (5.8–39.7)	7.2 (5.5–10.1)
AOC [$10^{-3} g$ acetate-C m^{-3}]	11.5 (5.7–32)	35.0 (16–99)	85.0 (40–199)	114.1 (42–265)

Notes water quality parameters: table values are average concentrations; values between parentheses are minimum and maximum concentrations; carbon dioxide concentrations calculated according (van Schagen et al., 2008).

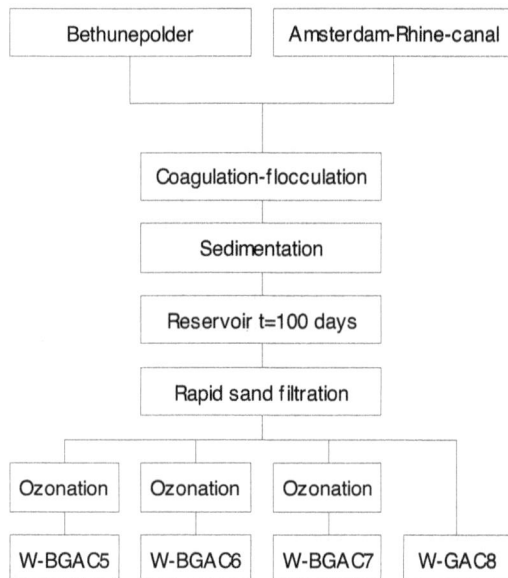

Figure 1. Flow scheme at the Weesperkarspel pilot plant.

2.3 Statistical methods

The Matlab® procedure "anova1" (Mathworks, 2007) was used to study differences between datasets. The datasets were considered significantly different from each other, when the p-value was smaller than 0.05, which means that the chance that the different datasets did not originate from the same distribution function is more than 95%.

The Matlab® procedure "robustfit" (Mathworks, 2007) was used to make linear regressions between different datasets. The procedure "robust" was used to assign lower weights to outliers. The input variables were considered to correlate significantly with the target variables, when the p-value was smaller than 0.05.

Multivariate linear regressions (MLRs) were used to determine which input variable(s) from a group of input variables influenced the target variable. The Matlab® procedure "robust" (Mathworks, 2007) was used to assign lower weights to outliers. For each target variable, MLRs were made for all possible combinations of the input variables. MLRs that included one or more input variables with no significant correlation to the target variable (p-value larger than 0.05) were rejected. The MLRs with the lowest root mean sigma error (RMSE) were considered the best MLRs.

3 Results and discussion

Figure 2 shows the concentrations of DOC, AOC and oxygen and the pH of the (B)GAC filters at the Weesperkarspel pilot plant at different ozone doses. The minimum, maximum and average influent concentrations are included in Table 1.

3.1 Effects of ozonation and water temperature on water quality

Based on the water quality after ozonation, MLRs were made for the following target parameters: DOC, AOC, oxygen, carbon dioxide and pH (see Table 2). The following input parameters were used: ozone dose, water temperature, DOC, AOC, oxygen, carbon dioxide and pH, all from the raw water. All target parameters after ozonation correlated significantly

Table 2. MLRs water quality parameters after ozonation.

MLR – Robust Input variables: O_3, temp, $[DOC]_{raw}$, $[AOC]_{raw}$, $[O_2]_{raw}$, $[CO_2]_{raw}$	R^2	RMSE
$[DOC]_{inf} = 0.51 + 0.92 [DOC]_{raw} - 0.070 O_3$	0.60	0.14
$[AOC]_{inf} = 1.0 \times 10^{-3} + 1.0 [AOC]_{raw} + 35 \times 10^{-3} O_3$	0.62	10×10^{-3}
$[O_2]_{inf} = 0.46 + 0.98 [O_2]_{raw} + 0.57 O_3$	0.86	0.49
$[CO_2]_{inf} = 0.44 + 0.94 [CO_2]_{raw} - 0.32 O_3$	0.15	0.35
$pH_{inf} = 0.17 + 0.98 pH_{raw} + 0.014 O_3$	0.48	0.013

Notes: $_{raw}$ indicates raw water (i.e., water before ozonation) and $_{inf}$ indicates (B)GAC influent (i.e., water after ozonation).

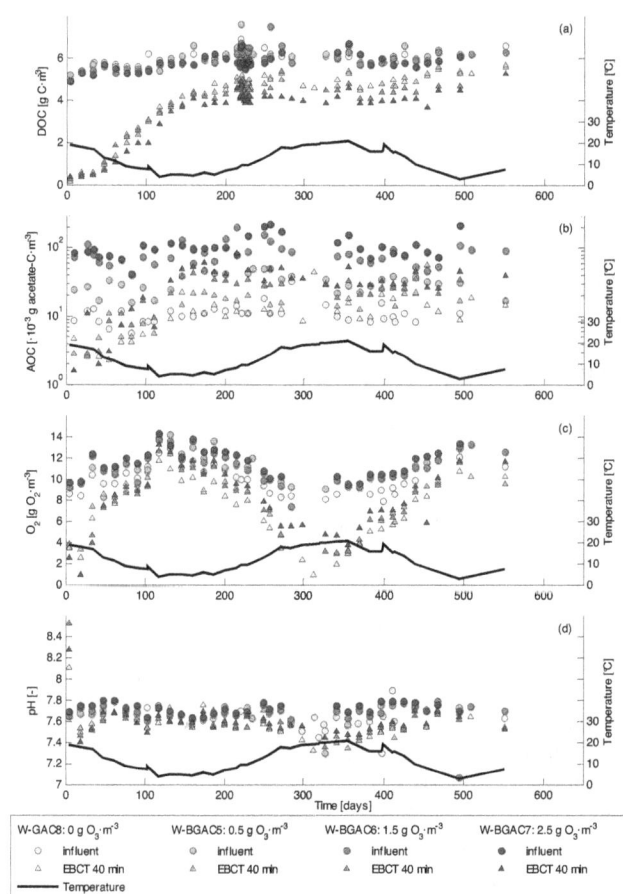

Figure 2. DOC (**a**), AOC (**b**), oxygen (**c**) and pH (**d**) in Weesperkarspel pilot (B)GAC filters.

with their value before ozonation. The linear coefficients were all between 0.92 and 1.0. All target parameters after ozonation also correlated significantly with the ozone dose, as seen in Table 2. The DOC concentration after ozonation mainly depended on the DOC concentration in the raw water and decreased slightly with increasing ozone doses. At the highest ozone dose of $2.5 \, g \, O_3 \, m^{-3}$, resulting in an average ozone/DOC ratio of $0.43 \, g \, O_3 \, g \, C^{-1}$, the average DOC

removal was $0.17 \, g \, C \, m^{-3}$ (3%), as seen in Table 1. Other studies reported NOM removal of 3% to 16%, depending on the ozone dose (Sontheimer et al., 1988). Apparently this DOC was completely oxidized to carbon dioxide. However, the carbon dioxide concentration did not increase with increasing ozone doses. On the contrary, it decreased with increasing ozone doses, although the correlation was weak (see Table 2). The ozone was dosed via a bubble column, using an ozone-air mixture. It was assumed that the produced carbon dioxide was stripped in the bubble column. No data about carbon dioxide concentrations in the off-gas were available to confirm this assumption. The oxygen concentration increased with the ozone dose. Apparently, additional oxygen dissolved from the gas mixture that was dosed in the bubble columns. Also, some of the ozone did not react with substances in the water and was converted to oxygen. The AOC concentration increased with the ozone dose. The increase in AOC of $35 \times 10^{-3} \, g \, C \, g \, O_3^{-1}$ was within the broad range of values between 20×10^{-3} and $250 \times 10^{-3} \, g \, C \, g \, O_3^{-1}$ found by other researchers (Hammes et al., 2006; van der Helm et al., 2007; van der Kooij et al., 1989).

There was no significant correlation between AOC concentration after ozonation and the DOC concentrations in the raw water. Other studies have shown that the formation of AOC and BDOC during ozonation does increase with the concentration of NOM in the raw water (Carlson and Amy, 1997; van der Helm, 2007). The limited variation in the DOC concentration in the raw water in this study could be a reason for not finding a significant correlation. It is know that the ozone decay rate increases with temperature. Therefore, at a constant ozone dose, the concentration multiplied by the contact time (CT) for ozone decreases with increasing temperatures. Neither the DOC nor the AOC concentration after ozonation depended on the water temperature. Apparently both concentrations depended more on the total amount of ozone dosed than on the CT-value. Van der Helm et al. (2007) also found that the reaction of ozone with NOM was linear with the ozone dose, and not with the CT-value. Hammes et al. (2006), on the other hand, found that the formation of AOC did increase with CT.

Table 3. Biological conversions in (B)GAC pilot filters.

Column	W-GAC8	W-BGAC5	W-BGAC6	W-BGAC7
Net ozone dose [$g\,O_3\,m^{-3}$]	0	0.5	1.5	2.5
ΔDOC [$g\,C\,m^{-3}$]	−1.2 (−20%)	−1.3 (−21%)	−1.5 (−25%)	−1.8 (−30%)
ΔAOC [$10^{-3}\,g$ acetate-$C\,m^{-3}$]	+2.7 (+21%)	−12.3 (−35%)	−68.3 (−73%)	−85.4 (−68%)
ΔO_2 [$g\,O_2\,m^{-3}$]	−3.3 (−54%)	−3.4 (−55%)	−3.6 (−59%)	−3.5 (−59%)
ΔCO_2 [$g\,CO_2\,m^{-3}$]	+2.6 (+33%)	+3.4 (+44%)	+0.9 (+10%)	+2.7 (+39%)
$\Delta O_2\,\Delta DOC^{-1}$ [$g\,O_2\,g\,C^{-1}$]	3.2	2.8	2.3	1.9
$\Delta CO_2\,\Delta DOC^{-1}$ [$g\,CO_2\,g\,C^{-1}$]	−2.4	−2.7	−0.4	−1.4

Notes: table values are the average difference in (B)GAC influent and effluent concentrations $C_{eff} - C_{inf}$ from day 194 to day 559; percentage values are the average relative removal percentages $(C_{eff} - C_{inf})/C_{inf}$ from day 194 to day 559.

3.2 Effects of ozonation and water temperature on biodegradation of NOM in (B)GAC filters

Figure 2 shows that in all (B)GAC filters initial NOM removal was high; at the first day of operation 92% to 96% of the DOC and 90% to 100% of the AOC was removed. After 40 days of operation, both DOC and AOC concentrations in the effluents increased rapidly according to a typical S-shaped adsorption breakthrough curve. However, complete breakthrough was not reached. Instead, after approximately half a year of operation, DOC and AOC removal in all (B)GAC filters reached more or less a steady-state. It was assumed that the main NOM removal process in this "steady-state" situation was biodegradation. It was also assumed that biodegradation of NOM was the only biodegradation process in the filters. From experiences in the full-scale plant, it was known that there was no microbial oxidation of inorganic compounds because ammonium, nitrite and iron were absent due to pre-treatment by rapid sand filtration (see Fig. 1).

To take seasonal effects into account, the biological removal of DOC and AOC, the consumption of oxygen, and the production of carbon dioxide in the (B)GAC filters were analyzed for a period of one year, during steady-state conditions from day 194 to day 559 (see Table 3). During this selected period, the average DOC removal was significantly greater at higher ozone doses, $1.2\,g\,C\,m^{-3}$ (20%) at $0\,g\,O_3\,m^{-3}$ and $1.8\,g\,C\,m^{-3}$ (30%) at $2.5\,g\,O_3\,m^{-3}$, at almost identical DOC influent concentrations. Also, the average AOC removal was significantly higher at higher ozone doses. Note that the AOC influent concentrations were increased due to ozonation. However, in none of the filters receiving ozonated water were AOC concentrations reduced to the concentrations before ozonation (see Table 1). The BGAC filters removed up to 70% of the produced AOC. The GAC effluent on average contained more AOC than the influent, as seen in Table 3. Apparently, the GAC filter produced AOC. Therefore, both GAC filtration and ozonation followed by BGAC filtration were expected to have a negative effect on biostability.

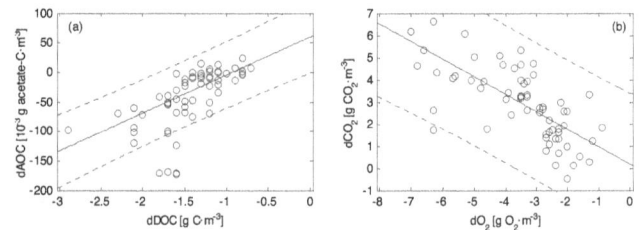

Figure 3. Linear regressions plus 95% confidence intervals for data from all 4 (B)GAC filters: $\Delta AOC = 64 \times 10^{-3}\,\Delta DOC + 60 \times 10^{-3}$ ($R^2 = 0.36$) (**a**) and $\Delta CO_2 = -0.79\,\Delta O_2 + 0.19$ ($R^2 = 0.13$) (**b**).

Neither average oxygen consumption nor the average carbon dioxide production depended on the ozone dose. The significant correlations between the biological removal of DOC, AOC and oxygen, and the production of carbon dioxide were $\Delta AOC = 64 \times 10^{-3}\Delta DOC + 60 \times 10^{-3}$ ($R^2 = 0.36$) and $\Delta CO_2 = -0.79\,\Delta O_2 + 0.19$ ($R^2 = 0.13$) (see Fig. 3).

MLRs were made for the biological removal of DOC and AOC, the consumption of oxygen and the production of carbon dioxide in the (B)GAC filters, again for the selected period of one year. Different sets of input variables were used: (a) water quality after ozonation (the ozone dose was not used), (b) water quality after ozonation plus the ozone dose, (c) raw water quality plus the ozone dose, and (d) water temperature plus the ozone dose. When only the water quality variables after ozonation were used as input parameters (set a), 3 MLRs showed negative values for R^2. This indicates that this set was severely influenced by outliers. Therefore, this set was excluded from further examination. Adding the ozone dose as an input variable (set b) reduced this problem and was therefore useful for predicting biodegradation in (B)GAC filters. Using the water quality variables from the raw water instead of those from the water after ozonation (set c) increased the RMSEs, meaning that the quality of the MLRs decreased. This is the result of additional noise in the input variables caused by the ozonation. The simplest

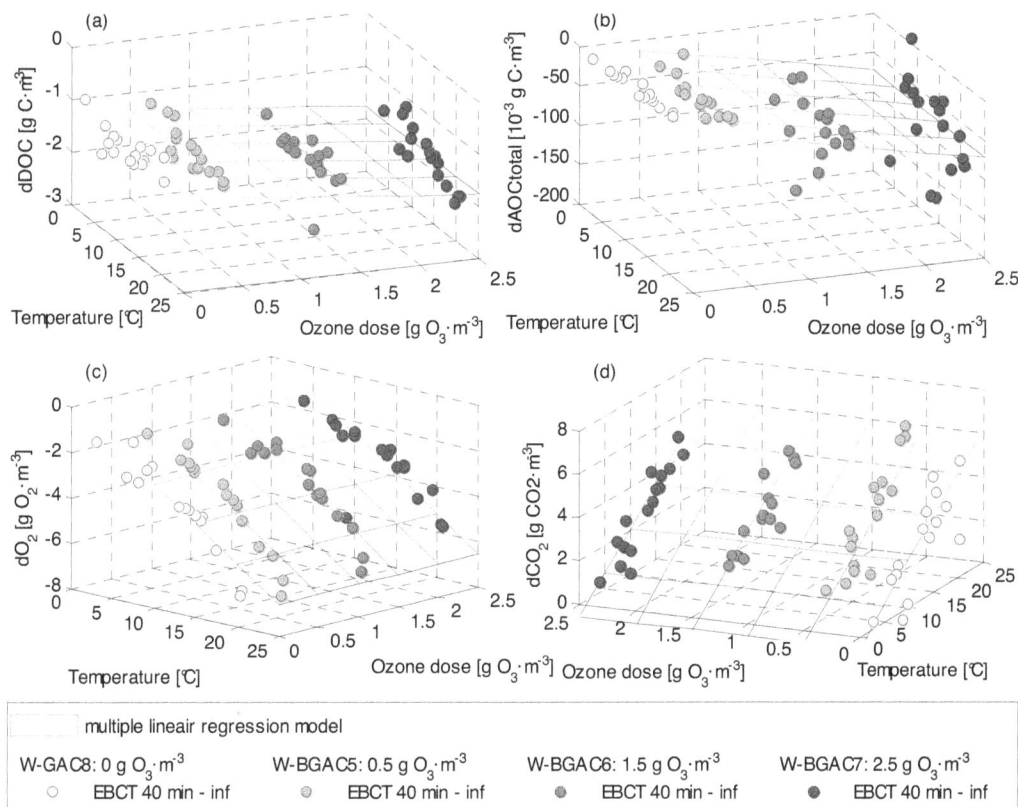

Figure 4. MLRs for biological conversions of DOC (**a**), AOC (**b**), oxygen (**c**) and carbon dioxide (**d**) as a function of the ozone dose and temperature from day 194 to 559 (set d).

set (d) only used the temperature and the ozone dose as input. Figure 4 illustrates the results of the MLRs for set (d). The results of the MLRs for sets b, c and d are shown in Table 4.

Figure 3, Table 4 and Fig. 4 illustrate that the removals of DOC and AOC correlated with each other and were greater at higher ozone doses. They were not significantly influenced by the water temperature. Oxygen consumption and carbon dioxide production correlated with each other and increased with temperatures. They were not significantly influenced by the ozone dose. The removal of DOC and AOC did not correlate significantly with the consumption of oxygen and the production of carbon dioxide.

It is known that different substrates require different amounts of oxygen for complete oxidation. For example, for complete oxidation of glucose ($C_6H_{12}O_6$) $2.7\,g\,O_2\,g\,C^{-1}$, for cellulose ($C_6H_{10}O_5$)$_n$ $2.7\,g\,O_2\,g\,C^{-1}$ and for ethanol (CH_3OH) $4.0\,g\,O_2\,g\,C^{-1}$ is needed. Perdue and Ritchie (2004) reviewed different studies on NOM and collected the compositions of NOM in 57 fresh waters. The average NOM of these 57 samples had a C:H:O ratio of 400:660:645. For complete oxidation of this average NOM $2.6\,g\,O_2\,g\,C^{-1}$ is needed. The oxygen consumption of all the NOM samples always was between $2.1\,g\,O_2\,g\,C^{-1}$ and $3.4\,g\,O_2\,g\,C^{-1}$. Complete oxidation of NOM always results in a carbon dioxide production of $3.7\,g\,CO_2\,g\,C^{-1}$.

Figure 5 shows that in the pilot (B)GAC filters both the ratio between oxygen consumption and DOC removal and the ratio between carbon dioxide production and DOC removal were higher at higher temperatures. In summer more oxygen was consumed and more carbon dioxide was produced than was needed for complete NOM oxidation (approximately 200% and 150% respectively). This was also observed by Jekel (1979) and Sontheimer et al. (1988).

It was assumed that in the pilot filters the type of NOM remained fairly constant during the year and that always $2.6\,g\,O_2\,g\,C^{-1}$ was needed for complete oxidation. When biodegradation of NOM from the water phase is assumed to be the only relevant process for run times longer than 194 days, the removal of DOC and AOC, the consumption of oxygen and the production of carbon dioxide: (a) should correlate with each other, (b) should all be higher at higher temperatures, (c) should all be higher at higher ozone doses, and (d) should not have oxygen consumption exceeding $2.6\,g\,O_2\,g\,C^{-1}$ and should not have carbon dioxide production exceeding $3.7\,g\,CO_2\,g\,C^{-1}$. None of these requirements were completely fulfilled. It was suggested that, besides biodegradation of NOM from the water phase, the following mechanisms could explain the observations from the pilot plant: a decrease in the NOM reduction degree (see Uhl, 2000) by ozonation, the degassing of (B)GAC filters,

Table 4. MLRs biological conversions in (B)GAC pilot filter.

MLR – Robust – set b Input variables: O_3, temp, $[DOC]_{inf}$, $[AOC]_{inf}$, $[O_2]_{inf}$, $[CO_2]_{inf}$	R^2	RMSE
$\Delta[DOC] = 1.2 - 0.39\,[DOC]_{inf} - 0.26\,O_3$	0.53	0.22
$\Delta[AOC] = 13 \times 10^{-3} - 0.90\,[AOC]_{inf} - 4.0 \times 10^{-3}\,O_3$	0.96	6.3×10^{-3}
$\Delta[O_2] = 173 - 0.57\,\text{temp} - 0.27\,[O_2]_{inf} - 20\,\text{pH}_{inf} - 1.5\,[CO_2]_{inf}$	−7.1	0.50
$\Delta[CO_2] = -5.4 + 0.36\,\text{temp} + 0.54\,[O_2]_{inf} - 0.18\,[CO_2]_{inf} - 0.57\,O_3$	0.46	0.94
MLR – Robust – set c Input parameters: O_3, temp, $[DOC]_{raw}$, $[AOC]_{raw}$, $[O_2]_{raw}$, $[CO_2]_{raw}$		
$\Delta[DOC] = -1.2 - 0.23\,O_3$	0.34	0.25
$\Delta[AOC] = 179 \times 10^{-3} - 29 \times 10^{-3}\,[DOC]_{raw} - 34 \times 10^{-3}\,O_3$	0.58	15×10^{-3}
$\Delta[O_2] = 7.3 - 0.38\,\text{temp} - 0.48\,[O_2]_{raw} - 0.13\,[CO_2]_{raw}$	0.88	0.53
$\Delta[CO_2] = -0.58 + 0.27\,\text{temp}$	0.20	1.0
MLR – Robust – set d Input parameters: O_3, temp		
$\Delta[DOC] = -1.2 - 0.23\,O_3$	0.34	0.25
$\Delta[AOC] = 0.92 \times 10^{-3} - 29 \times 10^{-3}\,O_3$	0.49	16×10^{-3}
$\Delta[O_2] = 0.20 - 0.27\,\text{temp}$	0.73	0.64
$\Delta[CO_2] = -0.58 + 0.27\,\text{temp}$	0.20	1.0

Notes: $_{raw}$ indicates raw water (i.e., water before ozonation); $_{inf}$ indicates (B)GAC influent (i.e., water after ozonation); Δ indicates the difference in (B)GAC influent and effluent concentrations $C_{eff} - C_{inf}$ from day 194 to day 559.

biomass production and losses, the temperature effects on NOM adsorption and bioregeneration. These hypotheses are discussed in the following sections.

3.2.1 Decrease of the NOM reduction degree by ozonation

The average ratio between oxygen consumption and DOC removal was lower at higher ozone doses, $3.2\,\text{g}\,O_2\,\text{g}\,C^{-1}$ at $0\,\text{g}\,O_3\,\text{m}^{-3}$ and $1.9\,\text{g}\,O_2\,\text{g}\,C^{-1}$ at $2.5\,\text{g}\,O_3\,\text{m}^{-3}$ (see Table 3). However, the significance was relatively low (p-value 0.06). The MLRs from sets d for DOC and oxygen consumption, and for DOC and carbon dioxide production (see Table 4) were combined into equations for the ratio between oxygen consumption and DOC removal and for the ratio between carbon dioxide production and DOC removal, as seen in Fig. 5. Both ratios were higher at higher temperatures and lower at higher ozone doses. Due to the reaction of NOM and ozone, more oxygen atoms are incorporated in the NOM molecules, resulting in a decrease in the reduction degree of the NOM. For biodegradation of NOM with a lower reduction degree, less oxygen is consumed (Uhl, 2000). From his data, Uhl derived that, due to this NOM oxidation, the reduction degree of NOM reduced by 0.26 to 0.48 in four different (B)GAC filters, corresponding to an oxygen uptake of 0.13 to $0.24\,\text{g}\,O_2\,\text{g}\,C^{-1}$. This is less than the observed difference of $3.2 - 1.9 = 1.3\,\text{g}\,O_2\,\text{g}\,C^{-1}$ that was found in this study. There-fore, the decrease of the NOM reduction degree by ozonation could not (completely) explain the observations from the pilot plant.

3.2.2 Degassing of (B)GAC filters

The average ratio between carbon dioxide production and DOC removal was between 0.4 and 2.7 for all (B)GAC filters and did not depend on the ozone dose (see Table 3). As 3.7 is the ratio between the molecular mass of carbon dioxide ($44\,\text{g}\,\text{mol}\,CO_2^{-1}$) and carbon ($12\,\text{g}\,\text{mol}\,C^{-1}$), this means that not all removed NOM carbon was found as carbon dioxide in the effluent. For example, in filter W-GAC8 only $2.4 \cdot 12/44 = 65\%$ of the biologically removed NOM carbon was found as carbon dioxide in the effluent. For filters W-BGAC5, W-BGAC6 and W-BGAC7 this was 11%, 38% and 74%, respectively. Ozone was dosed at a water depth of 5 m, while the supernatant water in the (B)GAC filters was at atmospheric pressure. The supernatant water was super-saturated with gas. In summer, gas bubbles frequently were observed in the (B)GAC filters. The escaping gas contained relatively high amounts of carbon dioxide (2%). No information on the amount of gas escaping was available to assess its effect on the carbon balance. However, the gas was escaping at atmospheric pressure, so the density was approximately $1.2 \times 10^3\,\text{g}\,\text{m}^{-3}$. The concentration of carbon dioxide as carbon was approximately $1.2 \times 10^{-3} \cdot 2\% \cdot 12/44 = 6.5\,\text{g}\,C\,\text{m}^{-3}$.

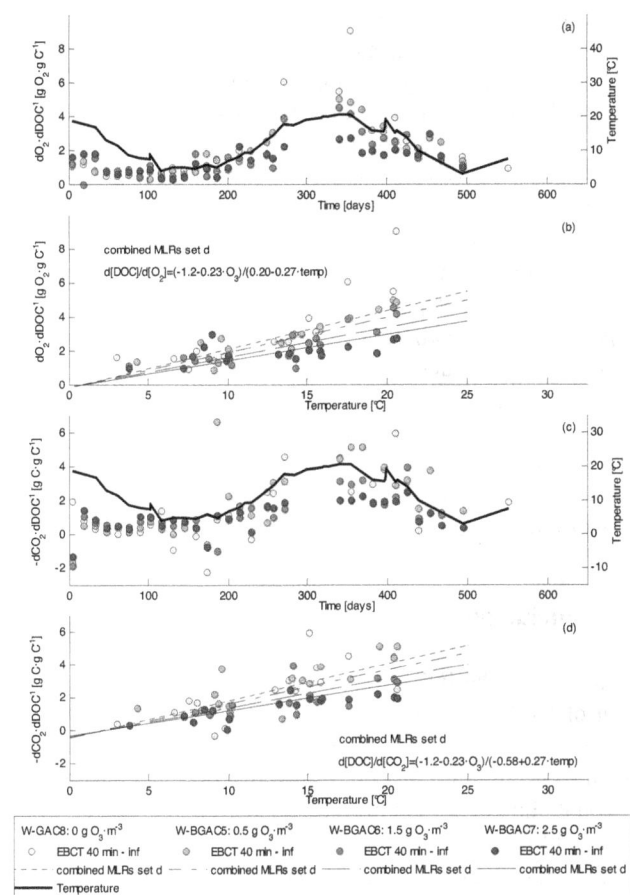

Figure 5. Oxygen consumption to DOC removal ratio (**a**), oxygen consumption to DOC removal ratio as function of temperature from day 194 to 559 (**b**), carbon dioxide production to DOC removal ratio (**c**) and carbon dioxide production to DOC removal ratio as function of temperature from day 194 to day 559 (**d**).

For example, for filter W-GAC8 it was demonstrated that 35% of the biologically removed NOM carbon was not found as carbon dioxide in the effluent. If this was caused by degassing of the filters, $35\% \cdot 1.2/6.5 = 0.065$ m^3 of gas per m^3 of treated water should have escaped from the filters. At a surface loading of 8.7×10^{-4} m s^{-1} this corresponded to 4.9 m^3 gas per m^2 filter bed per day, which is not realistic. Therefore, the degassing of the (B)GAC filters could not (completely) explain the observations from the pilot plant.

3.2.3 Biomass production and losses

A part of the biologically removed NOM carbon is converted into biomass. For aquatic bacteria the bacterial growth efficiency, or yield, varies from less than 0.05 to 0.6 g C g C^{-1}, depending on type and concentration of the substrate, type of bacteria, growth phase of bacteria, temperature and other water quality parameters (del Giorgio and Cole, 1998). For bacteria growing at a yield of 0.05 g C g C^{-1}, both the oxygen

consumption and the carbon dioxide production per amount of biodegraded NOM is about 2 times higher than for bacteria growing at a yield of 0.6 g C g C^{-1}. In the pilot (B)GAC filters these figures were more than 5 times higher in summer than in winter (see Fig. 5). In theory, this would only be possible at a maximum yield (in winter) of 0.8 g C g C^{-1} or more, which is not realistic. Variable yields also cannot explain why in summer more than 2.6 g O$_2$ g C^{-1} was consumed and more than 3.7 g CO$_2$ g C^{-1} was produced.

The produced biomass can either accumulate in the filter bed, or leave the filter with the effluent or with the backwash water. Besides, it is possible that the biomass dies, disintegrates (lysis) and serves as a carbon source for other microorganisms. Furthermore, bacteria in a starvation or limitation phase can utilize internally stored carbon. In this study, the concentrations of ATP, as a measure of active biomass, were determined in the filter influents, effluents and backwash water (data not shown). Assuming a biomass carbon/ATP ratio of 250 g C g ATP^{-1} (Magic-Knezev and van der Kooij, 2004), the amounts of biomass, expressed in carbon (biomass carbon), in the influents and effluents were less than 1% of the NOM that was biologically removed in the (B)GAC filters. During a backwash, 13% of the biomass was removed from pilot filter W-BGAC6 (1 observation). At a maximum frequency of 1 backwash per 4 days, the biomass carbon removed during backwashing was approximately 1% of the NOM that was biologically removed in the filter during 4 days. The observed changes in biomass activity were always between $+4 \times 10^{-9}$ g ATP g AC^{-1} day^{-1} (net biomass growth) and -4×10^{-9} g ATP g AC^{-1} day^{-1} (net biomass removal), which corresponded to less than 1% of the NOM that was biologically removed. Therefore, biomass growth, accumulation in the filter bed, lysis and the loss of carbon with biomass in the effluent and backwash water could not explain the discrepancy between biological NOM removal and carbon dioxide production in the BGAC filters.

3.2.4 Temperature effects on NOM adsorption

In theory, adsorption should be less at higher temperatures (Schreiber et al., 2007). This was shown experimentally for some individual compounds, e.g., phenol and o-cresol (Abuzaid and Nakhla, 1997). However, experiments showed more NOM adsorption at higher water temperatures (Schreiber et al., 2005 and 2007; Summers and Roberts, 1988). In this study, the adsorption isotherms were only determined at 12 °C. Therefore, the effect of temperature on the adsorption isotherm could not be assessed. It is possible that at higher temperatures adsorption decreases, or even completely stops, and simultaneously biodegradation of NOM increases. Because of the increased biodegradation, the oxygen consumption and carbon dioxide production typically increases in summer, while the DOC and AOC removal typically remains fairly constant. The phenomenon that NOM adsorbs in winter and is biodegraded in summer cannot explain why, in summer, the oxygen consumption exceeded $2.6\,\mathrm{g\,O_2\,g\,C^{-1}}$ and the carbon dioxide production exceeded $3.7\,\mathrm{g\,CO_2\,g\,C^{-1}}$. Temperature effects on NOM adsorption could only partly explain the discrepancy between DOC and AOC removal on one side and oxygen consumption and carbon dioxide production on the other.

3.2.5 Bioregeneration

Bioregeneration of AC is biodegradation of (previously) adsorbed NOM, which results in a decrease in the NOM loading on the AC (Sontheimer et al., 1988). Several authors described two possible mechanisms for bioregeneration. The first hypothesis is that biomass on the external AC surface takes up substrate. Therefore, the concentration of the substrate on the external AC surface becomes smaller than the internal equilibrium concentration. This causes diffusion of the substrate from the internal pores towards the external AC surface, where it is biodegraded. The concentration inside the pores decreases, which results in desorption. The AC is available for adsorption again: it has been bioregenerated. The second hypothesis is that the biomass releases extracellular enzymes that enter the meso-pores of the AC; micropores are believed to be too small for the exo-enzymes to enter. The exo-enzymes convert part of the adsorbed substrate into less adsorbable products. These products desorb and diffuse from the internal pores towards the external AC surface, where they are biodegraded. Again, the AC is bioregenerated. In both hypotheses, both desorption and biodegradation are conditions for bioregeneration (Aktas and Çeçen, 2007; Klimenko et al., 2003; Walker and Weatherley, 1998).

During bioregeneration, oxygen is consumed and carbon dioxide is produced. The NOM that is biodegraded originates from the adsorbed phase. An increase in oxygen consumption and carbon dioxide production is possible, without any effect on the measured DOC and AOC concentrations in the filter effluent. During complete oxidation of $1\,\mathrm{g\,C}$ $2.6\,\mathrm{g}$ oxygen is needed and $3.7\,\mathrm{g}$ carbon dioxide is produced. Assume that in winter, during a period of 6 months, $1.0\,\mathrm{g\,C\,m^{-3}}$ NOM adsorbed onto the AC and that $0.2\,\mathrm{g\,C\,m^{-3}}$ NOM was biodegraded. The oxygen consumption and the carbon dioxide production, per amount of NOM removed from the water phase, would have been $0.2 \cdot 2.6/1.2 = 0.4\,\mathrm{g\,O_2\,g\,C^{-1}}$ and $0.2 \cdot 3.7/1.2 = 0.6\,\mathrm{g\,CO_2\,g\,C^{-1}}$. Assume that in summer, during a period of 6 months, no NOM adsorbed, that $1.2\,\mathrm{g\,C\,m^{-3}}$ NOM was biodegraded from the water phase and that all adsorbed NOM from the previous winter period was biodegraded. The oxygen consumption and the carbon dioxide production, per amount of NOM removed from the water phase, would have been $(1.2+1) \cdot 2.6/1.2 = 4.8\,\mathrm{g\,O_2\,g\,C^{-1}}$ and $(1.2+1) \cdot 3.7/1.2 = 6.8\,\mathrm{g\,CO_2\,g\,C^{-1}}$. These figures correspond well to the measured results from the pilot experiment, as seen in Fig. 5.

AC bioregeneration was reported for different specific compounds in (industrial) waste waters (Aktas and Çeçen,

2007; Klimenko et al., 2003; Walker and Weatherley, 1998). Although no hard evidence was found, some researchers suggested that bioregeneration of NOM in drinking water is possible (Sontheimer et al., 1988). Both desorption and biodegradation of the compounds are conditional for bioregeneration. It is obvious that a part of the NOM in the pilot (B)GAC filters was biodegradable and adsorbable (Fig. 2 and Table 3). For batch experiments, desorption of 4% to 58% of previously adsorbed NOM was reported. The percentage of NOM desorption depended on the type of NOM and on the type of AC (Yapsakli et al., 2009), Therefore, in theory, the conditions for biodegradation and for desorption can be met. Because in summer the reported ratio between oxygen consumption and DOC removal exceeded $2.6\,\mathrm{g\,O_2\,g\,C^{-1}}$ and the ratio between carbon dioxide production and DOC removal exceeded $3.7\,\mathrm{g\,CO_2\,g\,C^{-1}}$, it was likely that bioregeneration of NOM did occur in the (B)GAC pilot filters.

4 Conclusions

In this study the influence of ozonation on the biodegradation of NOM in (B)GAC filters was assessed in four pilot (B)GAC filters. To study a period with biodegradation as the main NOM removal process and to take seasonal effects into account, the filters were judged for a period of one year, from day 194 to day 559.

The main effect of the ozonation was the increase in the AOC concentration in the (B)GAC influents by $35 \times 10^{-3}\,\mathrm{g\,acetate\text{-}C\,g\,O_3^{-1}}$. Ozonation also resulted in limited decreases in the DOC and oxygen concentrations, and in limited increases in the pH and (calculated) carbon dioxide concentration.

The BGAC filters receiving ozonated water removed up to 70% of the produced AOC. However, in none of the BGAC filters were AOC concentrations reduced to the concentration levels before ozonation. The GAC filter produced small amounts of AOC. Therefore, both GAC filtration and pre-oxidation followed by BGAC filtration were expected to have a negative effect on biostability. Ozonation also stimulated the removal of DOC in the (B)GAC filters. At the highest ozone dose of $2.5\,\mathrm{g\,O_3\,m^{-3}}$, which resulted in an average ozone/DOC ratio of $0.43\,\mathrm{g\,O_3\,g\,C^{-1}}$, the DOC removal was 30% versus 20% in the filter receiving non-ozonated water. The removals of DOC and AOC correlated with each other and were greater at higher ozone doses. They were not significantly influenced by the water temperature. Oxygen consumption and carbon dioxide production correlated with each other and were higher at higher temperatures. They were not significantly influenced by the ozone dose. The removal of DOC and AOC did not correlate significantly with the removal of oxygen and the production of carbon dioxide. This discrepancy indicated that, besides biodegradation, other processes were relevant for NOM removal in the filters. In summer, the ratio between oxygen consumption and DOC removal was approximately 2 times the theoretical maximum of $2.6\,g\,O_2\,g\,C^{-1}$ and the ratio between carbon dioxide production and DOC removal was approximately 1.5 times the theoretical maximum of $3.7\,g\,CO_2\,g\,C^{-1}$. Theoretical calculations demonstrated that the production and losses of biomass, the degassing of (B)GAC filters, the decrease in the NOM reduction degree and the temperature effects on NOM adsorption could only partly explain these excesses and the discrepancy between DOC and AOC removal and oxygen consumption and carbon dioxide production. Bioregeneration of NOM could explain the excesses and the non-correlation. Therefore, it was likely that bioregeneration of NOM did occur in the (B)GAC pilot filters.

It is recommended that adsorption, desorption and biodegradation experiments be performed with labeled ^{14}C-glucose (Servais et al., 1994), or if possible with larger (both biodegradable and non-biodegradable) ^{14}C-NOM molecules. This will make it possible to determine the fate of NOM and to quantify relevant processes in BGAC filtration. Possibly, hard evidence for bioregeneration of NOM will be found.

Acknowledgements. This research was conducted by Waternet, Vitens, KWR, Norit, Delft University of Technology and Wageningen and Research. The research was part of the project BAKF/SCWR, which was co-funded by SenterNovem (currently Agentschap NL), agency of the Dutch Ministry of Economic Affairs.

References

Abuzaid, N. S. and Nakhla, G. F.: Modeling of the temperature variation effects on the polymerization reactions of phenolics on granular activated carbon, Sep. Sci. Technol., 32, 1255–1272, 1997.

Aktas, Ö. and Çeçen, F.: Bioregeneration of activated carbon: A review, Int. Biodeter. Biodegr., 59, 257–272, 2007.

Baghoth, S. A., Maeng, S. K., Salinas Rodriguez, S. G., Ronteltap, M., Sharma, S., Kennedy, M., and Amy, G. L.: An urban water cycle perspective of natural organic matter (NOM): NOM in drinking water, wastewater effluent, storm water and seawater, Water Sci. Technol. Water Supply, 8, 701–707, 2008.

Bosklopper, T. G. J., Rietveld, L. C., Babuska, R., Smaal, B., and Timmer, J.: Integrated operation of drinking water treatment plant at Amsterdam water supply, Water Sci. Technol. Water Supply, 4, 263–270, 2004.

Carlson, K. and Amy, G.: The formation of filter-removable biodegradable organic matter during ozonation, Ozone Sci. Eng., 19, 179–199, 1997.

Carter, M. C. and Weber Jr., W. J.: Modeling adsorption of TCE by activated carbon preloaded by background organic matter, Environ. Sci. Technol., 28, 614–623, 1994.

Carter, M. C., Weber Jr., W. J., and Olmstead, K. P.: Effects of background dissolved organic matter on TCE adsorption by GAC, J. Am. Water Works Assn., 84, 81–91, 1992.

del Giorgio, P. A. and Cole, J. J.: Bacterial growth efficiency in natural aquatic systems, Annu. Rev. Ecol. Syst., 29, 503–541, 1998.

Escobar, I. C., Randall, A. A., and Taylor, J. S.: Bacterial growth in distribution systems: effect of assimilable organic carbon and biodegradable dissolved organic carbon, Environ. Sci. Technol., 35, 3442–3447, 2001.

Gilbert, E.: Photometrische Bestimmung niedriger Ozonkonzentrationen in Wasser mit Hilfe von Diaëthyl-p-phenylendiamin (DPD) (Photometric determination of low ozone concentrations in water using Diethyl-p-phenylenediamine (DPD)), GWF Wasser Abwasser, 122, 410–416, 1981.

Graham, N. J. D.: Removal of humic substances by oxidation/biofiltration processes – a review, Water Sci. Technol., 40, 141–148, 1999.

Graveland, A.: Application of biological activated carbon filtration at Amsterdam Water Supply, Water Supp., 14, 233–241, 1994.

Hammes, F., Salhi, E., Köster, O., Kaiser, H.-P., Egli, T., and von Gunten, U.: Mechanistic and kinetic evaluation of organic disinfection by-product and assimilable organic carbon (AOC) formation during the ozonation of drinking water, Water Res., 40, 2275–2286, 2006.

Jekel, M.: Experience with biological activated carbon filters, in: Oxidation Techniques in Drinking Water, edited by: Kuhn, W. and Sontheimer, H., US Environmental Protection Agency EPA-570/9-79-020, 715–726, 1979.

Juhna, T. and Rubulis, J.: Problem of DOC removal during biological treatment of surface water with a high amount of humic substances, Water Sci. Technol. Water Supply, 4, 183–187, 2004.

Klimenko, N., Smolin, S., Grechanyk, S., Kofanov, V., Nevynna, L., and Samoylenko, L.: Bioregeneration of activated carbons by bacterial degraders after adsorption of surfactants from aqueous solutions, Colloid. Surface. A, 230, 141–158, 2003.

Knappe, D. R. U., Snoeyink, V. L., Roche, P., Prados, M. J., and Bourbigot, M.-M.: The effect of preloading on rapid small-scale column test predictions of atrazine removal by GAC adsorbers, Water Res., 31, 2899–2909, 1997.

Kruithof, J. C. and Maschelein, W. J.: State-of-the-art of the application of ozonation in Benelux drinking water treatment, Ozone Sci. Eng., 21, 139–152, 1999.

Labouyrie, L., Bec, R. L., Mandon, F., Sorrento, L. J., and Merlet, N.: Comparaison de L'Activite Biologique de Differents Charbons Actifs en Grains Comparison of Biological Activity of Different Types of Granular Activated Carbons, Environ. Technol., 18, 151–159, 1997.

Leenheer, J. A. and Croué, J.-P.: Characterizing aquatic dissolved organic matter, Environ. Sci. Technol., 37, 18A–26A, 2003.

Magic-Knezev, A. and van der Kooij, D.: Optimisation and significance of ATP analysis for measuring active biomass in granular activated carbon filters used in water treatment, Water Res., 38, 3971–3979, 2004.

Mathworks: Matlab® The language of technical computing, version 7.5.0.342 (R2007b), The Mathworks Inc, 2007.

Orlandini, E.: Pesticide removal by combined ozonation and granular activated carbon filtration, Ph.D. thesis, Wageningen University and International Institute for Infrastructural, Hydraulic and Environmental Engineering, Delft, The Netherlands, 1999.

Perdue, E. M. and Ritchie, J. D.: Dissolved Organic Matter in Freshwaters, in: Treatise on Geochemistry, edited by: Drever, J. I., 273–318, 2004.

Rietveld, L., van der Helm, A., van Schagen, K., van der Aa, R., and van Dijk, H.: Integrated simulation of drinking water treatment, J. Water Supply Res. Technol. AQUA, 57, 133–141, 2008.

Schreiber, B., Brinkmann, T., Schmalz, V., and Worch, E.: Adsorption of dissolved organic matter onto activated carbon – the influence of temperature, absorption wavelength, and molecular size, Water Res., 39, 3449–3456, 2005.

Schreiber, B., Schmalz, V., Brinkmann, T., and Worch, E.: The effect of water temperature on the adsorption equilibrium of dissolved organic matter and atrazine on granular activated carbon, Environ. Sci. Technol., 41, 6448–6453, 2007.

Servais, P., Billen, G., and Bouillot, P.: Biological Colonization of Granular Activated Carbon Filters in Drinking-Water Treatment, J. Environ. Eng., 120, 888–899, 1994.

Servais, P., Anzil, A., Gatel, D., and Cavard, J.: Biofilm in the Parisian suburbs drinking water system, J. Water Supply Res. Technol. AQUA, 53, 313–323, 2004.

Siddiqui, M. S., Amy, G. L., and Murphy, B. D.: Ozone enhanced removal of natural organic matter from drinking water sources, Water Res., 31, 3098–3106, 1997.

Sontheimer, H., Crittenden, J. C., and Summers, R. S.: Activated carbon for water treatment, 2 Edn., AWWA – DVGW Forschungssstelle Engler Bunte Institut, Karlsruhe, Germany, 1988.

Standard Methods: Standard Methods for the examination of water & wastewater, 20 Edn., American Public Health Association/American Water Works Association/Water Environment Federation, Washington DC, USA, 1998.

Summers, R. S. and Roberts, P. V.: Activated carbon adsorption of humic substances: II. Size exclusion and electrostatic interactions, J. Colloid Interface Sci., 122, 382–397, 1988.

Uhl, W.: Einfluss von Schüttungsmaterial und Prozessparmetern auf die Leistung von Bioreaktoren bei der Trinkwasseraufbereitung (Influence of filter material and process parameters on the performance of bioreactors in the drinking water treatment), Ph.D. thesis, IWW Rheinisch-Westfällisches Institut für Wasserforschung gemeinnützige GmbH, Institut an der Gerhard-Mercator-Universität Duisburg, Mülheim an der Ruhr, Germany, 2000.

Urfer, D. and Huck, P. M.: Measurement of biomass activity in drinking water biofilters using a respirometric method, Water Res., 35, 1469–1477, 2001.

van der Aa, L. T. J.: Effects of pre-oxidation and filter run time on the performance of biological granular activated carbon filters, in preparation, 2011.

van der Helm, A. W. C.: Integrated modeling of ozonation for optimization of drinking water treatment, Ph.D. thesis, Water Management Academic Press, Delft, The Netherlands, 2007.

van der Helm, A. W. C., Smeets, P. W. M. H., Baars, E. T., Rietveld, L. C., and van Dijk, J. C.: Modeling of Ozonation for Dissolved Ozone Dosing, Ozone Sci. Eng., 29, 379–389, 2007.

van der Helm, A. W. C., Rietveld, L. C., Bosklopper, T. G. J., Kappelhof, J. W. N. M., and van Dijk, J. C.: Objectives for optimization and consequences for operation, design and concept of drinking water treatment plants, Water Sci. Technol. Water Supply, 8, 297–304, 2008.

van der Kooij, D., Oranje, J. P., and Hijnen, W. A. M.: Growth of *Pseudomonas aeruginosa* in tap water in relation to utilization of substrates at concentrations of a few micrograms per liter, Appl. Environ. Microbiol., 44, 1086–1095, 1982a.

van der Kooij, D., Visser, A., and Hijnen, W. A. M.: Determining the concentration of easily assimilable organic carbon in drinking water, J. Am. Water Works Assn., 74, 540–545, 1982b.

van der Kooij, D., Hijnen, W. A. M., and Kruithof, J. C.: The effects of ozonation, biological filtration and distribution on the concentration of easily assimilable organic carbon (AOC) in drinking water, Ozone Sci. Eng., 11, 297–311, 1989.

van der Kooij, D.: Assimilable organica carbon as indicator of bacterial regrowth, J. Am. Water Works Assn., 84, 57–65, 1992.

van Schagen, K. M., Rietveld, L. C., and Babuška, R.: Dynamic modelling for optimisation of pellet softening, J. Water Supply Res. Technol. AQUA, 57, 45–56, 2008.

Walker, G. M., and Weatherley, L. R.: Bacterial Regeneration in Biological Activated Carbon Systems, Process Saf. Environ., 76, 177–182, 1998.

Wang, H., Ho, L., Lewis, D. M., Brookes, J. D., and Newcombe, G.: Discriminating and assessing adsorption and biodegradation removal mechanisms during granular activated carbon filtration of microcystin toxins, Water Res., 41, 4262–4270, 2007.

Yapsakli, K., Çeçen, F., Aktas, Ö., and Can, Z. S.: Impact of Surface Properties of Granular Activated Carbon and Preozonation on Adsorption and Desorption of Natural Organic Matter, Environ. Eng. Sci., 26, 489–500, doi:10.1089/ees.2008.0005, 2009.

Yavich, A. A., Lee, K.-H., Chen, K.-C., Pape, L., and Masten, S. J.: Evaluation of biodegradability of NOM after ozonation, Water Res., 38, 2839–2846, 2004.

7

Status of organochlorine pesticides in Ganga river basin: anthropogenic or glacial?

P. K. Mutiyar and A. K. Mittal

Department of Civil Engineering, Indian Institute of Technology Delhi, 110016 New Delhi, India

Correspondence to: A. K. Mittal (akmittal@civil.iitd.ac.in)

Abstract. This study reports the occurrences of organochlorine pesticides (OCPs) in the Ganga river basin covering 3 states, i.e., Uttarakhand, Uttar Pradesh and Bihar comprising 72 % of total river stretch consisting of 82 sampling points covered through 3 sampling campaigns. Samples were monitored for 16 major OCPs, including hexachlorocyclohexanes (HCHs), Endosulfan group, Aldrin group, DDTs and Heptachlor group pesticides.

The results showed the $ng\,L^{-1}$ levels contamination of OCPs in all the stretches sampled during these campaigns. The results also revealed that different types of OCPs were dominating in different stretches in accordance with the land use practices and agricultural runoff generated from those stretches. HCHs were most frequently detected (detection rate = 75 %) in mountainous stretch; Endosulfans were prominent in UP (detection rate = 75 %) stretch while in BR stretch Aldrin group pesticides were paramount (detection rate = 34 %). Source apportionment of the OCP's revealed that in the upper reaches of the Ganges i.e., in the state of Uttarakhand, the glacial melt may be responsible for the presence of OCP's. In the lower reaches, intensive agriculture and industrial activities may be significantly contributing these pesticides. The samples from tributaries of Ganga river were found to contain higher numbers of pesticides as well as higher concentrations. The maximum total pesticide concentration in an individual sample during these sampling campaigns was found in the Son river sample ($0.17\,\mu g\,L^{-1}$, Location: Koilwar, Bhojpur, Bihar).

1 Introduction

River basin management plans in India have traditionally considered the point sources of water pollution. The non-point sources of pollution have largely missed out. Non-point pollution sources are of greater importance than point source pollution particularly in rural catchments, where agricultural runoff is the major pollution contributor, which brings nutrients and pesticides to the rivers (Duda, 1993; Jain, 2002). Similar conditions of agricultural practices and runoff, exists in Ganga basin and thus non point sources of pollution to the rivers are of serious concern as mostly pesticides entering river systems via diffuse sources (Holvoet et al., 2007). Trends of high levels of pesticide residues in agricultural runoff leading to river contamination have been reported from different parts of the world (Schulz, 2001a, b;

Varca, 2012; Oliver et al., 2011). In the Ganga river basin, where agriculture predominate the land use activities, pesticides used in agriculture could easily find their way into the river via runoff. Organochlorine pesticides (OCPs) have been extensively used in India for agricultural and public health purposes. OCPs in different environmental matrices are a matter of concern as the complete environmental fate of these chemicals is still an unexplored field.

The Indo-Gangetic alluvium plain, due to fertile soils, is the region of high agriculture and industrial activities with high population density, where pesticides may enter the water environment through runoff. As the OCPs are persistent in nature and could easily find their way in runoff after several years of their application (Kreuger, 1998). So, even after the recent ban on the use of these pesticides, monitoring of their residues in the rivers is required to assess the

Table 1. Various river stretches and the major rivers covered during the sampling campaigns.

Stretches	Utrakhand (UK)	Uttar Pradesh (UP)	Bihar (BR)
Sampling months	Dec 2010–Jan 2011	Jun–Jul 2011	Jul–Aug 2011
Distance (km)	193	1131	341
1st Sampling Point	Rudraprayag	Bijnor	Chhapra
Last Sampling Point	Haridwar	Ballia	Kohalgaon
No of sampling points	20	36	26
Major rivers covered during sampling campaign	Mandakini, Alaknanda, Bhagirathi, Ganga	Ramganga, Ghaghra, Yamuna, Varuna, Gomti, Rapti, Aami, Ganga	Son, Gandak, Ganga

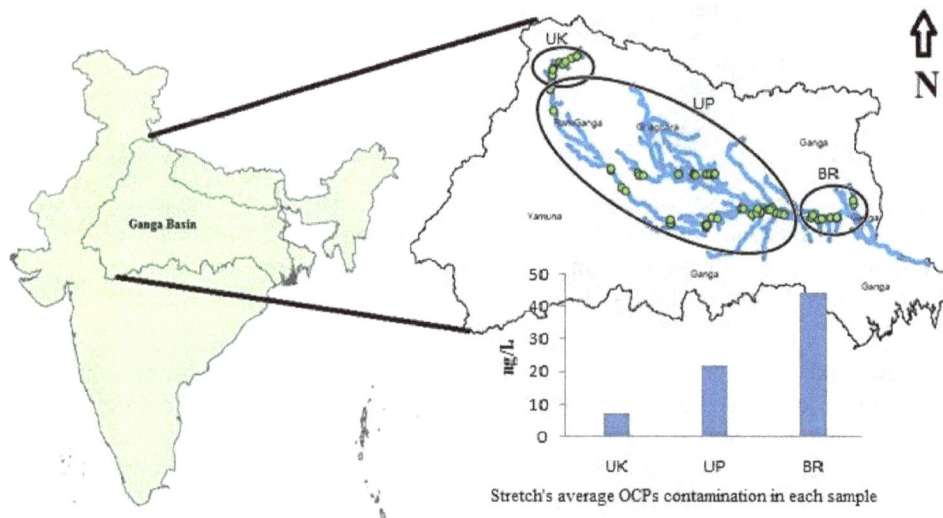

Figure 1. Sampling location of the river Ganga.

impact on human health and related ecological risks. Previously reported OCPs levels in river Ganga are either for specific tributaries or for a stretch of the river. There is no single study available reporting the levels of these contaminants across the Ganga basin. Source apportionment of OCPs in Ganga river is also of high importance. It is yet to be established whether glacial sources or anthropogenic activities contribute pesticides to various rivers of Ganga basin. Glacial melt could be another source of pesticide contamination in the Ganga basin as glacial melt contributes the major share of the Ganga and its tributaries. Rivers in the Ganga basin are the main source of freshwater for half the population of India and Bangladesh. Thus, an understanding of the fate of OCPs in Ganga basin rivers, and identification of their sources of origin is warranted. The present study reports the status of OCPs in the Ganga river and its major tributaries passing through three different states in India. The study area covers a 1805 km long stretch of river Ganges, covering 72 % of its entire length. Sampling campaigns were carried out in three states, i.e., Uttarakhand (UK), Uttar Pradesh (UP) and Bihar (BR), which represent a major part of the Ganga basin.

2 Materials and methods

2.1 Study area

The Ganga rises at 7010 m in Gangotri, Uttarakhand, India, on the southern slopes of the Himalayan range. It flows through five different states, Uttarakhand (UK), Uttar Pradesh (UP), Bihar (BR), Jharkhand (JK) and West Bengal (WB) covering a distance of 2525 km before it enters the Bay of Bengal. Ganga river and its major tributaries at Uttarakhand (UK), Uttar Pradesh (UP) and Bihar (BR) states represent the study area. The river Ganga was sub-divided into three stretches representing different watershed conditions as the UK, UP and BR stretch. The stretches were divided considering (1) different types of watershed, land-use activities, flow types and (2) state boundaries since states are responsible for managing the discharges to the river. The division of river stretches on the basis of states will help in understanding the health of the river in that particular state and their environmental awareness. Details on the sampling campaigns undertaken are presented in Table 1. Water samples were picked from 82 different points during these sampling campaigns from 3 different stretches/regions. The locations of the sampling points are shown in Fig. 1, while Fig. 2

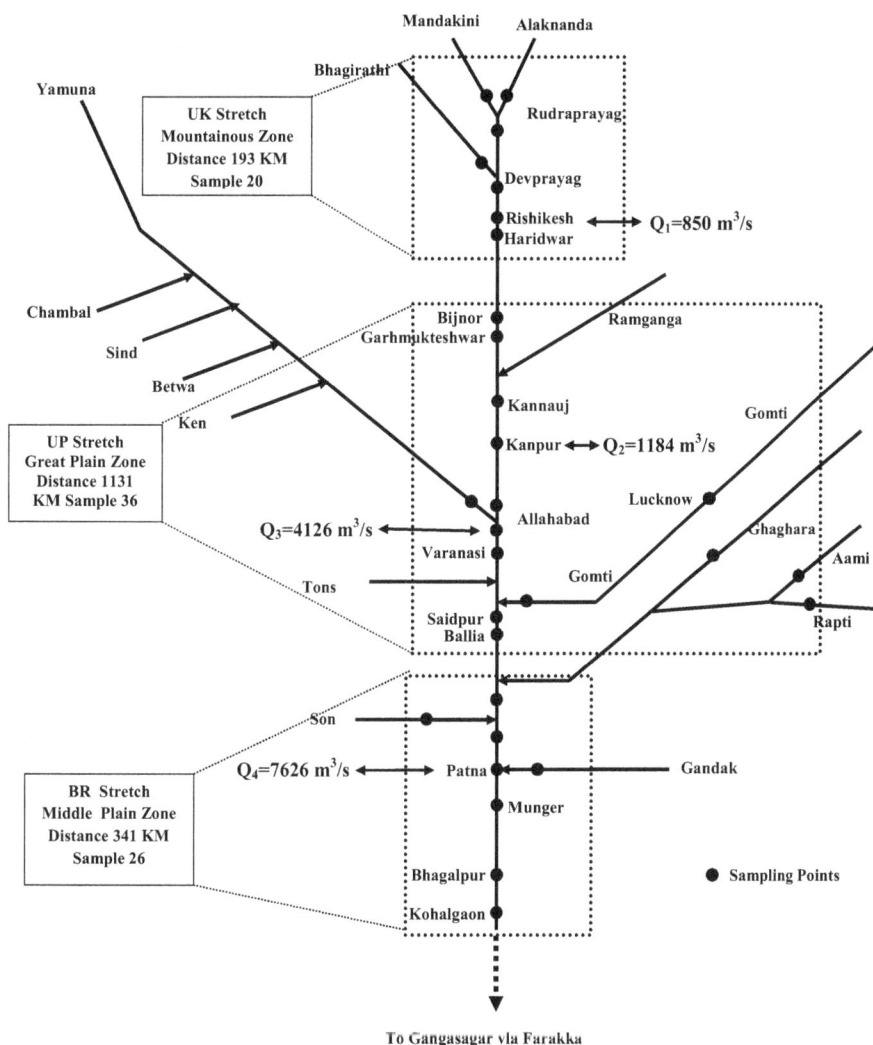

Figure 2. Sampling locations and space boundary of each sampling campaign along river Ganga (flow data were of CWC, Delhi taken from NGBRA, 2011 report as cross reference).

showed the sampling points and a flowchart of the rivers covered during each sampling campaign. The UK stretch was the smallest stretch, while UP stretch was the longest stretch of the sampling campaigns.

2.1.1   UK stretch

The State of Uttarakhand (UK) has three districts which fall in the mainstream of Ganga: Haridwar, Tehri Garhwal and Uttarkashi. Sampling campaigns of this stretch was carried out in December 2010. Sampling was started from upstream of Rudraprayag at Alaknanda and Mandakini rivers, and it went up to downstream of Haridwar (Table 1). This area is the hilly-mountainous zone of the river Ganga with a high bed slope (1 : 67) and mean flow rate of 856 m³ s⁻¹ (Fig. 2). Domestic sewage is the major source of pollution since there are no other major agricultural or industrial activities in this stretch.

2.1.2   UP stretch

It is the mid-stretch of the river and forms a part of the Great Plains of Ganga basin. It constitutes 17 districts of Uttar Pradesh (UP). Sampling campaign was carried out in June and July 2011. Sampling started from upstream of Bijnor and went up to downstream of Ballia, separating a distance of 1131 km (Fig. 2). Samples from Ramganga, Ghaghra, Yamuna, Gomti, Rapti and Aami; the major tributaries of Ganga, were also taken, to quantify the pesticide contamination contributed by the tributaries of river Ganges. Total 36 samples were taken from this stretch (Table 1). Rivers, in this stretch, receive pollution from highly diversified sources, including domestic, industrial and agricultural sources. Downstream to Haridwar, where the Ganga opens to the Gangetic Plains, major share of water is diverted by various barrages for irrigation and other purposes. The Ganga does not receive any major tributary until its tributary, Ramganga river joins at Kannauj, which is 460 km downstream from Haridwar (Fig. 2). River has relatively less flow upto Allahabad, where the Yamuna confluences the river Ganges. Downstream to Allahabad, river is joined by Tons and Gomti (Fig. 2). The subsequent upper plain section extends from Rishikesh to Allahabad at a slope of one in 4100 and a mean flow rate range between 850–1720 m^3 s^{-1} before its confluence with the Yamuna.

2.1.3   BR Stretch

There are 12 districts which fall within the Ganga basin in Bihar (BR), where agriculture and commercial fisheries in the river are the two most important source of livelihood for people. The river Ganga receives several major tributaries in this section, namely, Ghaghara, Son, Gandak, and Kosi. The flow in this stretch continually increases since major tributaries confluence to the Ganga river in this stretch. The average annual flow increased to 7626 m^3 s^{-1} at Patna (Fig. 2) from 4126 m^3 s^{-1} at downstream of Allahabad. The sampling campaign was carried up to Kohalgaon, just a few km before joining Kosi. The sampling campaign covered 341 km stretch (Fig. 2) of the river Ganges, from Chhapra to Kohalgaon, via Patna, Munger and Bhagalpur (Table 1). The samples were also collected from the Son and Gandak river, the major tributaries of Ganga in this stretch. This river stretch receives pollution from domestic, agricultural as well as industrial sectors. Raw sewage flows into the river in this stretch since sewage is not treated in Bihar due to various reasons as reported (CPCB report, 2009) (http://www.cpcb.nic.in/newitems/8.pdf).

2.2   Water sampling, collection and storage

During these sampling campaigns, which were carried out between December 2010 and August 2011, a total of 82 water samples were collected from different sites (Figs. 1 and 2). Two samples were taken from each site during each sampling campaign. One sample was collected in a 1000 mL HDPE bottles, was used to determine physico-chemical parameters and OCP analysis, while the second sample was taken in 100 mL HDPE bottle and preserved with acid. This acidified sample was used for TOC, NO_3^--N and NH_4^+-N analysis. Sampling bottles were rinsed with river water and were carefully filled to overflowing, without trapping air bubbles in sealed bottles. The samples were transported in cool-box with ice packs and subsequently stored in a refrigerator at 4 °C until further analysis. All the samples were transported on ice and kept under refrigeration until performance of laboratory analysis.

2.3   Reagents and standards

Analytical grade (AR) chemicals (Merck, Germany) were used throughout the study without any further purification.

Reagents and calibration standards for physico-chemical analysis were prepared using double glass distilled water. The glass-wares were washed with dilute nitric acid (1.15 N) followed by several portions of distilled water. EPA 502 Pesticide Standard Mix (49690-U) was procured from Sigma-Aldrich USA. The working standards of pesticides were prepared by diluting EPA pesticide mixture standard in n-Hexane and were stored at −20 °C. The samples were analysed within one week of sampling campaigns.

2.4   Physico-chemical parameters

Samples were analysed for different physico-chemical parameters: pH, electrical conductivity (EC), alkalinity, chloride, hardness, dissolved oxygen (DO), total organic carbon (TOC), nitrate, and ammonia as per APHA (1998). EC, pH, DO and total dissolved solids (TDS) were measured onsite using portable meters. Alkalinity, chloride and hardness were measured by titration method in the laboratory. Nitrate and ammonia were measured by selective ion electrode (Thermo and HACH, respectively), while TOC was analysed on TOC analyser (Shimadzu).

2.5   Extraction

The method prescribed by APHA (1998) with some modifications was used for the extraction of OCP residues from the water samples. A liquid liquid extraction (LLE) method, using n-hexane as solvent, was used for extraction of pesticide residues. Samples were prefiltered using 0.45 µm glass fiber filter to remove suspended impurity and were extracted without any pH adjustment. Sample containers were shaken and each 500 mL portion of filtered sample was transferred to a separating funnel (1000 mL cap.) fitted with glass-stopper. It was mixed with 30 g of NaCl and 50 mL of n-hexane. The sample was shacked properly for 30 min and the hexane layer was separated. Two further extractions with 30 mL n-hexane were done and the combined hexane extract was treated with 5 g anhydrous Na_2SO_4 to remove traces of water. The water-free extract was rotary vacuum evaporated to a small volume and transferred to a glass-stoppered test tube followed by evaporation of solvent under a mild stream of N_2 to 0.5 mL. The concentrated extracts were transferred to air-tight gas chromatograph (GC) vials and stored at −20 °C until their analysis.

2.6   OCPs analysis

The determination of OCPs was performed on a *Thermo Trace GC Ultra* gas chromatograph equipped with 63-Ni micro-electron capture detector (GC-ECD) and an autosampler. The column specifications and operating conditions are given in Table 2. Analysis was performed by EPA method 508, with slight modification. Identification of individual OCPs was based on comparison of retention

Table 2. Operating conditions used for the operation of GC-ECD (Double column confirmation).

	GC-ECD	
Column	DB-5, fused silica capillary column (30 m × 0.25 mm ID, film thickness 0.25 μm)	DB-1701, fused silica capillary column (30 m × 0.25 mm ID, film thickness 0.25 μm)
Purpose	Screening and Quantification	Cross confirmation by Retention Pattern
Injector Temperature	250 °C	
Injection Volume	2.5 μL (Splitless mode)	
Oven Programming	90 to 150 °C @ 15 °C min^{-1}, 150 to 220 °C @ 3 °C min^{-1} and 220 to 270 °C @ 5 °C min^{-1}	
Detector Temperature	280 °C	
Carrier gas	Helium @ 1.2 mL min^{-1}	
Makeup gas	Nitrogen @ 40 mL min^{-1}	

time between samples and the standard solution by double column chromatography under similar conditions. DB-5 (30 m × 0.25 mm i.d., and 0.25 μm film thickness) and DB-1701 (30 m × 0.25 mm i.d., and 0.25 μm film thickness) columns were used in the analysis. Tentative identifications of the pesticides were made on the basis of retention time obtained using DB-5. These were subsequently confirmed with second capillary column, DB-1701, having dissimilar liquid phase with different retention properties. The injection volume, column conditions, temperature programming, injector and detector temperature were kept the same for GC-ECD in both analyses. Helium was used as the carrier gas at a constant flow of 1.2 mL min^{-1} and high purity nitrogen was used as make-up gas (40 mL min^{-1}). Samples were injected using Thermo AS 3000 auto-sampler. Injection volume was 2.5 μL in splitless mode for each sample (Table 2). The instrument was operated by Xcaliur software (Thermo Finnigan). Quality of extraction and detection procedure was ensured by spiking 5 different concentrations of each OCP standard with distilled water, and extracting by the same method. Recovery was determined. Table 3 presents recovery efficiency (RE), retention time (RT) and pattern of compounds eluting on both the columns. The DB-5 column was used for quantification, while DB-1701 was used for compound cross confirmation by retention pattern. The important physico-chemical properties of investigated OCPs are expressed elsewhere (Mutiyar et al., 2011).

3 Result and discussion

3.1 Physico-chemical parameters (general water quality parameters)

Both, the quantity and quality of water get affected as the water from the river is either being diverted for various beneficial uses (canals for irrigation, industrial and drinking purposes) or by the sewage from the cities and agricultural runoff from other areas flowing into the river. The water quality of all the three stretches covered during different sampling campaigns is shown in Table 4. Water quality in mountainous stretch (UK stretch) is very good, with high DO levels (DO$_{avg}$ (mg L^{-1}) = 7.7 + 0.6), very low EC, TDS and TOC, indicating no significant pollution load in this stretch. The lower organic loading from the small cities of UK and high flow in the river keeps the stretch relatively clean. In this stretch, the major class 1 cities on the Ganges are Rishikesh and Haridwar, where 3 sewage treatment plants (STPs) are in operation. The STPs reduce 61–93 % organic loadings and half of trace contaminants present in the sewage (CPCB report, 2009; Mutiyar and Mittal, 2013). Domestic sewage is the major contributor of pollution in this stretch, which is more significant towards the end of this stretch where the last sampling point is situated, i.e. Haridwar and Rishikesh. Kumar et al. (2010) reported that the water quality of UK river stretch is of category A as per CPCB river classification, except for the stretch downstream of Haridwar, the last sampling point of the campaign 1. Results revealed similar situation in this study (Table 4). The UP stretch is the longest stretch of the sampling campaign, including many rivers and the sub-basins of the Ramganga, Ghaghra and Gomti river. The total discharge of wastewater from this zone to Ganga basin is second maximum after Delhi. The water quality in

Table 3. Standardisation of OCP's compounds using GC-ECD.

Compound	RT (min) (DB-5)	Recovery (%)	R^2	MQL (ng L^{-1})	Retention Pattern (DB-1701)
α-HCH (H1)	12.00	71.28	0.999	0.01	α-HCH (H1)
β-HCH (H2)	13.42	79.42	0.998	0.01	γ-HCH (H3)
γ-HCH (H3)	14.76	70.99	0.999	0.01	Heptachlor (He1)
δ-HCH (H4)	15.02	78.56	0.999	0.01	Aldrin (A1)
Heptachlor (He1)	16.70	87.53	0.996	0.01	β-HCH (H2)
Aldrin (A1)	18.41	146.78	0.999	0.01	δ-HCH (H4)
Hepta-Epoxide (He2)	20.56	59.55	0.999	0.01	Hepta-Epoxide (He2)
α-Endo (E1)	22.45	107.97	0.999	0.01	α-Endo (E1)
4,4'-DDE (D1)	23.87	123.87	0.999	0.01	4,4'-DDE (D1)
Dieldrin (A2)	24.09	86.16	0.999	0.01	Dieldrin (A2)
Endrin (A3)	25.01	87.70	0.998	0.01	Endrin (A3)
β-Endo (E2)	25.75	90.85	0.998	0.01	4,4'-DDD (D2)
4,4'-DDD (D2)	26.40	87.47	0.997	0.01	β-Endo (E2)
Endrin-aldehyde (A4)	28.05	133.12	0.995	0.01	4,4'-DDT (D3)
Endo-Sulfate (E3)	28.44	85.80	0.995	0.01	Endrin-aldehyde (A4)
4,4'-DDT (D3)	31.53	99.22	0.990	0.01	Endo-Sulfate (E3)

MQL = methods quantification limit

Table 4. River Water Quality of Ganga Basin covered during sampling campaigns (December 2010–August 2011).

	UK			UP			Bihar		
	Range	Avg	SD	Range	Avg	SD	Range	Avg	SD
pH	7.7–8.1	7.9	0.1	7.2–8.6	7.9	0.4	7.1–8.8	8.4	0.4
EC (μS cm^{-1})	19.5–42.9	32.2	7.4	118.8–381.0	214.2	75.5	186.7–452	345	74
TOC (mg L^{-1})	0.050–0.664	0.261	0.274	0.1–4.6	2.5	1.1	0.1–18.6	2.3	3.5
Nitrate (mg L^{-1})	NM	–	–	0.7–2.8	1.8	0.6	1.5–5.1	2.9	0.9
Ammonia (mg L^{-1})	NM	–	–	0.5–7.9	3.4	1.9	0–0.5	0.1	0.1
Chloride (mg L^{-1})	20–40	29.5	6.9	13.0–32.1	23.3	4.8	29–149	49.4	24
Hardness (mg L^{-1})	NM	–	–	65.2–143.2	98.9	18.6	100–603	190.3	95.5
Alkalinity (mg L^{-1})	50–110	83.5	15.7	71.2–164.4	109.4	21.4	102–401	178.4	89.6
TDS (mg L^{-1})	12.5–27.5	20.6	4.7	114.8–286	175.9	40.6	119.5–289.3	220	47.4
DO (mg L^{-1})	6.7–9.2	7.7	0.6	1–6.5	5.6	1.4	4.0–9.4	6.9	1.4

Avg = Average, SD = Standard deviation, NM = Not measured

the stretch is affected by domestic and industrial discharges, and agricultural runoff. The DO levels in all the samples were in the range of 1–6.5 mg L^{-1} (DO$_{avg}$ (mg L^{-1}) = 6.5 + 1.4), however this zone has some of the worst polluted stretches, including Kanpur and Allahabad regions. But due to high monsoonal flow, the river water quality appeared good from the water quality data obtained during this sampling campaign (Table 4). The minimum DO (1 mg L^{-1}) was reported from Varuna river, a small tributary of the Ganga, at Varanasi where this river has very less flow even during the monsoon season. The recent report on trends on water quality in the Ganga basin (CPCB, 2009) showed that river water quality of this stretch is fine except maximum biochemical oxygen demand (BOD$_5$) levels, in Kannauj-Gazipur segment. In Bihar, no sewage treatment plant (STP) was found working during the sampling campaigns. The total installed sewage treatment capacity is 84 mld against total discharge of 671 mld

in Ganga basin, and none of the STPs is functional in Bihar (CPCB report, 2009). Though, the sewage management in Bihar is very poor, but the water quality continually improved. It may be attributed to the dilution provided by the high flow from the major tributaries in this stretch. The DO levels were high (DO$_{avg}$ (mg L^{-1}) = 6.9 + 1.4) and Gangatic dolphins were seen during the sampling at various places, from Patna to Kohalgaon. The water quality was good and the report on trends in water quality in the Ganga basin (CPCB, 2009) supports the data as water quality in Bihar segment was well within permissible limits except for fecal coliforms (FC).

3.2 Organochlorine pesticides

Various types of pesticides are widely used in agricultural sector all over the Ganga basin and have been frequently

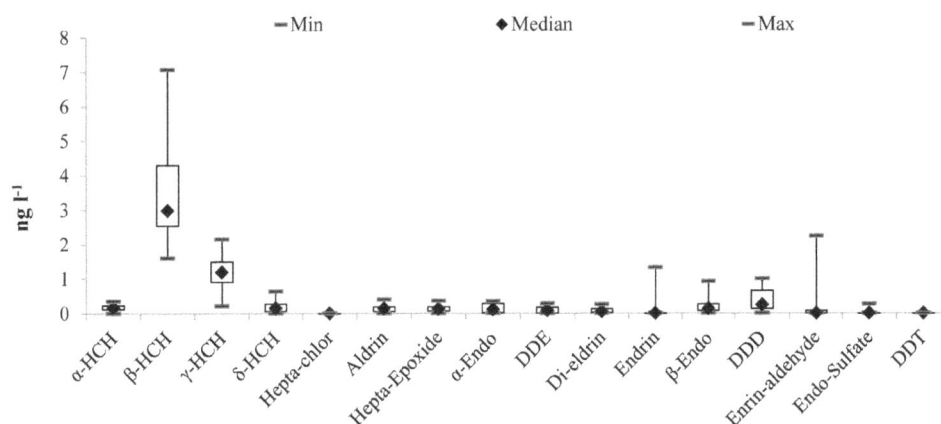

Figure 3. Organochlorine pesticide residues (ng L^{-1}) in river water samples of Uttarakhand (UK) area.

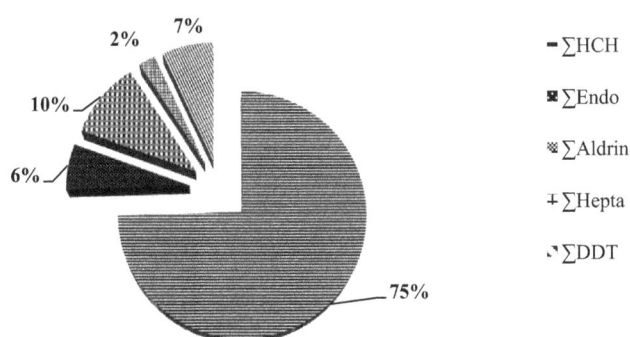

Figure 4. Detection frequency of individual OCPs across UK stretch of Ganga River.

reported in the water matrices from the basin (Rehana et al., 1995; Nayak et al., 1995; Sankararamakrishnan et al., 2005; Semwal and Akolkar, 2006; Malik et al., 2009; Singh et al., 2011). Beside the runoff from agricultural fields, the agricultural practices in the dry bed of the rivers, which are common in India (Hans et al., 1999) also, add pesticides to the river during monsoon. The OCPs levels in UK stretch are shown in Fig. 3. In this stretch concentration of all the targeted OCPs varies from not detected (ND) to 7.07 ng L^{-1}. Water sample from Rishikesh showed the maximum number of OCPs, i.e., 14 out of targeted 16, while the Ganga main river at Haridwar showed presence of least number; only 2 out of 16 were present. The occurrence frequency for the OCPs in this stretch varied from 0 to 100 %, as heptachlor and DDT were not detected in any of the samples (detection rate, 0 %), while β-HCH and γ-HCH were found in all the samples (detection rate, 100 %). The endosulfan sulfate and endrin were detected in one sample (detection rate, 5 %) while aldrin, endrin, dieldrin, heptachlor epoxide, α-Endo, β-Endo, DDE and DDD showed 75 % occurrence. The $\sum$HCH group was the most predominant in this stretch, accounting 75 % of relative abundance (Fig. 4). The low ratios of

α HCH/γ HCH (0.15) indicate that lindane may be an important source of HCHs in this stretch as technical HCH sources have high α HCH/γ HCH (Ridal et al., 1996). The lindane has been used extensively in the Indo-Gangetic plains for agriculture. Wang et al. (2008) reported that the OCPs used in Indo-Gangetic plains could reach the snow of Mount Everest via global circulation and cool deposition. Similar trends of deposition of OCPs in glacier via cold trap have also been reported by Valsechhi et al. (1999) and Kang et al. (2002). Blais et al. (2001) explained that melting glaciers supply up to 97 % of OCPs input while contributes 73 % of input water. In UK zone, the major share comes from melting ice from glaciers, thus high share of HCHs in glacial stream is expected. $\sum$Endo group's relative abundance was 10 % of the total indicates that this pesticide has limited use for agricultural purposes in this stretch. Very limited farming is done in this part of the Ganga basin, so trace levels of endosulfan residues could find their way into the river water from agricultural application via runoff. The heptachlor group formed 2 % of the total abundance, with no heptachlor been detected in any of the samples. Only heptachlor epoxide was detected in the samples suggested that this pesticide has been used in past in the basin.

The occurrence of OCPs in UP stretch is shown in Fig. 5. The trend of detection rate of OCPs was different in the UP stretch as the detection rate varies from 6–94 % (Fig. 6). All the samples were found to contain one or more pesticides. The minimum number OCPs detected in any sample was 3, while a maximum of 14 OCPs were present in one of the samples. The β-Endo OCP was frequently detected in many of the samples in relatively high concentration as the maximum concentration of β-Endo was 133.10 ng L^{-1} (Fig. 5). $\sum$Endo group pesticides contributed the maximum (75 %), while aldrin, DDT and HCH group contributed 11, 9 and 5 %, respectively (Fig. 6). The use of endosulfan is now banned in more than 60 countries but India has been the world's largest producer and consumer of endosulfan with a total use

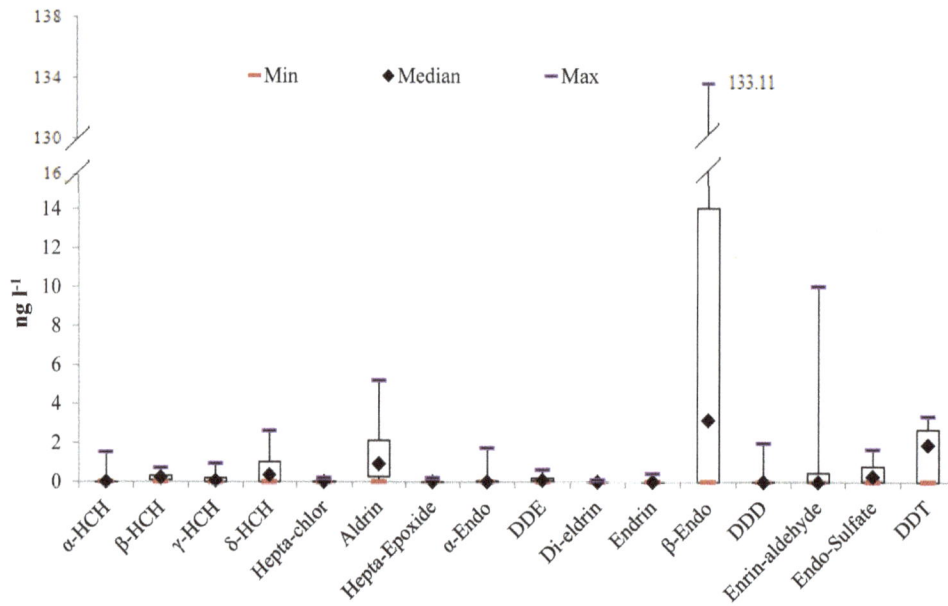

Figure 5. Organochlorine residues (ng L^{-1}) in water matrices of rivers of Uttar Pradesh (UP).

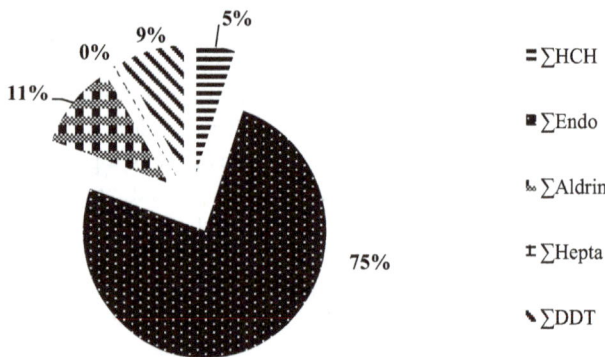

Figure 6. Detection frequency of individual OCPs across UP stretch of Ganga River.

of 113 000 tonnes from 1958 to 2000 (NGBRA, 2011). Recently, the supreme court of India has put a temporary ban on agricultural use of endosulfan pesticides (Writ petition 213/2011), but the impacts of this ban could only be noticed after a number of decades. The high concentration of Endo group pesticides is in conformity with its wide use in this area. The endosulfan was most widely used pesticide in Indo-Gangetic plains for agricultural purposes. Most studies suggest that α-endosulfan has a faster degradation than β-endosulfan, and that endosulfan sulfate is much more persistent (INIA, 1999–2004). Similar trends were observed for Endo group pesticides in UP stretch. β-endosulfan and endosulfan sulfate were more frequently detected in the water samples as compared to α-endosulfan. Endosulfan sulfate is the most persistent, but its reported concentrations are lower than its other isomers. It may be due to its lower share (< 1 %)

in technical endosulfan. As pesticides are used more sporadically, there are different reports on their occurrences in river Ganga. Higher levels of endosulfan (750 ng L^{-1}) have been previously reported in the Ganges river at Kannauj (Rehana et al., 1996) to the present level of 31.61 ng L^{-1} but another study has reported absence of endosulfan residues in Ganges at Kanpur (Sankararamakrishnan et al., 2005).

The occurrences of OCPs in Bihar stretch are shown in Fig. 7. The levels of OCPs varied from ND to 38.80 ng L^{-1} for aldrin in Son river, a major tributary of the river Ganga in Bihar. δ-HCH was not detected in any of the samples, while α-HCH was present in all the samples (Fig. 7). Frequency of appearance of the OCPs was higher for samples from tributaries of Ganga, than the parent river (Fig. 7). Sample from storm water drain near Ara (Bhojpur) which carried agricultural runoff was containing 15 out of 16 targeted compounds. Sultanganj and downstream to the Bhagalpur city area are located in the most downward stretch of the studied stretches. These stretches were found to contain 15 OCPs. Besides, the average concentration of monitored OCPs continually increased from UK to BR stretch (upstream to downstream) (Fig. 1). It may be due to the continual increase of contribution from the agricultural sector in the downstream stretches. In BR stretch, the occurrence patterns of OCPs were different from the UK and UP stretch. In UK stretch, HCHs group was more frequently detected, while in UP stretch the same was observed in Endo groups. High glacial water in the UK stretch may be a possible reason for this, since the glacial water streams reportedly have higher concentrations (1 log) of HCHs as compared to endosulfan and dieldrin pesticides in Bow lake in Canada (Blais et al., 2001) and Himalayan glaciers (Kang et al., 2009). Bizzotto et al. (2009)

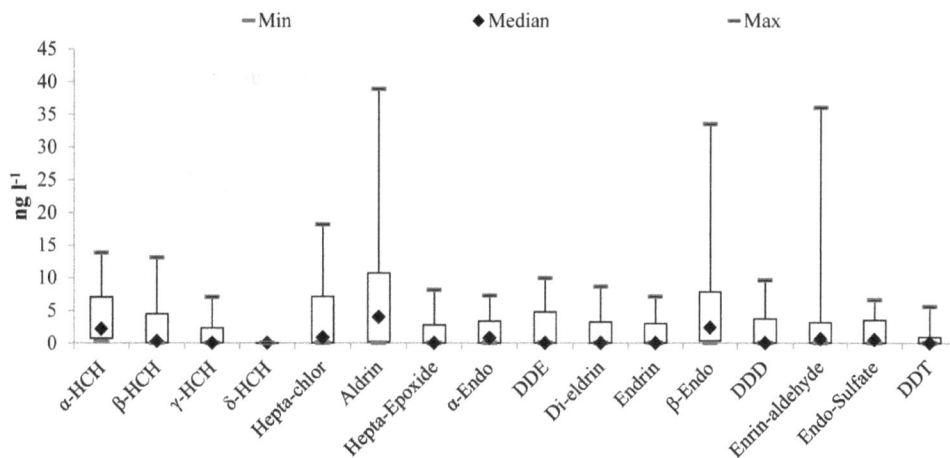

Figure 7. Organochlorine pesticide residues (ng L^{-1}) in water matrices of rivers of Bihar (BR) Area.

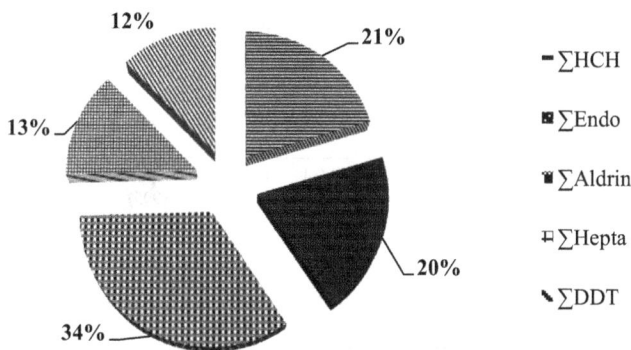

Figure 8. Detection frequency of individual OCPs across BR stretch of Ganga River.

compared the HCHs concentration in glacial and non-glacial streams from Alpine glaciers and found that glacial streams were always having a higher concentration (200–400 %) than non-glacial streams. Thus high frequencies of HCHs in the mountainous zone of Ganga were on the expected lines. The UP stretch was predominant with the Endosulfan group pesticides in connection with high previous use of this group of pesticides in the Indo-Gangetic plains for agriculture use (NGBRA, 2011). In Bihar stretch, none of these pesticide groups were dominant in occurrences. The maximum detection frequency was for the aldrin group (34 %), followed by HCHs (21 %), Endo group (20 %), heptachlor group (13 %) and DDTs group (12 %) (Fig. 8). It showed the mixed flow of glacial, domestic, agricultural and industrial discharge to the river. The high concentration of the heptachlor group in this stretch could be because of high previous use of this pesticide in parts of Bihar and West Bengal for termite control. As this stretch of the river receive flow from different river basins, having different agricultural practices and different pesticide uses. So the mixture of all OCPs was expected (Fig. 8). Table 5 presents the relative contamination levels of OCPs in

2011 along with the values reported in the literature. It shows reduction in OCPs levels in the Ganga water.

River waters are highly vulnerable to OCPs contaminations as OCPs once used in river basins are continually released and transported to the rivers and its tributaries via runoff from agricultural and urban areas, discharge from reservoirs and aquifers, and atmospheric deposition. The concentration of OCPs in river water is governed by several factors such as physiochemical properties of the pesticides, its usage, rainfall, and farming practices etc. and thus their concentration may have high spatial and temporal variations. People in the rural areas are directly using Ganges water for potable purposes and thus it is mandatory to compare the OCPs contamination with the regulatory standards. The maximum contaminant levels (MCLs) for pesticides in drinking water in India are described in BIS:10500 (Bureau of Indian Standards, Drinking Water Standard, 2003). The safe limit for pesticides in potable water is described as total pesticides concentration should be less than 1000 ng L^{-1} for drinking water, however this limit is quite higher compared to the world health organization (WHO) and European Union (EU) limits for potable water (500 ng L^{-1}). Variation in total OCPs contamination in individual sample across the monitoring stretches is described in Fig. 9. It's evident from the results (Fig. 9) that Ganges water is safe for potable purposes in terms of OCPs contamination; however other types of pesticides (organophosphates, synthetic pyrethroids etc.) may also be present in river water which was not monitored. As the maximum concentration of total OCPs in individual sample (0.17 µg L^{-1}, Fig. 9) was within the safe limits and pesticides are also removed during conventional water treatment process (partially to substantially) (Stackelberg et al., 2007) which will further reduce the pesticides levels in finished drinking water. Thus production of water through a water treatment plant is advantageous and minimizes the risk exposed by pesticide contamination as well. Treated water is supplied for potable purposes in most of the cities on the

Table 5. Comparison of reported OCPs levels in rivers of Ganga Basin to the present study at different locations.

River	Sampling Site	Compound Reported (ng L^{-1})										*Reference
		∑HCH		∑DDT		∑Endo		∑Aldrin		∑Hepta		
		Literature Levels*	Present Levels (2011)	Literature Levels*	Present Levels (2011)	Literature Levels*	Present Levels (2011)	Literature Levels*	Present Levels (2011)	Literature Levels*	Present Levels (2011)	
Ganga	Devprayag	ND (2006)	7.24	ND-365 (2006)	ND	ND-66 (2006)	ND	ND-46 (2006)	2.3	NM	0.07	Semwal and Akolkar (2006)
	Rishikesh	6–124 (1992)	5.5	4–98 (1992)	1.01		0.92		1.89		0.32	ITRC annual report (1992)
	Haridwar	4–153 (1992)	5.2	2–113 (1992)	0.19		0.16		0.12		0.06	ITRC annual report (1992)
	Kannauj	3010 (1995), 8–154 (1992)	0.1–1.0	7740 (1995), 3–150 (1992)	0.05–0.12	750 (1995)	0.8–31.6	3340 (1995)	1.2–1.3	NM	0.14–0.2	Rehana et al. (1995) ITRC annual report (1992)
	Kanpur	450 (2005), 14–359 (1992)	0.1–0.36	ND (2005), 8–174 (1992)	0.2	ND (2005)	9.7–11.6	ND (2005)	ND-1.1	NM	ND-0.08	Sankararamakrishnan et al. (2005) ITRC annual report (1992)
	Allahabad	7–270 (1992)	1.23–3.5	2–136 (1992)	0.08–2.21		ND-0.15		ND-0.4		ND	ITRC annual report (1992)
	Varanasi	9–156 (1992), 105–99517 (1995)	0.2–0.7	3–84 (1992), 64–143 226 (1995)	0.1–1.9	83–66 516 (1995)	ND-85.4	NM	0.5–2.2		ND-0.1	ITRC annual report (1992) Nayak et al. (1995)
	Patna	11–131 (1992)	0.3–5.0	5–385 (1992)	ND		ND-5.03		ND-1.17		ND	ITRC annual report (1992)
	Bhagalpur	ND-74.04 (2011)	12.4–17.6	NM (2011)	11.6–12.3	ND-208 (2011)	13.8–17.9	ND-489 (2011)	8.8–16.4	NM (2011)	3.2–11.8	Singh et al. (2011)
Gomti	Lucknow	ND-507 (2009)	0.72	ND-108 (2009)	2.75	ND-186 (2009)	1.16	ND-82 (2009)	0.6	ND-91 (2009)	ND	Malik et al. (2009)

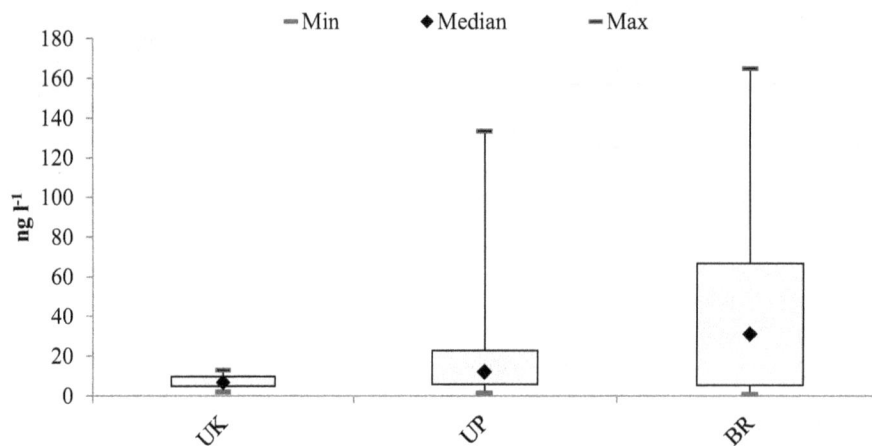

Figure 9. Variation in total OCPs residues in Ganges water (BIS Safe limit = 1000 ng L^{-1}).

bank of the rivers, but in rural areas it is still consumed without treatment as well. Thus, it is the need of the hour to upgrade the existing water practices and should be moved for treated potable water.

4 Conclusions

OCPs, 16 in number were monitored from 82 sampling sites in the Ganga river basin. These included mountainous and the great plains of the Ganga basin. The results revealed that different types of OCPs predominate in different regions depending upon the land use pattern and differential past use of the OCPs for agricultural and public health programme. Since, there is not even a single report on the levels of OCPs representing the entire Ganga stretch; the present findings could be effectively used in understanding the present status of the river. The comparative analysis of present study to the previous reports, showed the decline trend in OCPs contamination in the river water, which is a good sign at environmental and ecological front. The ban on the use of various OCPs has shown a positive sign for river health, but as these compounds are highly lipophillic, thus it becomes quite necessary to monitor these compounds continuously in the river stretch. There is a wide gap in the timeline of the continuous reporting of OCPs levels and thus, it is recommended to frequently monitor the river quality for OCPs contamination with changing land-use pattern, pesticide formulation, climatic conditions and ecological and environmental sense of the society. The decreasing trends in the OCPs contamination levels in Ganges water were also confirmed. The banned or restricted use of OCPs and increasing environmental awareness regarding pesticide application in farmers may be the possible contributor to this declining trend. The maximum total pesticide concentration in individual sample $(0.17 \, \mu g \, L^{-1}$, Son river sample from Koilwar) was less than $0.2 \, \mu g \, L^{-1}$ for all the samples against a safe drinking water limit of $0.5 \, \mu g \, L^{-1}$ by European Union (EU) and $1 \, \mu g \, L^{-1}$ of Bureau of Indian Standards (BIS). Lower concentration $(ng \, L^{-1})$ of pesticides in rivers indicates the wise use of pesticide in the area but the higher detection rate of endosulfan group pesticide in the UP stretch which receives agricultural runoff from Indo-Gangetic plains are of significant concern. The recent temporary ban on the use of endosulfan for agricultural use by the Supreme Court of India (Writ petition no 213/2011), is a precautionary and an appreciable step towards conserving the water resources from the further POP contamination.

Acknowledgements. Authors are thankful to D. P. Singh, Former Chief Engineer UP Jal Nigam, S. Z. S. Tabish, SE CPWD, S. K. Gupta, N. S. Maurya and Pooja Srivastava for their help during sampling campaigns. One of the authors (Pravin K. Mutiyar) is also thankful to University Grants Commission (UGC), New Delhi (India) for providing financial assistance in the form of Junior Research Fellowship (JRF) to conduct this work.

References

APHA, AWWA: Standard Methods for Examination of Water and Wastewater Investigations, APHA, AWWA, Washington, DC, 1998.

Bizzotto, E., Villa, S., Vaj, C., and Vighi, M.: Comparison of glacial and non-glacial-fed streams to evaluate the loading of persistent organic pollutants through seasonal snow/ice melt, Chemosphere, 74, 924–930, 2009.

Blais, J., Schindler, K., Muir, D., Donald, D., Sharp, M., Lafreniere, M., Braekevelt, E., and Strachan, W. M. J.: Melting glaciers dominate sources of persistent organochlorines to subalpine Bow Lake in Banff National Park, Canada, Ambio, 30, 410–415, 2001.

Bureau of Indian Standard (BIS): BIS10500-1991, Indian Standard Drinking Water-Specification, 2003.

CPCB: Ganga water quality trends, Monitoring of Indian Aquatic Resources, MINARS/31/2009–2010 available at: http://cpcb.nic.in/upload/NewItems/NewItem_168_CPCB-Ganga_Trend%20Report-Final.pdf, 2009.

CPCB report: Status of sewage treatment plants in Ganga basin, Central Pollution Control Board, available at: http://www.cpcb.nic.in/newitems/8.pdf, 2009.

Duda, A. M.: Addressing non-point sources of water pollution must become an international priority, Water Sci. Technol., 28, 1–11, 1993.

Hans, R. K., Farooq, M., Suresh Babu, G., Srivastava, S. P., Joshi, P. C., and Viswanathan, P. N.: Agricultural produce in the dry bed of the River Ganga in Kanpur, India – a new source of pesticide contamination in human diets, Food Chem. Toxicol., 37, 847–852, 1999.

Holvoet, K. M. A., Seuntjens, P., and Vanrolleghem, P. A.: Monitoring and modeling pesticide fate in surface waters at the catchment scale, Ecol. Modell., 209, 53–64, 2007.

INIA: Instituto Nacional de Investigación y Tecnología Agraria y Alimentaria (I.N.I.A.) including addenda available at: http://chm.pops.int/Convention/POPsReviewCommittee/Meetings/POPRC4/Convention/tabid/359/Default.aspx, 1999–2004.

ITRC: Industrial Toxicological Research Centre, Lucknow, 6th Annual Progress Report (July 1991–June 1992), Measurements on Ganga water quality – Heavy metal and Pesticides, http://www.itrc.org, 1992.

Jain, C. K.: Hydro-chemical study of a mountainous watershed: the Ganga, India, Water Res. 36, 1262–1274, 2002.

Kang, J. H., Choi, S. D., Park, H., Baek, S. Y., Hong, S., and Chang, Y. S.: Atmospheric deposition of persistent organic pollutants to the East Rongbuk Glacier in the Himalayas, Sci. Total Environ., 408, 57–63, 2009.

Kang, S. C., Mayewski, P. A., Qin, D. H., Yan, Y., Hou, S., Zhang, D., Ren, J., and Kruetz, K.: Glaciochemical records from a Mt. Everest ice core: relationship to atmospheric circulation over Asia, Atmos. Environ., 36, 3351–3361, 2002.

Kreuger, J.: Pesticides in stream water within an agricultural catchment in southern Sweden, 1990–1996, Sci. Total Environ., 216, 227–251, 1998.

Kumar, A., Bisth, B. S., Joshi, V. D., Singh, A. K., and Talwar, A.: Physical, Chemical and Bacteriological Study of Water from

Rivers of Uttarakhand, J. Hum. Ecol., 32, 169–173, 2010.

Malik, A., Ojha, P., and Singh, K. P.: Levels and distribution of persistent organochlorine pesticide residues in water and sediments of Gomti River (India) – a tributary of the Ganges River, Environ. Monit. Assess., 148, 421–435, 2009.

Mutiyar, P. K., Mittal, A. K., and Pekdeger, A.: Status of organochlorine pesticides in the drinking water well-field located in the Delhi region of the flood plains of river Yamuna, Drink. Water Eng. Sci., 4, 51–60, doi:10.5194/dwes-4-51-2011, 2011.

Mutiyar, P. K. and Mittal, A. K.: Occurrences and fate of an antibiotic amoxicillin in extended aeration-based sewage treatment plant in Delhi, India: a case study of emerging pollutant, Desalin. Water Treat., doi:10.1080/19443994.2013.770199, in press, 2013.

Nayak, A. K., Raha, R., and Das, A. K.:. Organochlorine pesticide residues in middle stream of the Ganga river, India, Bull. Environ. Contam. Toxicol., 54, 68–75, 1995.

NGBRA: National Ganga River Basin Authority (NGRBA), Draft, Environmental and Social Management Framework (ESMF), Volume I – Environmental and Social Analysis, 2011, available at: http://moef.nic.in/downloads/public-information/Draft%20ESA%20Volume%20I.pdf, 2011.

Oliver, D. P., Kookana, R. S., Anderson, J. S., Cox, J. W., Fleming, N., Waller, N., and Smith, L.: Off-site transport of pesticides from two horticultural land uses in the Mt. Lofty Ranges, South Australia, Agric. Water Manage., 106, 60–69, 2012.

Rehana, Z., Malik, A., and Ahmad, M.: Mutagenic activity of the Ganges water with special reference to the pesticide pollution in the river between Kachla to Kannauj (U.P.), India, Mutat. Res., 343, 137–144, 1995.

Ridal, J. J., Kerman, B., Durham, L., and Fox, M. E.: Seasonality of air-water fluxes of hexachlorocyclohexanes in Lake Ontario, Environ. Sci. Technol., 30, 852–858, 1996.

Sankararamakrishnan, N., Kumar Sharma, A., and Sanghi, R.: Organochlorine and organophosphorous pesticide residues in ground water and surface waters of Kanpur, Uttar Pradesh, India, Environ. Int., 31, 113–120, 2005.

Schulz, R.: Comparison of spray drift- and runoff-related input of azinphos-methyl and endosulfan from fruit orchards into the Lourens River, South Africa, Chemosphere, 45, 543–551, 2001a.

Schulz, R.: Rainfall-induced sediment and pesticide input from orchards into the Lourens River, Western Cape, South Africa: Importance of a single event, Water Res., 35, 1869–1876, 2001b.

Semwal, N. and Akolkar, P.: Water quality assessment of sacred Himalayan rivers of Uttaranchal, Curr. Sci. India, 91, 486–496, 2006.

Singh, L., Choudhary, S. K., and Singh, P. K.: Organochlorine and Organophosphorous pesticides residues in Water of River Ganga at Bhagalpur, Bihar, India, Int. J. Res. Chem. Environ., 1, 77–84, 2011.

Stackelberg, P. E., Gibs, J., Furlong, E. T., Meyer, M. T., Zaugg, S. D., and Lippincott, R. L.: Efficiency of conventional drinking-water-treatment processes in removal of pharmaceuticals and other organic compounds, Sci. Total Environ., 377, 255–272, 2007.

Valsecchi, S., Smiraglia, C., Tartari, G., and Polesello, S.: Chemical composition of Monsoon deposition in the Everest region, Sci. Total Environ., 226, 187–199, 1999.

Varca, L. M.: Pesticide residues in surface waters of Pagsanjan-Lumban catchment of Laguna de Bay, Agric. Water Manage., 106, 35–41, 2012.

Wang, X., Xu, B., Kang, S., Cong, Z., and Yao, T.: The historical residue trends of DDT, hexachlorocyclohexanes and polycyclic aromatic hydrocarbons in an ice core from Mt. Everest, central Himalayas, China, Atmos. Environ., 42, 6699–6709, 2008.

Writ Petition 213/2011: Supreme Court of India, Record of Proceeding, available at: http://supremecourtofindia.nic.in/outtoday/wc21311p.pdf, 2011.

8

Abnormal quality detection and isolation in water distribution networks using simulation models

F. Nejjari[1], **R. Pérez**[1], **V. Puig**[1], **J. Quevedo**[1], **R. Sarrate**[1], **M. A. Cugueró**[1], **G. Sanz**[1], and **J. M. Mirats**[2]

[1]Technical University of Catalonia (UPC), 10 Rambla Sant Nebridi, 08222 Terrassa, Spain
[2]CETaqua Water Technology Centre, Crta. d'Esplugues n.75, 08940 Cornellà de Llobregat, Spain

Correspondence to: F. Nejjari (fatiha.nejjari@upc.edu)

Abstract. This paper proposes a model based detection and localisation method to deal with abnormal quality levels based on the chlorine measurements and chlorine sensitivity analysis in a water distribution network. A fault isolation algorithm which correlates on line the residuals (generated by comparing the available chlorine measurements with their estimations using a model) with the fault sensitivity matrix is used. The proposed methodology has been applied to a District Metered Area (DMA) in the Barcelona network.

1 Introduction

Drinking water distribution networks are complex large-scale systems that are designed to supply clean water to consumers at any time based on demand. To guarantee the high quality of water, chlorine is certainly the most common disinfectant used in these systems. The maintenance of chlorine residual is needed at all the points in the network. Its propagation and level are affected by both bulk and pipe wall reactions. Water quality fault will result in reduction of residual oxidant and can be manifested in added variability and lower chlorine concentrations at sensor locations in the network. This variability of chlorine concentration in the network would allow differentiating normal from anomalous water quality conditions.

The water distribution network can be perceived as a complex chemical reactor in which various processes occur simultaneously (Sadiq et al., 2003). Some of these processes take place in the bulk phase and others on the pipe's wall, and all of them can degrade water quality. Water quality faults can be caused directly or indirectly by an internal corrosion due to an oxidation/reduction reaction, by detachment and leaching of pipe material or biofilm formation (LeChevallier et al., 1987), by regrowth of microorganisms on the internal surface and/or by a loss of disinfectant and formation of byproducts (DBPs), or an intrusion of contaminants (Lindley, 2001).

This paper proposes a detection and localisation method for abnormal water quality levels based on chlorine measurements and chlorine sensitivity analysis of the nodes in a network. Simulations of the water quality of the network performed both for realistic and abnormal bulk decay at a certain time in each pipe segment (link) provides an approximation of this sensitivity. A fault isolation algorithm that correlates the residuals (generated by comparing the available chlorine concentration measurements with their estimated values using a model) with the fault sensitivity matrix is used. The correlation between the observed residual fault signature and each column of the sensitivity matrix is a measure of the similarity of the residual effect concerning pipe bulk decay fault due to pipe material detachment. The proposed approach is illustrated by a simulation example, based on a real water distribution network. Water quality data were obtained by simulation using EPANET 2.0 software, which can track the chlorine decay evolution through the water distribution system. The same methodology has been applied to detect and isolate leakages in nodes of the same network (Quevedo et al., 2011). In addition, a comparative study of this technique based on correlation analysis and the one based on a binary matrix has been done in Pérez et al. (2011). Correlation analysis is also an important tool for data validation. Further good results, using the same methodology to determine flow demand patterns given some metered data from

Barcelona water distribution network, are shown in Quevedo et al. (2010). Nevertheless, the application to abnormal water quality detection introduces irrelevant dynamics in leakage detection achieved using quasi-static models.

This paper is organised as follows: in Sect. 2, the methodology used for water quality diagnosis is presented. In Sect. 3, the proposed approach is demonstrated by its application on a DMA of the Barcelona water distribution network. Finally, in Sect. 4 some conclusions are given.

2 Water quality fault diagnosis

2.1 Fault diagnosis procedure

Model based fault diagnosis consists in the process of detecting and isolating a fault by using analytical redundancy techniques to monitor the changes in the dynamics of a system (Gertler, 1998). The consistency check is based on residuals $r(k)$ computation, obtained from measured input $u(k)$ and output $y(k)$ signals and the analytical relationships which are obtained by system modelling that provide an estimated output $\hat{y}(k)$:

$$r(k) = y(k) - \hat{y}(k) \tag{1}$$

At each time step k, the residual is compared with a threshold value (zero in the ideal case and almost zero in a real case). The threshold value is typically determined using statistical or set-based methods that take into account the effect of noise and model uncertainty (Blanke et al., 2006). When a residual is bigger than the threshold, it is determined that there is a fault in the system; otherwise, it is considered that the system is working properly. In practice, because of input/output noise, nuisance inputs and modelling errors affecting the model considered, robust residual generators must be used. Robustness can be achieved at the residual generation phase (active) or at the evaluation phase (passive). Robust residual evaluation allows obtaining a set of observed fault signatures

$$\Phi(k) = \left[\phi_1(k),\ \phi_2(k),\ \dots \phi_{n\phi}(k) \right] \tag{2}$$

where each fault indicator is obtained as follows:

$$\phi_i(k) = \begin{cases} 0 \text{ if } |r_i(k)| \leq \tau_i(k) \\ 1 \text{ if } |r_i(k)| > \tau_i(k) \end{cases}$$

and where $\tau_i(k)$ is the threshold associated to the residual $r_i(k)$.

2.2 Fault localisation using the correlation method

Fault isolation is carried out on the basis of observed fault signatures, $\phi_i(k)$, generated by the detection module and its relation with all the considered faults, $f(k) = \left[f_1(k),\ f_2(k),\ \dots f_{nf}(k) \right]$, which are compared with the theoretical fault signature matrix.

The use of the information associated with the relationship between the residuals and faults, by means of the residual fault sensitivity, allows improving the isolation results. Sensitivity Matrix $\mathbf{S}$ (Eq. 3) provides the sensitivity of the chlorine sensor residuals for each different fault f_j affecting the system:

$$\mathbf{S} = \begin{array}{c} \\ r_1 \\ r_2 \\ \vdots \\ r_n \end{array} \begin{array}{c} f_1 \quad f_2 \quad \cdots \quad f_m \\ \begin{bmatrix} s_{11} & s_{12} & \cdots & s_{1m} \\ s_{21} & s_{22} & \cdots & s_{2m} \\ \vdots & \vdots & \ddots & \vdots \\ s_{n1} & s_{n2} & \cdots & s_{nm} \end{bmatrix} \end{array} \tag{3}$$

where $s_{ij} = \frac{C_{1fj} - C_{1nf}}{f_j}$ with C_{1fj} and C_{1nf} being the mean values of the chlorine concentration C_1 in node j at a certain time, when the bulk decay is abnormal or normal, respectively.

The main idea of correlation-based fault diagnosis consists in comparing the columns (faults) of the sensitivity matrix $\mathbf{S}$ (Eq. 3) with the corresponding residual vector at time k by using the correlation function. The correlation coefficient $\rho_{r,S_{fj}}(k)$ between $r(k)$ and each column j of S (i.e. $S_{fj}(k)$) is computed as the Pearson's correlation coefficient, that is defined as:

$$\rho_{r,S_{fj}} = \frac{\text{cov}(r, S_{fj})}{\sqrt{\text{cov}(r,r)\text{cov}(S_{fj}, S_{fj})}} \tag{4}$$

where $\text{cov}(r, S_{fj}) = E\left[(r - \bar{r})\left(S_{fj} - \overline{S}_{fj} \right) \right]$ is the covariance function between r and S_{fj} being $\bar{r} = E(r)$ and $\overline{S}_{fj} = E(S_{fj})$, respectively.

The columns of $\mathbf{S}$ having higher correlation values with the residual vector at time k are the most probable elements to have a fault. The correlation between the observed residual fault signature (i.e. $r(k)$) and each column of the matrix S is a measure of the similarity between the real fault residual effect (with unknown magnitude) and the faults considered in matrix $\mathbf{S}$ (Eq. 3) (with known magnitude) that allows discovering which is the column of this matrix (fault) having the same behaviour. For faults with similar size to those used to obtain the sensitivity matrix $\mathbf{S}$, the correlation function obtains the maximum similarity (shape and form), i.e. $\rho = 1$ in the element having the fault, for any magnitude of the real fault. Because of the non-linearity of the water network system, if the real magnitude of the fault is far from the fault size used to compute matrix $\mathbf{S}$, the similarity of the correlation function decreases but a high correlation between the residuals corresponding to a particular fault and the corresponding column of the sensitivity matrix still exists.

The vector obtained is the decision vector that will be used to figure out which is the fault occurring in the system. More concretely, the maximum correlation value in this vector will point out the corresponding $\mathbf{S}$ column (fault) as the most probable element to be faulty:

$$\max_j \left(\rho_{r,S_{fj}}(k) \right) \tag{5}$$

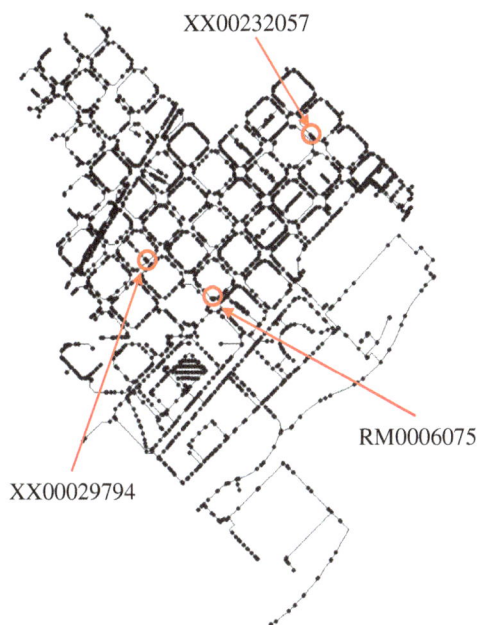

Figure 1. The layout of Nova Icària Network with the 3 installed chlorine sensor.

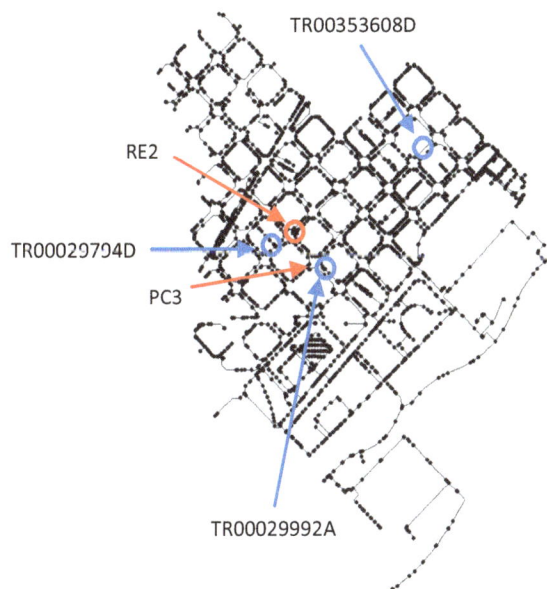

Figure 2. Location of chlorine injectors (red) and links with chlorine decay fault (blue).

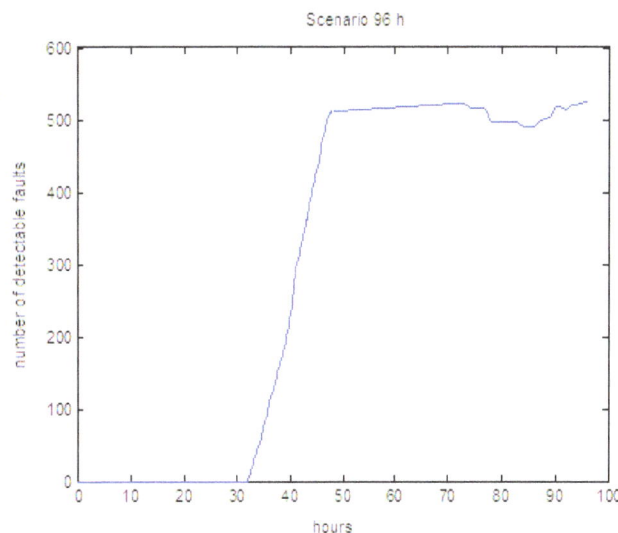

Figure 3. Number of bulk chlorine decay faults using 3 pre-installed sensors.

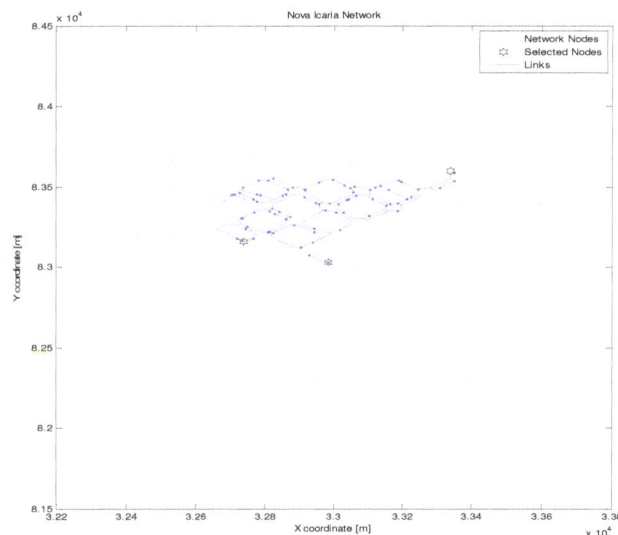

Figure 4. Geographic distribution of faults using 3 pre-installed sensors.

3 Application

A DMA of a real water distribution network (Fig. 1) was used to verify the proposed method for abnormal water quality detection and location. This DMA, located in Barcelona's Nova Icària area, is included in the 55th pressure level within the city network and has 1996 nodes and 3442 pipes. The DMA has two inputs, called Alaba and Llull, and three chlorine sensors, which are also shown in Fig. 1.

Scenarios have been generated using EPANET simulation software, considering these three sensors already installed in the network (see Fig. 1). The events have been generated setting the *Initial Quality* and *Source Quality* parameters of the injectors Alaba (RE) and Llull (PC3) in Fig. 2. The scenarios have been generated changing the K_{bulk} parameter at the corresponding links of the network at 08:00 a.m of day 2 (32nd hour of the episode). The values of faulty/non-faulty behaviours on these links are $K_{bulk} = -2.306415 \, 1/d$ for the non-faulty mode and $K_{bulk} = -50 \, 1/d$ for the faulty mode.

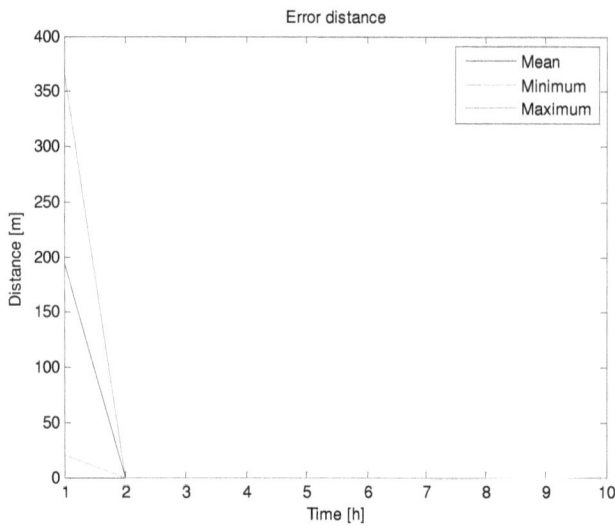

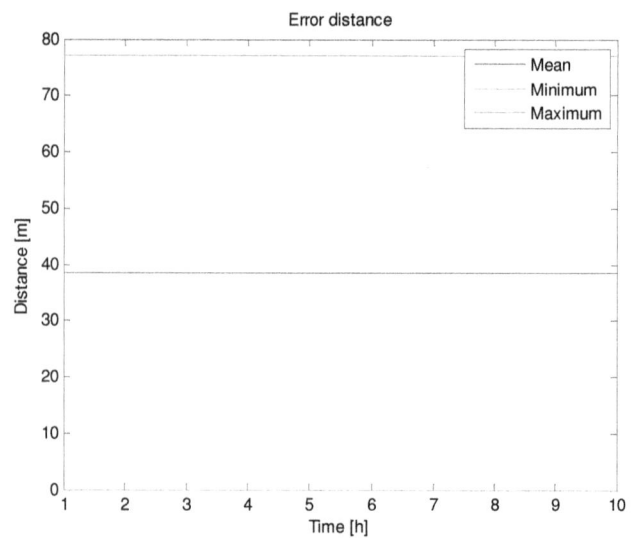

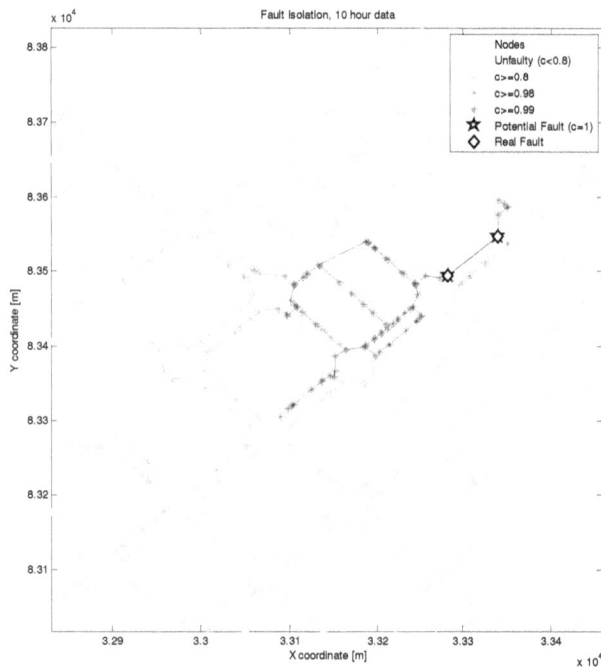

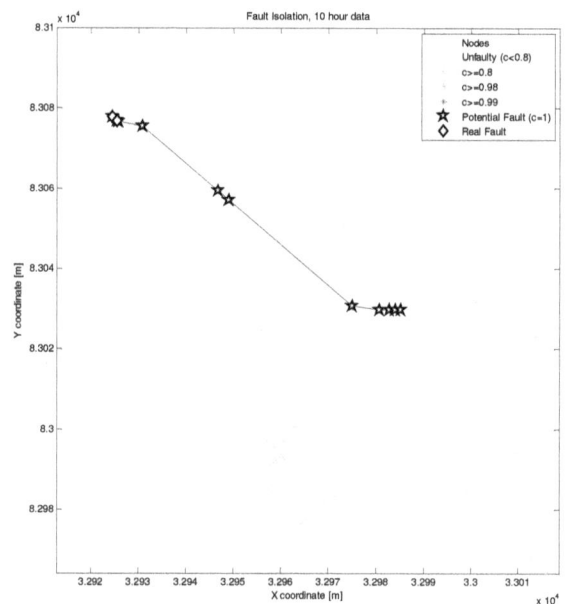

Figure 5. Error distance and geographic detail of chlorine bulk decay fault detection for the link TR00353608D at hour 48.

Figure 6. Error distance and geographic detail of chlorine bulk decay fault detection for the link TR00029992A at hour 48.

The scenario has been run for 48 h and data sets have been created according to these scenarios.

Figure 3 shows the number of bulk chlorine decay faults that may be detected using sensors located in Fig. 1 after 96 h of simulation. It may be seen how, with the sensors already installed, the number of faults that may be detected is about 500. These links are distributed geographically as shown in Fig. 4.

Three different links of the network, included in the set of detectable faults in Fig. 3, are considered to generate the faulty scenarios. These links labelled "TR00353608D",

"TR00029992A" and "TR00029794D", respectively are shown in Fig. 2. The simulated data are obtained using the methodology presented. Figures 8–10 show the fault detection results obtained at hour 48 for every chlorine bulk decay fault. In these figures, the link with the actual fault is represented with a diamond, and the starred nodes are the potential faulty nodes suggested by the correlation method. The rest of the nodes are divided in grey-scaled areas, depending on the correlation value they have: the most correlated with the fault signature, the darkest the area they are located. The evolution of the error distance through the period of results generation (15 to 24 h) and a detail of the detection for each

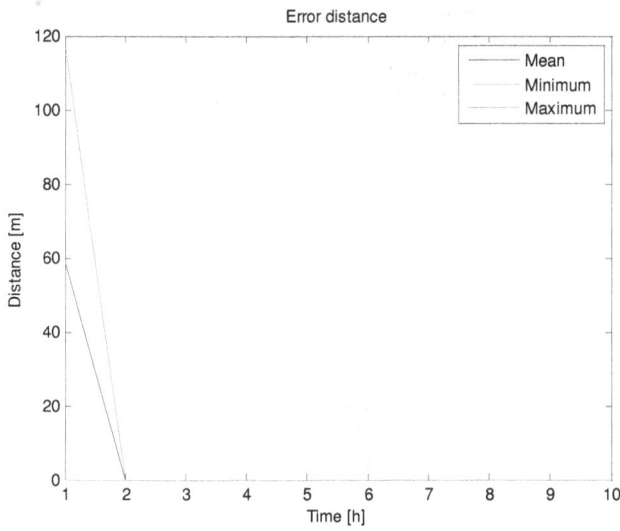

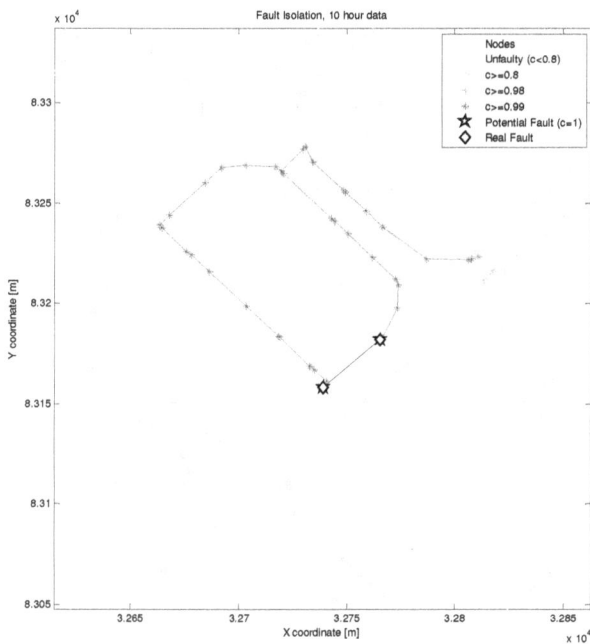

Figure 7. Error distance and geographic detail of chlorine bulk decay fault detection for the link TR00029794D at hour 48.

fault are depicted in Figs. 5–7. In the latters, the distance between the potential faulty links (starred in Figs. 5–7) and the actual faulty link (diamonds in Figs. 5–7) is represented. If more than one potential faulty link is obtained at a particular time, minimum, maximum and mean distances among potential faulty links set and actual faulty link are represented. For all cases, a good match between the actual and the identified fault has been achieved.

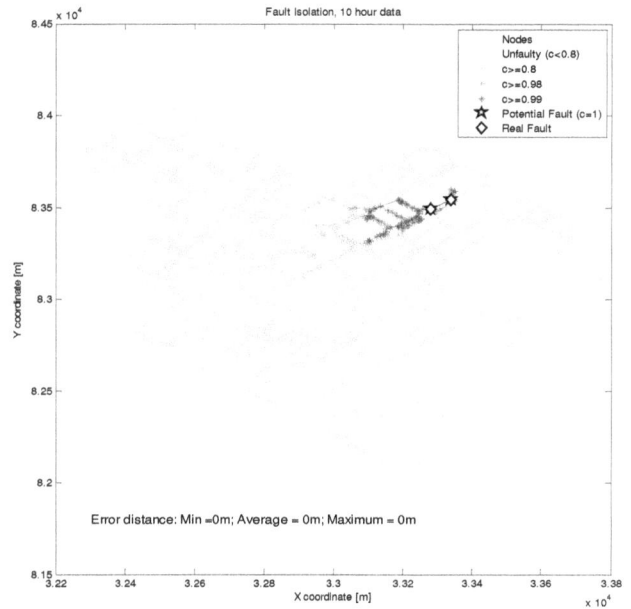

Figure 8. Chlorine bulk decay fault detection in link TR00353608D at hour 48.

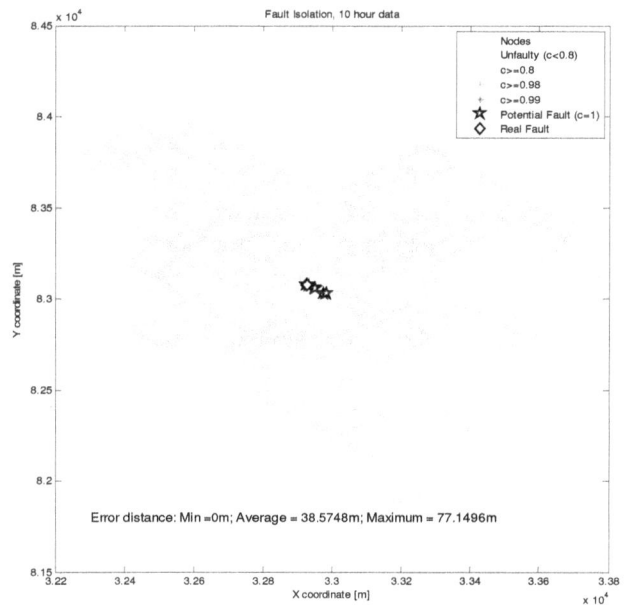

Figure 9. Chlorine bulk decay fault detection in link TR00029992A at hour 48.

4 Conclusions

In this work, the problem of detection and localisation of water quality anomalies has been addressed. The abnormal water quality localisation method is based on the chlorine measurements and chlorine sensitivity analysis of nodes in a water distribution network. A fault localisation algorithm which correlates on line the residuals (generated by comparing the

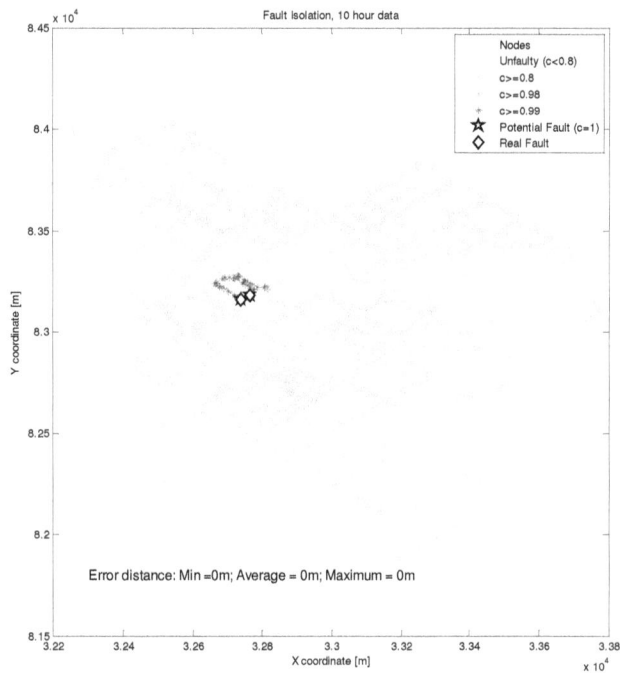

Figure 10. Chlorine bulk decay fault detection in link TR00029794D at hour 48.

available chlorine measurement with their estimation using a model) with the fault sensitivity matrix is used. The proposed algorithm has been applied in simulation to a DMA (Nova Icària) in the Barcelona network using EPANET software. The effectiveness of the method ensures the ability for a rapid-response to an abnormal event and, consequently, minimizes exposure risks to water consumers. The solution should help water companies to estimate the presence and the location of a bad chlorine concentration through a simple field data analysis.

Acknowledgements. This work was supported in part by the project RTNM AM0901 of Alliance, by the European Commission through contract i-Sense (ref. FP7-ICT-2009-6-270428) and the projects DPI2011-26243 (SHERECS) and DPI2009-13744 (WAT-MAN) of the Spanish Ministry of Economy and Competitiveness (MINECO).

References

Blanke, M., Kinnaert, M., Lunze, J., and Staroswiecki, M.: Diagnosis and Fault-tolerant Control, 2nd Edn., Springer, 2006.

Gertler, J. J.: Fault Detection and Diagnosis in Engineering Systems, Marcel Dekker, 1998.

LeChevallier, M. W., Babcock, T. M., and Lee, R. G.: Examination and characterization of distribution system biofilms, Appl. Environ. Microbiol., 53, 2714–2724, 1987.

Lindley, T. R.: A framework to protect water distribution systems against potential intrusions, MS thesis, University of Cincinnati, Cincinnati, 2001.

Sadiq, R., Kleiner, Y., and Rajani, B.: Forensics of water quality failure in distribution system – a conceptual framework, Journal of Indian Water Works Association, 35, 267–278, 2003.

Pérez, R., Quevedo, J., Puig, V., Nejjar, F., Cugueró, M. A., Sanz, G., and Mirats, J. M.: Leakage Isolation in Water Distribution Networks: a Comparative Study of Two Methodologies on a Real Case Study, The 19th Mediterranean Conference on Control and Automation, 138–143, Corfu, Greece, 20–23 June 2011.

Quevedo, J., Puig, V., Cembrano, G., Blanch, J., Aguilar, J., Saporta, D., Benito, G., Hedo, M., and Molina, A.: Validation and reconstruction of flow meter data in the Barcelona water distribution network, Control Eng. Pract., 18, 640–651, 2010.

Quevedo, J., Cuguero, M. A., Pérez, R., Nejjari, F., Puig, V., and Mirats, J. M.: Leakage Location in Water Distribution Networks based on Correlation Measurement of Pressure Sensors, 8th IWA Symposium on Systems Analysis and Integrated Assessment, Watermatex 2011, San Sebastián, Spain, 20–22 June 2011.

9

The large-scale impact of climate change to Mississippi flood hazard in New Orleans

T. L. A. Driessen[1] **and M. van Ledden**[2]

[1]Royal HaskoningDHV, Rivers, Deltas & Coasts, Nijmegen, the Netherlands
[2]Royal HaskoningDHV, Rivers, Deltas & Coasts, Rotterdam, the Netherlands

Correspondence to: T. L. A. Driessen (tjeerd.driessen@rhdhv.com)

Abstract. The objective of this paper was to describe the impact of climate change on the Mississippi River flood hazard in the New Orleans area. This city has a unique flood risk management challenge, heavily influenced by climate change, since it faces flood hazards from multiple geographical locations (e.g. Lake Pontchartrain and Mississippi River) and multiple sources (hurricane, river, rainfall). Also the low elevation and significant subsidence rate of the Greater New Orleans area poses a high risk and challenges the water management of this urban area. Its vulnerability to flooding became dramatically apparent during Hurricane Katrina in 2005 with huge economic losses and a large number of casualties.

A SOBEK Rural 1DFLOW model was set up to simulate the general hydrodynamics. This model included the two important spillways that are operated during high flow conditions. A weighted multi-criteria calibration procedure was performed to calibrate the model for high flows. Validation for floods in 2011 indicated a reasonable performance for high flows and clearly demonstrated the influence of the spillways.

32 different scenarios were defined which included the relatively large sea level rise and the changing discharge regime that is expected due to climate change. The impact of these scenarios on the water levels near New Orleans were analysed by the hydrodynamic model. Results showed that during high flows New Orleans will not be affected by varying discharge regimes, since the presence of the spillways ensures a constant discharge through the city. In contrary, sea level rise is expected to push water levels upwards. The effect of sea level rise will be noticeable even more than 470 km upstream. Climate change impacts necessitate a more frequent use of the spillways and opening strategies that are based on stages.

1 Introduction

The Mississippi River has the third largest river basin of the world and drains 41 % of the area of the United States. The flood risk of the Mississippi River imposed on urban areas, like Baton Rouge and New Orleans, became apparent during the great flood in 1927 which was the most destructive river flood in the history of the United States (Barry, 1998). The river, discharging more than $85\,000\,\mathrm{m}^3\,\mathrm{s}^{-1}$, caused levee failures at numerous places, displaced over $600\,000$ people and flooded almost $70\,000\,\mathrm{km}^2$ causing damage for \$400 million.

Looking ahead, climate change is expected to affect the river discharge due to alterations in the water cycle. Ericson et al. (2006) found that the relative vulnerability of the Mississippi delta, compared to 39 other deltas, is extreme and is demonstrated by the population potentially being displaced by current sea level trends to 2050. USGCRP (2000) showed significant increases in the heaviest precipitation events and noted that changes in streamflow follow these changes in precipitation, but are amplified by about a factor three. Additionally, climate change induced sea level rise in combination with the local subsidence of the Mississippi River Delta causes the projected local relative sea level rise in 50 yr to range between +0.5 m and +0.9 m for different rates of sea level rise (USACE, 2009; NOAA, 2011). The uncertainty in these projections is large, since the local subsidence rates vary strongly throughout the delta. Also, river morphological changes play a role, but they were neglected in this paper. Altogether, relative sea level rise is expected to affect river stages further upstream in the future. In combination with the alterations in streamflow the future flood water levels in the Mississippi River are subject to change.

In view of future changes there are various management challenges in the Mississippi River Delta to be dealt with in order to sustain the economic benefits for society, reduce the flood risk of urban areas, such as New Orleans, and to preserve the environmental value of the wetlands. Many of these issues are interrelated which each other. As an example, the high river levees prevent the river overflowing into the historic flood plains and therefore withhold fertile river sediment being deposited in the floodplains. This sediment starvation induces coastal marsh deterioration. Since the 1930s some 4900 km^2 of coastal wetlands have already disappeared in the Gulf of Mexico and this process continues at a rate of 65 km^2 per year (Bourne Jr., 2004). Marsh deterioration is also enhanced by the strong subsidence in the area which increases the potential for inundation (Nicholls et al., 2007). Subsidence enables more intrusion of salty water which destructs the fresh/brackish marshlands and allows sea water to penetrate further inland. Another water management challenge is more closely related to flood hazard in City of New Orleans. The Mississippi River levees downstream of Baton Rouge, LA, do not have large floodplains to overflow when the bankfull discharge is exceeded. This results in a steep discharge-stage relation compared to a river system that has large floodplains available on the riverside of the levees. Near New Orleans a river stage of +5.18 m above the North American Vertical Datum of 1988 epoch 2004.65 (NAVD88) is used as the safe water elevation. In order to prevent exceedance of this river stage near New Orleans two important spillways have been constructed further upstream: the Bonnet Carre Spillway and the Morganza Spillway (see Fig. 1). The Bonnet Carre Spillway is closest to New Orleans and consists of 350 bays that includes wooden beams. Its inlet structure is opened when the upstream discharge at Tarbert Landing exceeds ca. 35 400 m^3 s^{-1}. The spillway is capable of discharging almost 7080 m^3 s^{-1}. The Morganza Spillway is opened when the upstream discharge at Red River landing exceeds 42 475 m^3 s^{-1} and consists of 125 gates which are able to divert a maximum flow of almost 17 000 m^3 s^{-1}. The Morganza Spillway diverts water to the Atchafalya River basin which is located to the west of the Mississippi River, and was opened for the second time in spring 2011. The main question that is addressed in this article is how this system of levees and spillways can be adapted in view of future changes.

The objective of this paper is to address the future changes in the Mississippi River stages in the context of climate change and sea level rise. For this purpose, a hydraulic model and an overview of modelling scenarios are presented in the

following section of this paper. Next, the model results are discussed in the context of Mississippi flood hazard.

2 Methods

2.1 Hydrodynamic model

An improved one-dimensional hydraulic model was set up to describe the general hydrodynamics of the river and to simulate stages for the lower reach of the Mississippi River, which includes the cities Baton Rouge and New Orleans. This included a model of the last 492 km of the river before it enters the Gulf of Mexico (see Fig. 1). This model was subsequently used to perform a sensitivity analysis on the response of the river stage to climate change.

A HEC-RAS geometry was used that was provided by the Hydraulics and Hydrologic Branch of the U.S. Army Corps of Engineers (USACE) New Orleans District containing survey data obtained in 2002 and 2003. This geometry was imported in the SOBEK Rural Advanced Version 2.11.002c software (Deltares, 2010) and provided 164 cross-sections between the upstream boundary at Tarbert Landing and the downstream boundary at Venice (see Fig. 1). The modelled reach receives little lateral inflow due to precipitation which justifies the use of the hydrodynamic 1DFLOW module within SOBEK Rural and the exclusion of a rainfall-runoff module.

Daily discharge at Tarbert Landing was used for setting the upstream boundary condition for unsteady flow computations. A stage-discharge relation was used to compute the daily discharge based on observed daily stages. A fixed water level, representing the yearly averaged stage, was applied at the downstream boundary condition at Venice, since no hourly data was available to incorporate the tidal signal. Although the tidal signal may have a pronounced effect on the stages in the Mississippi River during low discharges, a fixed boundary was justified, since the study focused on high flows wherein tidal influence on river stages was negligible. Storm surges on the other hand can have a considerable effect. However, the study focused on peak flows of the Mississippi River, which occur mostly during the first six months of the year. Storm surges are related to hurricanes that mostly occur from June to November. Hence, the effect of storm surges was not within the scope of this study.

2.2 Calibration

Calibration of the model was performed by a weighted multi-criteria procedure for the main channel friction (Manning) parameter. The model was calibrated with a particular interest for high flow conditions. In order to include a realistic opening of the Bonnet Carre Spillway stage a discharge and water level time series originating from 1997 was used for the calibration procedure. The discharge into the spillway is schematised as a lateral outflow that is the surplus of

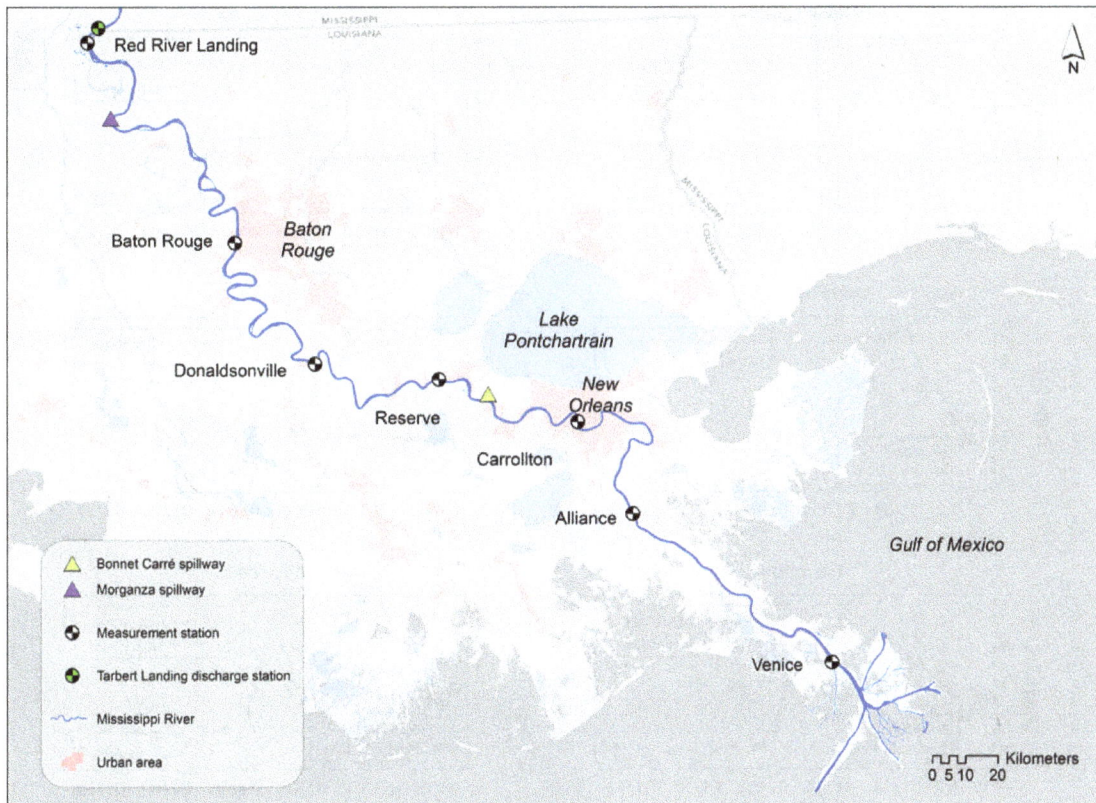

Figure 1. Map of the study area of the Mississippi River.

discharge relative to the discharge criterion of the spillway. Ten discharge bins were defined based on the minimum and maximum discharge of that year. Based on the frequency of days that features each bin a certain weighting factor was defined for the daily discharges in each bin, so that each bin had an equal weight (10 %) in the calculation of the statistical parameters. These weighting factors were higher for higher discharges, since their occurrence was lower. A multi-criteria procedure was used, since one parameter does not indicate every aspect of a time series. An optimal friction coefficient was expected by compromising between multiple calibration parameters, such as the Nash-Sutcliffe coefficient (Nash and Sutcliffe, 1970), the weighted linear regression coefficient, the weighted absolute error and the weighted root square error. For six different segments along the Mississippi River the Manning friction value in the main channel was calibrated based on the simulated stages.

2.3 Validation

After completing the calibration procedure a validation of the model needed to be performed by running the model for a different time period. Validation of the model was performed using the Mississippi River flood of 2011. In this year there was an exceptionally high discharge of $45\,845\,\mathrm{m^3\,s^{-1}}$ measured at Tarbert Landing (USACE, 2011) that required the opening of the Bonnet Carre and Morganza Spillway for several days. The validation run simulated stages for the time period 1 January to 14 June 2011.

2.4 Climate change projections

The climate change impact on the Mississippi river stages was investigated by using different scenarios of sea level rise and upstream discharges. Based on the projections and the practical application of sea level rise, four situations were used: no sea level rise, +0.45 m, +0.70 m and +1.00 m sea level rise. These scenarios were within the limits of the appropriate USACE directive (USACE, 2009). A study from USGCRP (2000) describes the general effects of climate change to the American catchments. These effects are used to simulate potential changes in river discharges. Hence, four projections of the upstream discharges were formulated by shifting the discharge hydrograph by a factor 0.95, 1.05, 1.10 and 1.15. Both the mean and standard deviation changed according to this factor. Three other projections were formulated by multiplying the difference to the mean with a factor 1.1, 1.2 and 1.3. Hereby, the stages above and below the mean were exaggerated. It did not affect the mean, but increased the standard deviation by the used factors. Tweaking the hydrograph in aforementioned way finally resulted in four sea level rise projections and eight upstream discharge

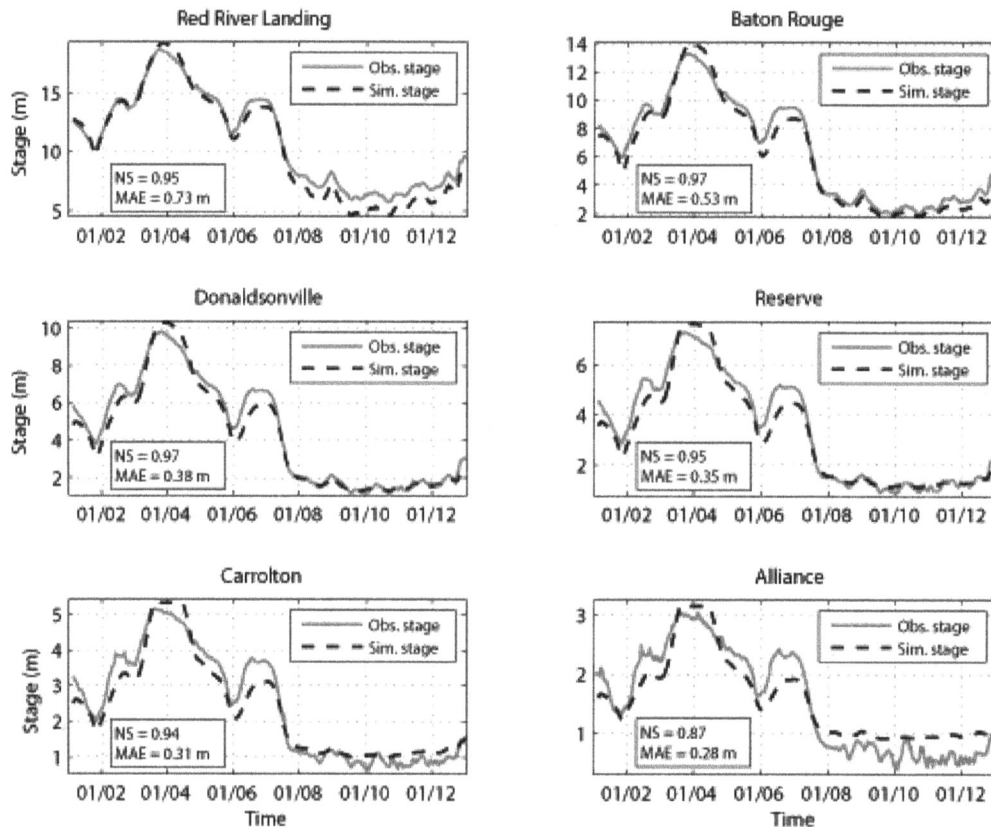

Figure 2. Hydrographs of calibrated simulation run 1997.

projections including the reference situations. Hence, 32 scenarios were used to assess the sensitivity of the Mississippi River stages to climate change. These scenarios are all under the assumption that there is an unchanged river bed.

3 Results

Figure 2 presents the results of the calibration procedure. The Nash-Sutcliffe coefficient and the mean absolute error are given for each location. The Nash-Sutcliffe coefficient assesses the predictive power of a hydrological model and its optimum value is 1. It is calculated by one minus the residual variance divided by the data variance. The model results show an overestimation of the maximum discharge peak for all six measurement stations that were used. For the calibration period, the performance of the model to simulate water levels at Carrolton (New Orleans) is described with an 10th and 90th percentile of respectively −0.6 m and +0.3 m. At Alliance, the model errors were relatively large and were partly caused by the tidal signal that was dominant at this location under low flow conditions. The error became relatively larger in downstream direction.

The validation results are given in Fig. 3 and show a good representation of the flood that occurred in May. Both the Bonnet Carre Spillway and the Morganza Spillway were opened during this flood. The latter flooded an area as large as 7700 km^2 and caused the displacement of ca. 25 000 people. At Carrolton (New Orleans) the mean absolute error of the model simulation was 0.31 m. At Red River Landing and Carrolton the difference of maximum stage was less than 0.01 m, while at Baton Rouge the maximum stage was underestimated with 0.09 m. At Alliance the maximum stage was overestimated with 0.25 m. The simulated hydrographs had a shift with respect to the observed hydrographs, probably because the velocities in the model were underestimated. However, the model errors described above were relatively small considering the stage range of the Carrolton gage. Its average annual lowest and highest stages are respectively NAVD88 −0.01 m and NAVD88 +4.34 m. The model demonstrated to be capable of simulating the hydrodynamics and, therefore, allowed a legitimate use of the model for a comparative analysis.

The model was also used to simulate the stages when the spillways would not have been opened (see Fig. 4). The impact on the stage at Carrolton (New Orleans) was especially of interest. The peak stage for this situation was projected to be NAVD88 +7.17 m, while the levee heights at Carrolton are ca. NAVD88 +7.30 m. This clearly demonstrates the benefit that the spillways have in reducing flood hazard for the urban area of New Orleans.

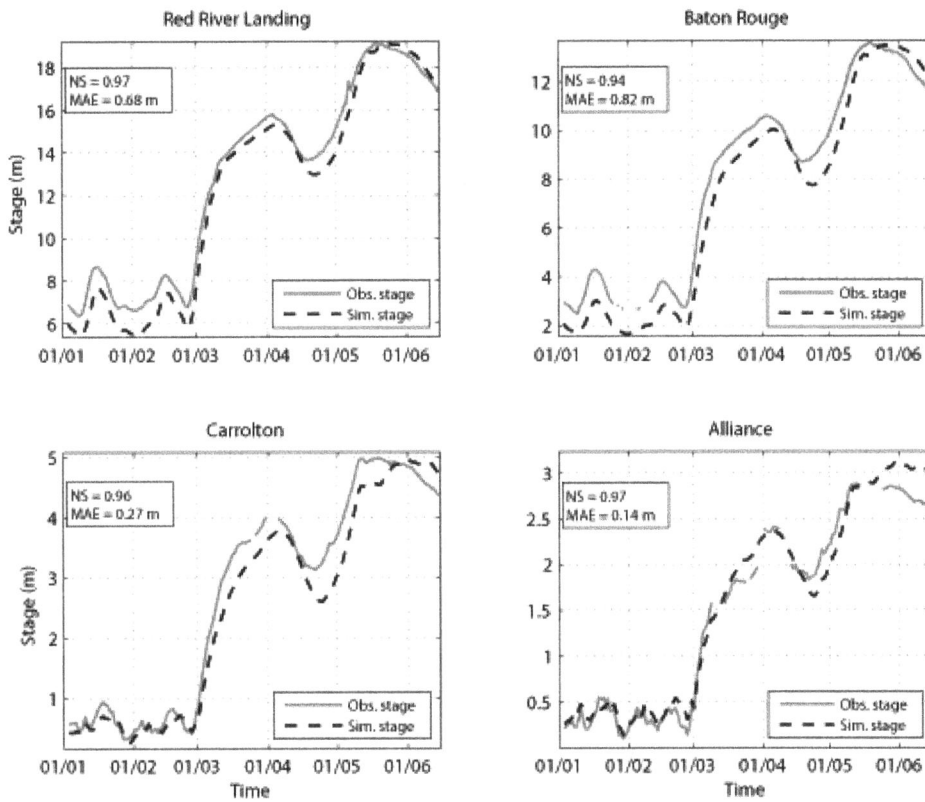

Figure 3. Validation of the flood in 2011.

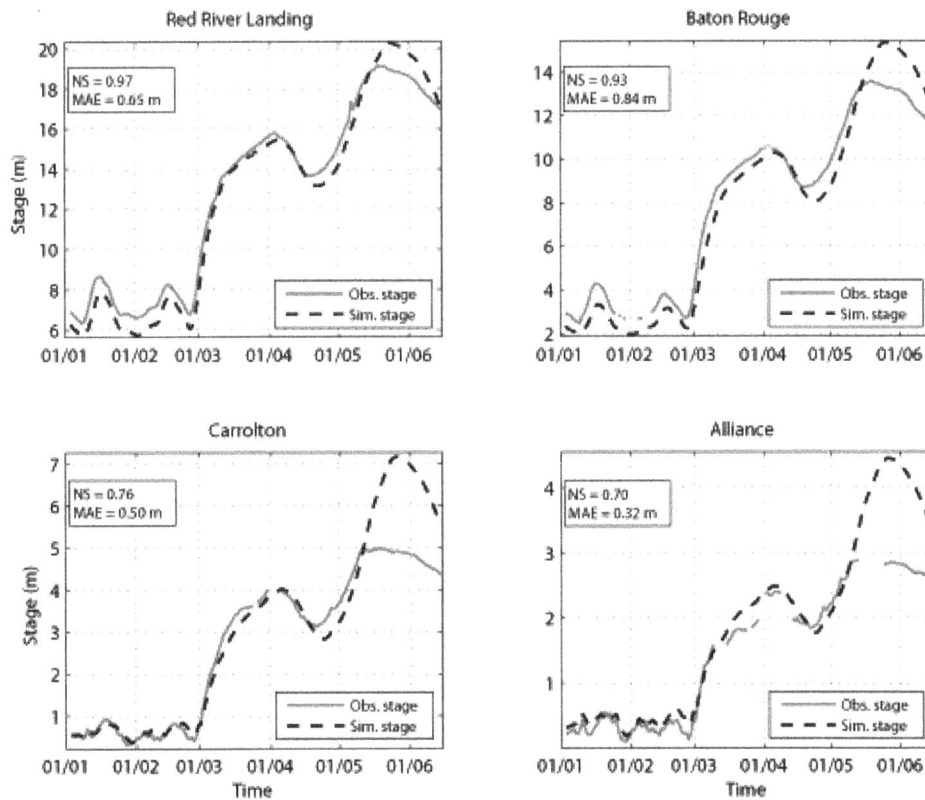

Figure 4. The validation of the flood in 2011 when no spillways are opened.

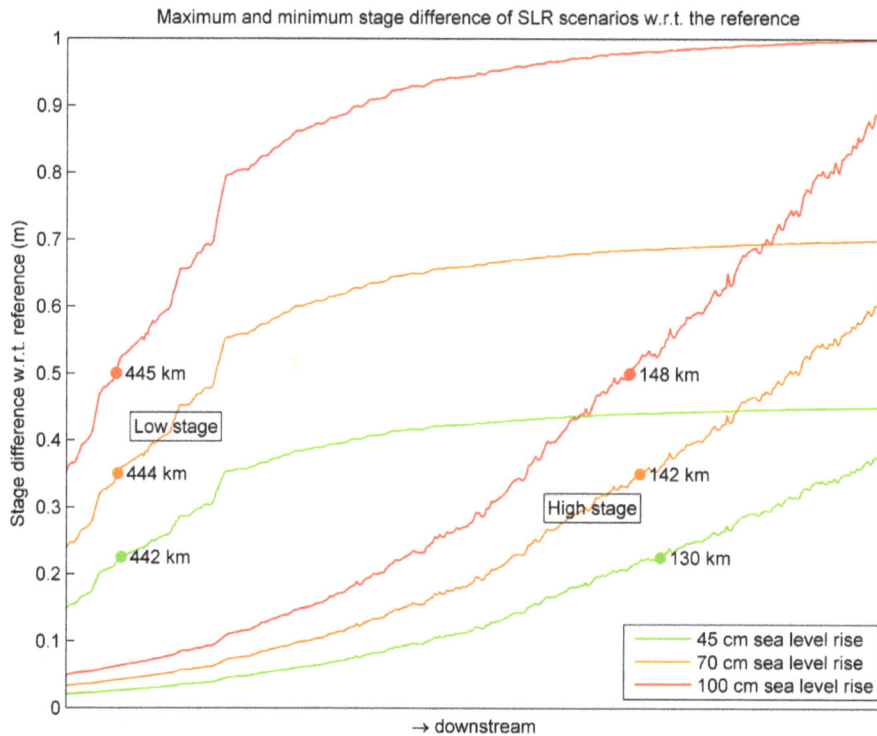

Figure 5. Minimum and maximum stage differences with respect to the reference for three sea level rise projections.

The defined scenarios were used to get an insight in the potential future behaviour of the Lower Mississippi system. By looking at the sea level rise scenarios where no discharge change is expected it can be concluded that the distance to the Gulf of Mexico, the position where the effect of sea level rise is halved, becomes larger with increasing sea level rise. The minimum stage difference with respect to the reference scenario occurs during high flow conditions when the stage is predominantly dependent on the discharge. Then, the sea level rise is halved at 130 km, 142 km and 148 km from the downstream boundary for respectively +0.45 m, +0.7 m and +1.0 m sea level rise as can be seen in Fig. 5. The minimum increase of stage at Carrolton becomes +0.21 m, +0.34 m and +0.50 m for all discharge regimes except the "−5 %"-scenario. During low flow conditions the sea level affects the river stages stronger. In this case the sea level rise is halved at 442 km, 444 km and 445 km for respectively the same three sea level rise projections. It was found that the stages along the entire modelled reach are affected by sea level rise for the three projections. The maximum stage increase at Carrolton is dependent on the type of discharge regime as well as the strength of sea level rise.

The relative steep stage-discharge relation due to the confined levees downstream of Baton Rouge causes a strong increase in stages when upstream discharges increase. At the same time changes in discharge characteristics did not affect the stages downstream of Bonnet Carre Spillway when it was closed, since a constant discharge passed through. However,

changes in the upstream discharge would affect the frequency and duration for opening of the spillways. Their effect on the water level near Carrolton is demonstrated in Figs. 3 and 4. The increase of stages in the downstream section is predominantly caused by the sea level rise. Currently, the threshold for a spillway opening is defined by the upstream discharge at Tarbert Landing. Sea level rise causes an increase in stage that changes the stage-discharge relationship. Thus in order to not exceed the NAVD88 +5.18 m threshold at Carrolton it is likely that a lower discharge threshold is needed for opening the spillways.

4 Conclusions

The paper addressed the future changes in the Mississippi River stages regarding "climate change"-induced changes of upstream discharge and sea level. The weighted multi-criteria calibration procedure provided a hydrodynamic model that was capable of simulating high flows in the Mississippi River. During validation a mean absolute error at Carrolton (New Orleans) of 0.31 m was found and a difference during maximum stage that was less than 0.01 m. It was concluded that the model could be used for a comparative analysis.

The presence of the two spillways proved to be of vital importance to reduce the flood hazard risk of the city New Orleans. In the future their role will even be more important. The challenges of the Mississippi River delta become even more apparent with expected climate change impacts.

All of them increase the flood hazard in the Greater New Orleans area, which is already situated in a flood-prone area. This study showed that sea level rise is likely to be dominant on the downstream end of the Bonnet Carre Spillway. During high flow conditions the stages near New Orleans could be increase with a range between 0.21 m and 0.50 m. At the same time the increased upstream discharge increases the frequency and duration of spillway openings.

Appropriate measures to face these future water management challenges and reduce the flood hazard are not discussed in this paper. Nevertheless, the Mississippi River model that is presented in this paper can be of added value in the analyses of these flood risk reduction measures. In that case it is recommended to update the bathymetry and include a morphological module to further enhance the model's performance.

References

Barry, J.: Rising Tide – The Great Mississippi Flood of 1927 and How it Changed America, Simon & Schuster, New York, 1998.

Bourne Jr., J. K.: Gone with the water, National Geographic Magazine, available at: http://ngm.nationalgeographic.com/ngm/0410/feature5/ (last access: May 2009), 2004.

Deltares: Design and analysis tools – SOBEK suite, available at: http://www.deltaressystems.com/hydro/product/108282/sobek-suite (last access: October 2011), 2010.

Ericson, J., Vörösmarty, C., Dingman, S., Ward, L., and Meybeck, M.: Effective sea-level rise and deltas – Causes of change and human dimension implications, Global Planet. Change, 50, 63–82, 2006.

Nash, J. I. and Sutcliffe, I. V.: River flow forecasting through conceptual models: Part I – A discussion of principles, J. Hydrol., 10, 282–290, 1970.

Nicholls, R. J., Wong, P. P., Burkett, V. R., Codignotto, J. O., Hay, J. E., McLean, R. F., Ragoonaden, S., and Woodroffe, C. D.: Coastal systems and low-lying areas. Climate Change 2007: Impacts, Adaptation and Vulnerability. Contribution of Working Group II to the Fourth Assessment Report of the Intergovernmental Panel on Climate Change, Cambridge University Press, Cambridge, UK, 315–356, 2007.

NOAA: Sea Levels Online, available at: http://tidesandcurrents.noaa.gov/sltrends/sltrends.shtml, last access: 28 August 2011.

U.S. Army Corps of Engineers Washington: Water Resource Policies and Authorities Incorporating Sea-level Change Considerations in Civil Works Programs, available at: http://140.194.76.129/publications/eng-circulars/ec1165-2-211/entire.pdf (last access: 28 August 2011), 2009.

U.S. Army Corps of Engineers: Discharge Data; Mississippi River at Tarbert Landing, MS, available at: http://www2.mvn.usace.army.mil/cgi-bin/wcmanual.pl?01100, last access: 11 October 2011.

USGCRP: Climate change impacts on the United States – The potential consequences of climate variability and change, US Global Change Research Program, 2000.

10

Assessing variable speed pump efficiency in water distribution systems

A. Marchi[1], A. R. Simpson[1], and N. Ertugrul[2]

[1]School of Civil, Environmental and Mining Engineering, The University of Adelaide, SA 5005, Australia
[2]School of Electrical and Electronic Engineering, The University of Adelaide, SA 5005, Australia

Correspondence to: A. Marchi (amarchi@civeng.adelaide.edu.au)

Abstract. Energy savings and greenhouse gas emission reductions are increasingly becoming important design targets in many industrial systems where fossil fuel based electrical energy is heavily utilised. In water distribution systems (WDSs) a significant portion of operational cost is related to pumping. Recent studies have considered variable speed pumps (VSPs) which aim to vary the operating point of the pump to match demand to pumping rate. Depending on the system characteristics, this approach can lead to considerable savings in operational costs. In particular, cost reductions can take advantage of the demand variability and can decrease energy consumption significantly. One of the issues in using variable speed pumping systems, however, is the total efficiency of the electric motor/pump arrangement under a given operating condition. This paper aims to provide a comprehensive discussion about the components of WDS that incorporate variable speed pumps (including electric motors, variable frequency drives and the pumps themselves) to provide an insight of ways of increasing the system efficiency and hence to reduce energy consumption. In addition, specific attention is given to selection of motor types, sizing, duty cycle of pump (ratio of on-time and time period), losses due to installation and motor faults. All these factors affect the efficiency of motor drive/pump system.

1 Introduction

Attention towards efficient and eco-friendly management has been growing in recent years, especially in systems characterized by a large consumption of non-renewable energy. Most water distribution systems (WDSs) require the operation of pumps to deliver the necessary quantity of water with the adequate pressure to the final consumers. As reported by Lingireddy and Wood (1998) and Bene et al. (2010), the electrical energy used to pump water is a significant portion of the total operational costs in WDSs.

While sustainability has been considered in the problem only recently (Wu et al., 2010a, b), a vast literature regarding the minimization of energy costs in WDSs exists. In most of the cases, optimization of operations has only considered fixed speed pumps (FSPs) and the cost savings that may be obtained by exploiting a multi-pattern electric tariff (McCormick and Powell, 2003; van Zyl et al., 2004; Salomons et al., 2007; López-Ibáñez et al., 2008; Broad et al., 2010).

As a result the energy consumption by the fixed speed pump remains approximately the same.

As an alternative, the cost related to pump operation can be reduced by decreasing energy consumption. An attractive alternative to reach this target is the use of variable speed pumps (VSPs) instead of FSPs. VSPs are pumps coupled with a motor that is controlled by a variable frequency drive (VFD). The principal duty of the variable frequency drive is to alter the mains supply to vary the speed of the motor while delivering the required torque at higher efficiency. As a result, as the pump speed changes, the pump curve is adjusted for different operating conditions. The main advantages of VSPs are exploited when the operating conditions in the system are characterized by a high variability. In this case, while an FSP's operating point is forced to remain on the pump curve for a constant pump speed, with a possible decrease in the pump efficiency, VSPs have the possibility of maintaining

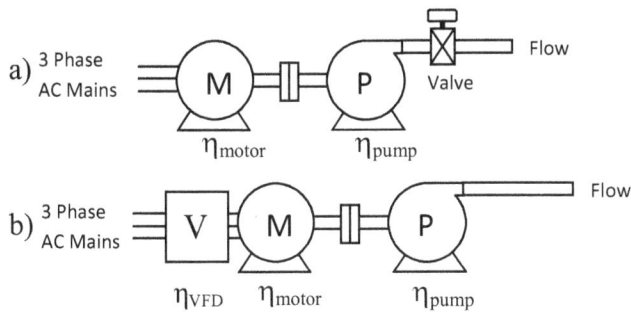

Figure 1. Simplified representation of pump systems: (a) fixed speed pumps and (b) variable speed pumps. M: motor, P: pump, V: Variable frequency drive, η: efficiency.

high efficiency by effectively scaling the pump curve as the pump speed is varied.

Although good efficiencies can be maintained by considering proper options, such as using control valves or selecting the correct number and type of parallel pumps, VSPs can still be advantageous in reducing energy consumption, water losses and pipe wear (Lingireddy and Wood, 1998). However, VSP convenience is not always guaranteed: manuals such as Hydraulic Institute et al. (2004) or BPMA et al. (2003) describe the WDSs characteristics for which VSP are not as attractive. In addition, prediction of the total efficiency of VSP system is not clear specifically at light loads. Although it is known that the efficiency decreases when the speed rate is reduced (Lingireddy and Wood, 1998; Sârbu and Borza, 1998), Walski et al. (2003) showed that the efficiency can be considerably lower than expected also for small reduction of the pump speed rate (although the pump that was tested in the laboratory was quite small).

The overall efficiency of VSP, called wire to water efficiency, is a result of various factors since every component in the pump system has losses. In the simplified scheme shown in Fig. 1, the main difference between VSPs and FSPs is the introduction of the variable frequency drive and the absence of the valve, which have an impact on the total efficiency of the pump system. As the wire to water efficiency depends on the particular type and size of components used and on the system requirements, the effectiveness of VSPs has to be assessed for each specific case. However, as VSPs require knowledge in different fields, this task can be complicated.

This paper aims to provide a comprehensive discussion about the components of pumping systems in WDSs (variable frequency drives, electric motors and pumps) to provide insight into the assessment of efficiency and energy consumption of variable speed pumps applied water distribution systems. The next section explains why VSPs can be advantageous compared to FSPs. The following sections describe the interaction between pump, motor and VFD characteristics, and provide insights how these characteristics are modified when the pump speed rate is changed. Further discus-

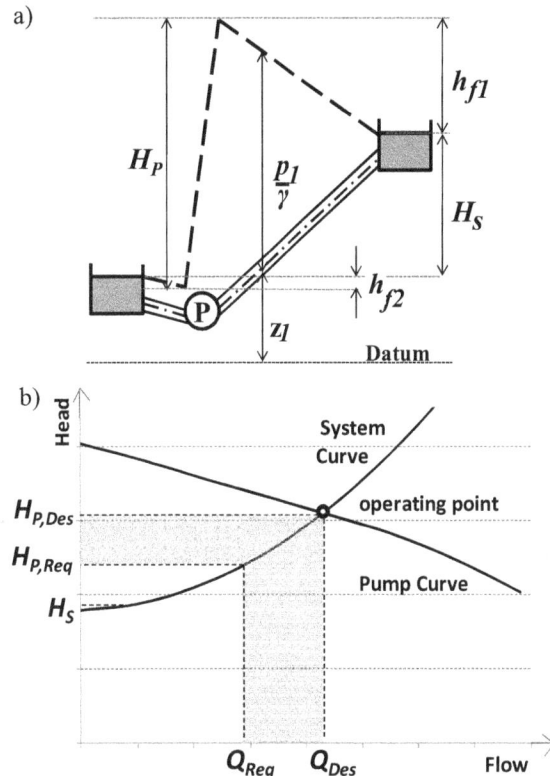

Figure 2. (a) Hydraulic grade line (dashed line) when the pump transfers water between two reservoirs. (b) Additional energy consumption in case of water transmission systems (in grey). (Figure adapted from ITP and Hydraulic Institute, 2006.)

sions are also provided to assess potential motor types and motor management issues in practical systems.

2 Possibilities for energy saving in WDSs with VSPs

The energy used by a pump in a water distribution system, E (kWh), is the product of the specific weight of water, γ (N m^{-3}), the flow delivered, Q (m^3 s^{-1}), the increase in head across the pump, H_P (m), and the operating time t (h) divided by the wire to water system efficiency, η_{ww}.

The pump head H_P is the sum of the static head H_s, i.e. the difference in water levels between the two reservoirs, and the friction losses before and after and the pump P, shown as h_{f2} and h_{f1} respectively in Fig. 2a. As the velocity head can normally be neglected in WDSs, H_P is representative of the increase of energy in the flow: part of this energy will be dissipated due to the friction losses, while the remaining part will be converted from pressure to potential energy associated with the increase of elevation. The friction losses h_f (the sum of h_{f1} and h_{f2}) are an approximately quadratic function of the fluid velocity and therefore of flow. For the same diameter pipe, if a different pump P$_1$ is used to raise a flow Q_1 that is smaller than Q, the friction losses are reduced and therefore the new head H_{P1} will be smaller than H_P. Thus, the energy consumed by the new pump P_1 will be smaller than the energy used by P because of the smaller flow and head.

As pumps are often sized considering the system conditions at the end of its service life, the actual operating points can be different from the design criterion. In addition, the daily and seasonal variability in water demand can further move away the actual pump operating point from the expected one.

If the pump used is a fixed speed pump, the operating point is forced to move along the pump curve corresponding to the fixed nominal speed. If the system is not modified, the system curve can possibly intersect the FSP curve in a region of low efficiency. A common way to avoid this drawback is the modification of the system curve by the insertion of valves. Although this leads to an increase in pump efficiency, it is a waste of energy because of the additional losses introduced by the valve.

VSPs can take advantage of the different required operating conditions in WDSs. In particular, the reduction in energy consumption exploits the possibility of reducing head or flow in the system. Figure 2b shows the case in which the required flow is smaller than the actual operating point. This can be the case of a water transmission system, where pumps are used to move water from a lower to an upper tank. Since the pumping has been at a particular time of the year may be sized for a peak-day demand, the pump will deliver a flow associated with the operating point despite the fact that the required demand is decreased (for example because of the seasonal variability). Given that smaller flows imply smaller friction losses, the grey area in Fig. 2b can be seen as energy wasted. However, this operating strategy can be economically convenient if the system has a sufficient storage volume so that pumps can be switched on only during the off-peak-tariff period.

The energy used by the pumping system depends also on its efficiency and the adjustment of the pump curve when VSPs are used can affect the wire to water efficiency, η_{ww}. The general expression of the wire to water efficiency is represented by Eq. (1), where η_{VFD}, η_{mot} and η_{pump} are the efficiencies of the variable frequency drive, of the motor and of the pump, respectively. If the pump is a fixed speed pump, there is no variable frequency drive and thus η_{VFD} may be taken as equal to 1.

$$\eta_{ww} = \eta_{VFD} \cdot \eta_{mot} \cdot \eta_{pump} \tag{1}$$

The following section of the paper will consider the impact of changing the pump speed rate on the pump efficiency, η_{pump}, which will be followed by the discussions about types of motors and VFDs to obtain a variable speed pump system.

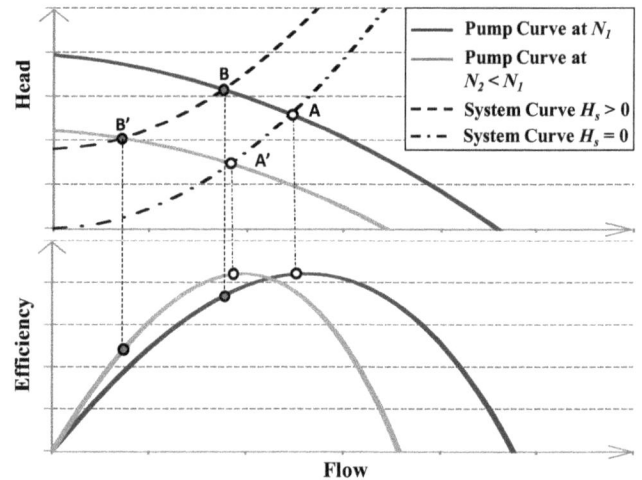

Figure 3. Operating points and efficiencies with varying speeds: in case of system with zero static head, the operating point A at the rotational speed N_1 moves to A' when the speed is reduced to N_2, but A and A' maintain the same efficiency; in case of system with considerable static head the operating point at the reduced speed, B', has an efficiency lower than the operating point B at the nominal speed N_1.

3 Pump operation at variable speeds

Pumps used in WDSs are often centrifugal pumps characterized by the fact that the flow-head curve remains sufficiently flat for a wide range of flows. The relationships that link the pump characteristics (flow, Q, head, H, power, P) operating at different speeds (N_1, N_2) are described by the affinity laws: the ratio of the pump flow rates, pump heads and pump powers is a linear, quadratic and cubic function, respectively, of the pump speed ratio.

The approximation introduced in the power-speed relation implies that the efficiency will remain constant for speed N_1 and N_2, i.e. that the efficiency curve will be only shifted to the left in case of speed reduction. Equation (2) is given by Sârbu and Borza (1998), which provides an analytical relationship between two different speeds (N_1 and N_2) and the corresponding efficiencies (η_1, η_2).

$$\eta_2 = 1 - (1 - \eta_1) \cdot \left(\frac{N_1}{N_2}\right)^{0.1} \tag{2}$$

Therefore, as indicated in Sârbu and Borza (1998), the changes in efficiency can be neglected if the changes in speed rate do not exceed the 33 % of the nominal pump speed and this approximation is particularly justified for large pumps. However, this approximation only states that the pump curve points will maintain the same efficiency, e.g. the pump will have the same best efficiency point (BEP), but it does not imply that the pump will operate with the same efficiency once inserted in the system. In fact, the operating point is defined by the intersection of the pump curve with the system curve.

Figure 3 shows that, in a system with zero static head, the pump can operate with the same efficiency of the nominal speed. In contrast, when the static head of the system is a significant portion of the total head delivered by the pump, the efficiency of the pump operating at a reduced speed can rapidly decrease below an acceptable level.

In systems where there is a static head component, the allowable decrease in pump speed rate is limited by several factors. Firstly, the minimum pump speed rate is defined by the speed at which the shutoff head of the pump corresponds to the static head of the system. Secondly, the efficiency of the new operating point should be evaluated as it is the most restrictive condition in case of system with large static head relative to the total pump head. Finally, the speed should not be reduced beyond the 70 % of the nominal speed, so to consider that the pump maintains the same BEP. However, as it can be observed in Fig. 1, the highest efficiency in variable speed pump systems can be achieved if all the system components (included motor and variable frequency drive) have higher efficiency at the operating points, which will be discussed in the following sections.

4 Motor options

Reducing losses in motors and associated motor drives can result in substantial cost savings. In fact, pump operations account for about 20 % of the total electric energy used in industry and more than 95 % of an electric motor's life cycle cost is related to energy consumption (Fuchsloch et al., 2008).

The majority of electric motors used in pump applications are induction (asynchronous) types as they operate directly from AC mains supply and are self starting. In addition, they are low cost machines, are reliable and have relatively high efficiency, which can be considered relatively constant if motor operates above 50 % of rated load (Ulanicki et al., 2008). The rated speeds of induction motor (IM) is usually very close (typically 90–99 % depending on machine size) to synchronous speed that is determined by the supply frequency and can be altered by a variable frequency drive. The operating speed range of these motors are normally between synchronous speed (no-load) and rated speed (full-load). In IMs, the ratio of the speed difference between synchronous and rated speed relative to the synchronous speed is known as slip, s. The name plate of an IM always gives a rated speed that is slightly less than the synchronous speed, and such motors are designed to deliver rated torque (that is, rated output power) at rated speed. To a first approximation the torque versus speed curve of an IM can be approximated as a straight line between synchronous speed and rated speed. Although the motor characteristics can differ slightly, their efficiency varies significantly with the load and the highest possible efficiency of an induction motor operating at a slip s, is equal to $1-s$. For instance at a slip $= 0.1$, the maximum possible ef-

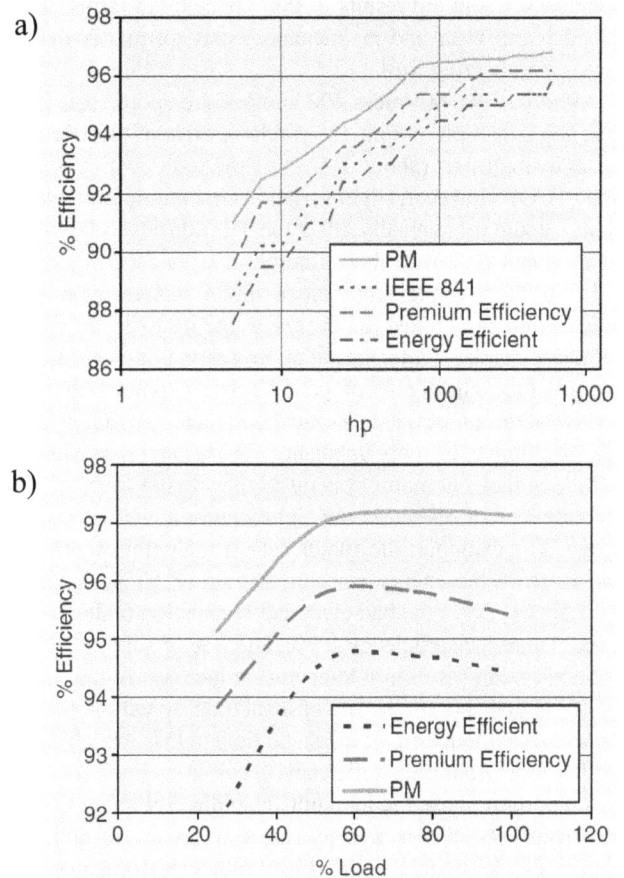

Figure 4. The improved efficiency at rated load (**a**) and typical partial load efficiencies (**b**) for PM motors and energy efficient IMs (Melfi et al., 2009).

ficiency is 90 %. In practice the actual efficiency is somewhat lower than this due to other losses in the machine.

A detailed loss breakdown of induction motor as function of power rating has been studied in Fuchsloch et al. (2008). It was reported that typically 50 % of IM losses can be removed by keeping the stator the same and replacing the rotor with a permanent magnet (PM) rotor, which results in a motor named brushless PM motor. These types of motors show opportunity to improve efficiency (Fig. 4a), and they generally have better low-speed and light-load efficiency (Fig. 4b) because of lack of magnetising current (which is required in induction motors). Figure 4 also illustrates that brushless PM motors can provide an opportunity to extend efficiency to a level beyond that defined by National Electric Manufacturing Association (NEMA), Premium Efficiency. Furthermore, brushless PM motors can provide a very high power factor (about 0.95) that is noticeable specifically at light loads (which typically vary between 0.55–0.75 at light loads in IMs) (Melfi et al., 2009). Note that the power factor indicates the effective utilisation of the supply power and reduces the reactive power consumption. In addition, low speed operation of a pump results in some benefits in terms of reduced pump wear and maintenance costs (primarily due to increase in bearing life).

Although the brushless PM motors are more expensive than the induction motors, the purchase price of electric motor is a small part (about 3–5 %) of the total cost of operation (TCO). However, slight variations in their operating efficiency could dramatically affect the TCO through electricity costs (about 70 % over their lifetime).

In addition, implementation of motor management program alone can reduce the energy usage up to 15 %, which should consider motor installation related issues including soft-foot, misalignment, poor electrical connection, poor power factor, undersized electrical conductor as well as supply variation and supply unbalance. These can be considered as hidden faults in motor systems as they do not cause an immediate loss of operation but significantly affect the energy usage. For example, the motor set-up fault, misalignment, can increase the energy consumption up to 10 %, the extra resistances causes the bad electrical connection (called a hot joint) can reduce efficiency 1.5 %, and 2 % of supply voltage imbalance can result in a temperature increase in the motor by 10 °C that shorten the life of insulation by half as well as cause reverse sequence of magnetic field in IMs, which cause further reduction of the efficiency.

In addition, these hidden faults also directly affect the reliability of the system as they can develop severe faults in motors as they reduce bearing life, motor performance and efficiency as well as causing noise and vibration.

5   Variable frequency drives

Variable frequency drives (or variable speed controllers) can simply control motor speed by varying effective voltage and frequency, which is obtained from a fixed voltage and fixed frequency three-phase mains supply. Brushless PM motors cannot operate without a variable frequency drive due to their structure. However, from a cost and complexity point of view, there is not much difference between the variable frequency drives used in IMs or brushless PM motors for the same power ratings. Although there are different types of VFDs and control schemes, they should be assessed together with the choice of a motor.

There have been difficulties in estimating and measuring the efficiency of a motor (IM or brushless PM motor) controlled by a variable speed controller. Internationally approved standards or testing procedures of motor/VFD combination are not available at present (Stockman et al., 2010). Since the variable frequency drive can operate the motor over a wide range of speeds, it is important to know efficiency values of VFD/motor combination at each operating point to accurately determine the operating point of a pump, which is known as iso-efficiency contours. Stockman et al. (2010) have reported a successful measurement procedures how to

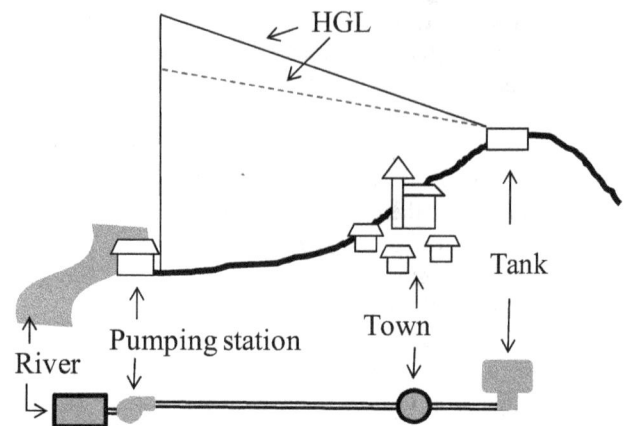

Figure 5. Case study layout and its EPANET representation. The hydraulic grade line (HGL) of the pump run at reduced speed (dashed line) has a flatter slope than the HGL at full speed (continuous line).

obtain the iso-efficiency contours of VFDs connected to both IMs and brushless PM motors. It was demonstrated that the brushless PM motor has far better efficiency values compared to IM.

It should be noted here that in variable drive systems, there are additional losses generated in the motor by the variable frequency drive. In addition, the VFDs perform rapid switching of DC voltages across the motor windings which can cause damage to the motor insulation (resulting reduced motor life) which is normally only designed for sinusoidal voltages. Therefore, in variable speed pump applications it is important to use motors that have specially designed insulation to allow them to handle with VFD operation.

6   Case study

A case study is presented to show how the efficiencies of variable frequency drive, motor and pump impact the VPS power. The case study consists of a pump that lifts water from a reservoir to a tank while supplying customers (Fig. 5). If the pump can be run at a reduced speed, the hydraulic grade line will have a flatter slope than the one at full speed, because of the reduced flow and hence reduced head losses.

It is assumed that during the low demand periods (e.g. at night) the pump speed can be reduced to 75 % of the full speed, while still meeting the water supply requirement, i.e. the same amount of water is pumped during the day, often resulting in the VSP being run continuously.

The hydraulic model of the case study (lower part of Fig. 5) has been implemented in EPANET 2 (Rossman, 2000). EPANET uses the affinity laws to compute the pump characteristics at the pump reduced speed: the simulation results show that pump flow and head are $51.91 \, \text{s}^{-1}$ and $61.03 \, \text{m}$, respectively. The pump efficiency computed using the affinity laws is 83.03 %. As a result, the input power to the pump is 37.40 kW. However, the motor and variable frequency drive efficiencies also need to be considered to compute an accurate estimate of the total power.

The efficiency of the motor is related to motor type, size and load as shown in Fig. 4. At full speed the pump power is 86.84 kW (flow = 65.3 l s^{-1}, head = 110.59 m and pump efficiency = 81.48 %). However, considering that the maximum power of the pump is 104 kW, a motor size of 110 kW was chosen. Therefore, during normal operation at full speed, the motor works at 79 % (86.84/110×100) of the full load; at the reduced speed the motor load is reduced to 34 %. Assuming that the motor characteristics are those presented in Fig. 4b for an energy efficient motor, the motor efficiency is about 94.7 and 93 % for the pump run at full speed and at the reduced speed of 75 %, respectively.

The variable frequency drive efficiency depends on the VFD type, size and on the speed reduction. If the data reported in the Industrial Technologies Program (2008) are used, the VFD efficiency for our case study is 97 %. However, to be on the safe side, data from Bernier and Bourret (1999) are used, and the VFD efficiency is assumed to be equal to 92 %. Therefore, when the pump is run at 75 % of the full speed, the power required is 43.7 kW as shown by Eq. (3) when $Q = 51.86 l s^{-1}$, $H = 61.03$ m, $\eta_{VFD} = 0.92$, $\eta_{mot} = 0.93$ and $\eta_{pump} = 0.83$. This value is about half of the power required at full speed, i.e. 95.2 kW ($Q = 65.27 l s^{-1}$, $H = 110.59$ m, $\eta_{VFD} = 0.97$, $\eta_{mot} = 0.947$ and $\eta_{pump} = 0.81$).

$$P \text{ (kW)} = \frac{9810 \left(\frac{N}{m^3}\right) \cdot Q \text{ (m}^3\text{s}^{-1}) \cdot H \text{ (m)}}{\eta_{VFD} \cdot \eta_{mot} \cdot \eta_{pump}} \cdot \frac{1}{1000} \quad (3)$$

Note that the power required for the VSP to run at full speed is larger than the power required for a FSP (95.2 vs. 92.1 kW) because of the VFD efficiency (0.97 in this case) that decreases the wire-to-water efficiency also when the pump is run at full speed. Therefore to assess the effectiveness of VSPs the duty cycle of the pump has to be known. Assuming that in normal operations the FSP is switched on for 19 h per day, the pump operation consumes 1754 kWh day^{-1} to pump 4465 m^3 day^{-1} (corresponding to 0.393 kWh m^{-3}). We also assume that the VSP can be run at the relative speed of 75 % for 24 h. If VSPs are used, the energy consumption is 24 × 43.7 = 1049 kWh day^{-1} (4481 m^3 day^{-1} pumped corresponding to 0.234 kWh m^{-3}). Assuming that a cooling system is not required for the pump in the case study, VSPs save energy. Note that computations are here simplified for ease of discussion, but in reality the VSP speed is likely to vary during the day to comply with the network requirements. For example, depending on the size of the tank and on the demand profile, the pump could be forced to work at full speed during the high demand periods, increasing its energy consumption.

Finally, to assess the effectiveness of VSPs, capital costs of the investment have to also be evaluated and compared to the energy saving during the lifetime of the project. VSPs are usually more effective in existing systems where the actual pumps may be operating in response to conditions different from the ones they were designed for. However, it is worth noting that VSPs are only one option among others and therefore also the life cycle analysis of the other options has to be considered so as to indentify the best solution.

7 Conclusions

This paper considers the energy saving that may be obtained by the introduction of variable speed pumps in water distribution systems. This assessment should evaluate pump and pipeline characteristics, load cycle and efficiency characteristics of the motor.

Variable speed pumps can provide significant economies if the WDS is characterized by a high variability of operating conditions, so that the advantages of adjusting the pump operating point can be fully exploited. In this perspective, systems characterized by a large static head are penalised because of the limited speed range the pump can operate and because of the substantial decrease in pump efficiency. In addition, the introduction of the variable frequency drive lowers the wire to water efficiency of the system. Moreover the efficiency of the motor and the pump are also lower when operated at lighter loads and lower speeds. Although induction motors are commonly used in WDSs with FSPs, brushless PM motors offer significantly higher efficiency and higher power factor specifically at low speeds and light loads. In addition, motor installation related issues play a significant role in the total efficiency of the system.

Acknowledgements. We would like to thank the Editor Guangtao Fu, Tom Walski and the other anonymous reviewer for their comments, which improved the quality of this paper.

References

Bene, J. G., Selek, I., and Hös, C.: Neutral search technique for short-term pump schedule optimization, J. Water Res. Pl.-ASCE, 136, 133–137, 2010.

Bernier, M. A. and Bourret, B.: Pumping energy and variable frequency drives, ASHRAE J., 41, 37-40, 1999.

British Pump Manufacturers' Association (BPMA), Gambica's Variable Speed Drive group and Electric Motor industry: Variable speed driven pumps – Best practice guide, 2003.

Broad, D. R., Maier, H. R., and Dandy, G. C.: Optimal operation of complex water distribution systems using metamodels, J. Water Res. Pl.-ASCE, 136, 433–443, 2010.

Fuchsloch, J. F., Finley, W. R., and Walter, R. W.: The Next Generation Motor, IEEE Ind. Appl. Mag., 14, 37–43, 2008.

Hydraulic Institute, Europump and the US Department of Energy's (DOE) Industrial Technologies Program: Variable speed pumping – A guide to successful applications, Executive Summary, 2004.

Industrial Technologies Program: Adjustable speed drive part – load efficiency, Motor Tip #11, US Department of Energy, Energy Efficiency and Renewable Energy, June 2008.

Lingireddy, S. and Wood, D. J.: Improved operation of water distribution systems using variable-speed pumps, J. Energ. Eng.-ASCE, 124, 90–102, 1998.

López-Ibáñez, M., Prasad, T. D., and Paechter, B.: Ant Colony Optimization for Optimal Control of Pumps in Water Distribution Networks, J. Water Res. Pl.-ASCE, 134, 337–346, 2008.

McCormick, G. and Powell, R. S.: Optimal pump scheduling in water supply systems with maximum demand charges, J. Water Res. Pl.-ASCE, 129, 372–379, 2003.

Melfi, M., Evon, S., and McElveen, R.: Induction versus Permanent Magnet Motors, IEEE Ind. Appl. Mag., 15, 28–35, 2009.

Rossman, L. A.: EPANET2. EPA/600/R-00/057, Water Supply and Water Resources Division, National Risk Management Research Laboratory, Office of research and Development, USEPA, Cincinnati, Ohio, USA, 2000.

Salomons, E., Goryashko, A., Shamir, U., Rao, Z., and Alvisi, S.: Optimizing the operation of the Haifa-A water distribution network, J. Hydroinform., 9, 51–64, 2007.

Sârbu, I. and Borza, I.: Energetic optimization of water pumping in distribution systems, Period. Polytech. Mech., 42, 141–152, 1998.

Stockman, K., Dereyne, S., Vanhooydonck, D., Symens, W., Lemmens, J., and Deprez, W.: Iso Efficiency Contour Measurement results for Variable Speed Drives, in: Proceedings of the XIX International Conference on Electrical Machines, ICEM, Rome, Italy, 6–8 September, 2010.

The US Department of Energy's Industrial Technologies Program (ITP) and the Hydraulic Institute (HI): Improving pumping system performance: A sourcebook for industry, The United States Department of Energy Office of Energy Efficiency and Renewable Energy, 2006.

Ulanicki, B., Kahler, J., and Coulbeck, B.: Modelling the efficiency and power characteristics of a pump group, J. Water Res. Pl.-ASCE, 134, 88–93, 2008.

van Zyl, J. E., Savic, D. A., and Walters, G. A.: Operational optimization of water distribution systems using a hybrid genetic algorithm, J. Water Res. Pl.-ASCE, 130, 160–170, 2004.

Walski, T., Zimmerman, K., Dudinyak, M., and Dileepkumar, P.: Some surprises in estimating the efficiency of variable speed pumps with the pump affinity laws, in: Proceedings of the World Water and Environmental Resources Congress 2003, Philadelphia, PA: ASCE, 1–10, 2003.

Wu, W., Maier, H. R., and Simpson, A. R.: Single-objective versus multiobjective optimization of water distribution systems accounting for greenhouse gas emissions by carbon price, J. Water Res. Pl.-ASCE, 136, 555–565, 2010a.

Wu, W., Simpson, A. R., and Maier, H. R.: Accounting for greenhouse gas emissions in multiobjective genetic algorithm optimization of water distribution systems, J. Water Res. Pl.-ASCE, 136, 146–155, 2010b.

11

Water supply project feasibilities in fringe areas of Kolkata, India

K. Dutta Roy[1], B. Thakur[2], T. S. Konar[3], and S. N. Chakrabarty[4]

[1]Kolkata Metropolitan Water and Sanitation Authority, Kolkata, India
[2]Meghnad Saha Institute of Technology, Kolkata, India
[3]Civil Engineering Dept., Jadavpur University, Kolkata, India
[4]Civil Engineering Dept., Jadavpur University, Kolkata, India

Abstract. Water supply management to the peri-urban areas of the developing world is a complex task due to migration, infrastructure and paucity of fund. A cost-benefit methodology particularly suitable for the peri-urban areas has been developed for the city of Kolkata, India. The costs are estimated based on a neural network estimate. The water quality of the area is estimated from samples and a water quality index has been prepared. A questionnaire survey in the area has been conducted for relevant information like income, awareness and willingness to pay for safe drinking water. A factor analysis has been conducted for distinguishing the important factors of the survey and subsequent multiple regressions have been conducted for finding the relationships for the willingness to pay. A system dynamics model has been conducted to estimate the trend of increase of willingness to pay with the urbanizations in the peri-urban areas. A cost benefit analysis with the impact of time value of money has been executed. The risk and uncertainty of the project is investigated by Monte Carlos simulation and tornado diagrams. It has been found that the projects that are normally rejected in standard cost benefit analysis would be accepted if the impacts of urbanizations in the peri-urban areas are considered.

1 Introduction

1.1 Background

The urban population is increasing at a rapid rate. In Asia, about 37% of the population was living in urban space in the year 2000. By 2030, the percentage is expected to reach about 54% which is equivalent to about 0.58% increase in each year. This is a significant departure from the trend of the same population group in 1950 when the annual average increase was only about 0.29% (Cohen, 2006). The trend is more pronounced in Kolkata (earlier called Calcutta). The population has increased from 11.02 million in 1991 to 13 million in 2001 indicating about 1.8% annual growth rate (Bannerjee and Das, 2006). Such an urbanization rate generated a tremendous pressure on civic facilities like water supply.

World Health Organization, in a half a century old study, found that only 12% of urban population in India had house connections and 18% used public stand post for water. The remaining 70% had no facility for piped water (Dietrich and Henderson, 1963). The same study also found that in all most all of the 75 developing countries that were within the study scope, the water works developments are too slow to match the future needs. In a decade later, International Bank for Reconstruction and International Development Association (IBRD) started financing for the water supply system. Improvements like construction of mains, reservoirs, booster stations and small groundwater based systems in the fringe areas of Kolkata have been initiated (IBRD, 1973). Such ground water based small systems have become a norm for fringe areas of Kolkata for many years mainly because of lesser investments and smaller gestation periods. The maintenance cost for small ground water supply is relatively high (Dutta Roy and Chakravarti, 2006). In Kolkata, the ground water was found to be bacteria free but showed high concentrations of chlorides and other chemicals. Hardness and Iron (Fe), Chlorine (Cl), and sulfates increase from north to south

Table 1. Methodology tools.

Steps	Tools	Input	Output
Technical analysis	CCME Water Quality Index	1. Chemical data of water 2. Indian standard limits	Quality index for potable water
Cost analysis	Present value analysis	1. Fund flows 2. Service life 3. Interest rate	Present value of cost
	ANN model estimate	1. Land rent 2. Population 3. Distance from plant	Cost estimate of the booster stations
Benefit analysis	Factor analysis for variable selections	1. Variables affecting WTP 2. WTP survey data	Ten nos. selected based on actual impact on WTP
	Multiple regression for WTP estimate	Selected variables values from WTP survey	Relationships among WTP and variables
	Dynamic simulation for WTP forecast in fringes	1. Increase of built-up area 2. WTP and income survey	Forecast of WTP increase with time
Cost benefit analysis	Monte Carlo's simulation for risk analysis	Input variations like family size, inflation, interest	Distributions of the net present value
	Sensitivity analysis for uncertainty	Net present value relationship data	Tornado diagram for sensitivity of variables

side of Kolkata. Recent studies also show that the groundwater in some areas contains arsenic, lead, and cadmium in excess of the levels prescribed by the WHO for drinking water (KMC, 2006). Experience and awareness have increased the demand of piped surface water in the fringe area. The policy makers now shun away from ground water sources because of its detrimental impacts. For example, Asian Development bank (ADB) has recently sanctioned a 230 Mil US$ loan for environmental improvement in Kolkata which includes construction of 8500 public stand posts but ground water based small water supply systems have not been financed (KMC, 2007).

Many of the users in the fringe areas find the ground water unpalatable and regularly purchase surface water for drinking from unorganized sector. Llorente and Zerah (2003) found that in New Delhi, India the water suppliers in such areas enjoyed a sizable market demand. Gessler and Brighu (2008) also found that in Jaipur, India the alternative providers supplied good quality water in tankers or camel-powered cart in peripheral areas of the city.

Water is traditionally supplied free of cost in Kolkata. Costs for the water supply are clubbed to the property tax (Ruet, 2002). Roy et al. (2003) reported that the Kolkata Municipal Corporation (KMC) grants about 75% subsidy to water supply. As a result, the water supply department is resource starved and soft international loans are availed for water supply infrastructures. However, international funding agencies require cost benefit analysis (CBA) and the author-

ity is obligated to prepare an explicit CBA as per best management practice. The development of such a CBA for the funding agencies has been found to be daunting for want of data and interdisciplinary nature of applications.

1.2 Methodology

The objective of this study is to develop a methodology that can be effectively used for development of a CBA for water supply projects in the peri-urban areas of the cities in India. The CBA should have sufficient details for catering the requirements of the international funding agencies. In typical CBA studies in India, the benefits are estimated from willingness to pay (WTP) studies conducted in the service area. It seems that the values obtained from such a WTP study in the peri-urban area would be inaccurate in the long run. The effect of rapid urbanization in the fringe areas should be integrated to the CBA study. In this paper, a CBA methodology with special reference to fringe area of the city that is under transition to piped water supply has been investigated.

In order to develop an appropriate CBA a number of steps from varities of technical fields have been used as shown in the block diagram in Fig. 1. The steps essential for the CBA are indicated in blocks in Fig. 1. A number of tools as shown in Table 1 have been used to complete the steps presented in Fig. 1. The explanation of these steps as well as the tools used in CBA are presented herein.

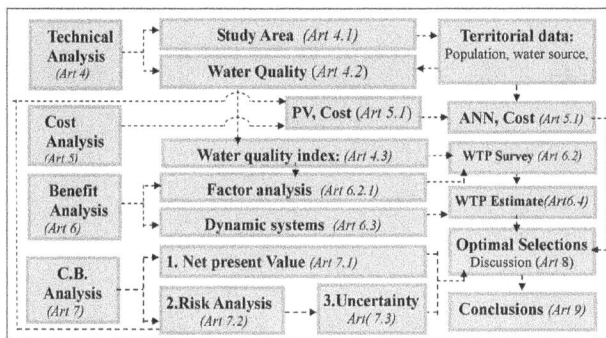

Figure 1. Block Diagram for the CBA.

2 Existing status

2.1 Previous study

The present authors investigated about the water supply network feasibilities in Kolkata in an earlier study (Dutta Roy et al., 2010). In the study, artificial neural network (ANN) have been trained for estimation of booster pumping station costs based on size, land rent and distance from the treatment plant. The benefit of the project is estimated from the willingness to pay survey conducted among beneficiaries within the Kolkata city. The costs and benefits occurred during the project life have been reduced to the present value for comparison. The variability of data has been studied with the help of Monte Carlo's simulation, uncertainty analysis and tornado diagram. The study has resulted in two key observations. Firstly, the net present value (NPV) of a booster pumping station is reduced to about zero at a distance of 20 km from the treatment plant. Secondly, the NPV is heavily dependent on the social discount and inflation rate. Accurate predictions of long term discount and inflation rate are really in the domain of the economists. Apart from it the study seems to have other latent issues in reference to its application to the fringe areas as explained in the following.

2.2 Willingness to pay

The monetary benefit from the project is based on WTP survey conducted on the beneficiaries. A number of WTP surveys have been conducted for water supply benefits in Kolkata (World Bank, 2001; Roy et al., 2003; Guha, 2007; Majumdar and Gupta, 2009). In addition, the authors also conducted a WTP survey in connection with the previous study (Dutta Roy et al., 2010). Each of these studies was conducted among the residents who had access to piped surface water. The dwellers living in the fringe areas of the city does not have access of the piped surface water. They use ground water either from bore/tube wells or use the public stand posts served by ground water based small systems. Users suffer from the disadvantages like hardness and iron contents. Many of these users buy surface water for drinking/cooking.

Ground water in some of the fringe areas contains harmful chemicals like arsenic (Bhattacharya et al., 1997; Chatterjee et al., 1995). The user groups who suffer from water quality have already revealed their economic choices by purchasing water. The negative experience from hardness and iron content in ground water is likely to alter the WTP among those living in the fringe areas. The rising awareness about the harmful chemical in ground water may also change the WTP with time. The WTP of the previous studies were all conducted among the dwellers enjoying the piped surface water. The WTP in the fringe areas might be different which should be used in the CBA for water supply projects in these areas. Such issues are investigated in this study.

2.3 Issues

Efforts have been made in many countries for standardization of CBA. For example, the Office of Management and Budget (OMB) in US issued guidelines for CBA and risk assessment (Kopp et al., 1997). European Union started TECH-NEAU, an integrated project funded by the European Commission that had a mandate to develop CBA for water supply system (Baffoe-Bonnie et al., 2006) and published guides (Florio, 2006). These studies do not specifically concentrate into urban-rural borderline where urbanization is taking place rapidly. The dwellers in fringe areas who purchase surface water reveal economic preferences that are not discernable in benefits assessed by willingness to pay (WTP) method conducted within Kolkata where water is supplied free of cost. The present paper has investigated the issues related to CBA of supplying piped surface water to these peri-urban users.

3 Technical analysis

3.1 Study area

The urbanization is taking place rapidly in the study area. Bhatta, Saraswati et al. (2010) estimated the degree-of-freedom and degree-of-sprawl towards the analysis of urban growth in Kolkata from satellite images of 15 years. Bhatta (December 2009) analyzed the urban growth boundary for the city of Kolkata. The satellite image data were used to model the growth boundary of Kolkata for the years 2020 and 2035. The proposed model discouraged scattered development and increase in urban growth rate. They proposed a plan with polycentric urban blobs into a monocentric tract. In order to monitor the urban growth from remote sensing, researchers have attempted the concept of Shannon's entropy for determining the dispersion of built-up land growth. The concept is applied to several Indian cities like Pune (Sekhar, 2010), Hyderabad (Lata et al., 2010), Mangalore (Sudhira, Ramachandra et al., 2004) and Indore (Antony, Kumar et al., 2007) with encouraging results. Bhatta (September 2009) applied Shannon's entropy model to Kolkata's urban growth.

Table 2. Water pumping stations in Kolkata and southern fringes.

Supply	Sl	Station Name	Tributary pop. (000)	Tank (MG)	Supply (MGD)
GRWW	1	GRWW Ph1			
	2	GRWW Ph2	579.6	NA	60.00
	3	Behala	550.9	3.5	9.00
	4	Mahestala	236.1	5.0	10.00
	5	Ranikuthi	241.3	3.5	7.00
	6	Garfa	185.9	3.5	7.00
	7	Kalighat	319.7	4.0	8.00
	8	Bansdroni	122.7	2.0	4.00
	9	Pujali	34.7	0.70	2.10
		Total	2630.9	22.20	107.10
DRWW		DRWW Direct	757.8	NA	21
	10	Kamagazi	341.8	1.0	3
	11	Begampur	419.7	1.0	3
	12	Krishnamohanpur	47.4	1.0	3
		Total	1993.9		30

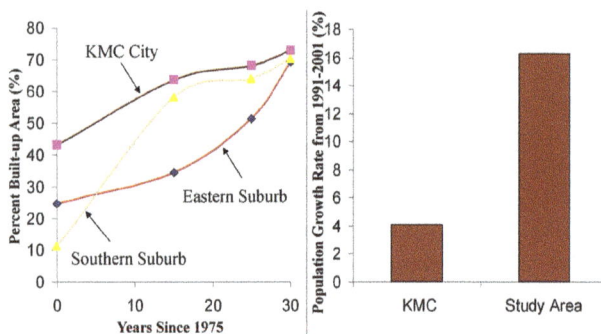

Figure 2. Urban growths in fringes of Kolkata.

He found higher growth rate of the built up area in the present study location that is the southern and eastern fringes of the city. The increase in built up area derived from satellite data by Bhatta (September 2009) is presented in Fig. 2. The population growth in Kolkata Municipal Corporation (KMC) and areas within the present study area as obtained from census records are also presented in Fig. 2. The Fig. 2 would indicate that population growth in the study area is increasing at a higher rate than that of the Kolkata city limits exerting a greater demand for drinking water services in the study area.

The water supply scheme based on ground water based small systems in the peri-urban area of Kolkata was initiated in about three decades earlier (IBRD, 1973). The divisions of ground and surface water distribution as observed in 1995 were presented by Basu and Main (2001). The population in the study area either uses ground water based small systems or individual bore wells. They are subjected to the detrimental effects of the ground water. Many of them purchase

drinking water from vendors. Some of these areas are now in a transition stage for piped surface water from the two water works namely the Garden Reach Water Works (GRWW) and the Dakshin Roypur Water Works (DRWW). GRWW is administratively designated only for the city of Kolkata and DRWW serves the peri-urban areas. The details of the water works are presented in Table 2.

3.2 Water quality

Water quality standards are not a fixed world wide standard. It varies significantly even in a country with the passage of time. In the setting of standards, competent agencies make political, economical, technical and scientific decisions about how the water will be used. For example, the US act requires reviewing the water quality standards in every three years (US EPA, 2010). Indian standard (CPHEEO, 1999) has specified a series of requirements for potable water standards.

The Septic/Imhoff tank is the common form of sewage disposal method in the study area. Effluents are generally dispersed to ground through soak pits. Ground water up to a depth of about 15M is not potable for contamination (Sahu and Sikder, 2009). In Kolkata, a Quaternary aquifer is sandwiched between two clay sequences (Sahu and Sikder, 2009). It starts from a depth of about 50 m. The depth of aquifer is about 300 m below which the bottom clay bed of Pliocene age occurs (Sikdar et al., 2001). The ground water is usually drawn from a depth of about 100 to 300 m depths. The Central Ground Water Board (CGWB, 2010) in India indicated salinity and arsenic as the major problem for ground water in Kolkata. Users in the study area depend on ground water because it is mostly safe from biological contamination which at least prevents the epidemics. However, ground

Table 3. Water quality observed in southern fringes of Kolkata.

Source	Area	pH	Chloride (mg/l)	Hardness (mg/l₃)	Iron (mg/l)	Arsenic (mg/l)	Fluoride (mg/l)	Coliform (N/100 ml)	WQI
Piped surface water	W. Fringe Behala	7.04 (±0.07)	15.65 (±1.55)	4.80 (±2.77)	BDL	BDL	BDL	0	91.10
	E.Fringe Mukunda.	7.03 (±0.05)	15.98 (±1.20)	5.00 (±1.58)	BDL	BDL	BDL	0	91.10 *Good*
Pond	W. Fringe Behala	7.02 (±0.07)	58.60 (±6.02)	20.20 (±1.64)	BDL	BDL	BDL	3900 (±291.6)	39.00
	E.Fringe Mukunda.	7.03 (±0.09)	90.60 (±5.77)	29.20 (±2.39)	BDL	BDL	BDL	3800 (±158.1)	38.60 *Poor*
Open Well	W. Fringe Behala	7.01 (±0.02)	23.41 (±3.30)	108.00 (±8.37)	0.14 (±0.05)	BDL (±)	BDL (±)	1760 (±167.3)	35.70
	E.Fringe Mukunda.	7.06 (±0.02)	44.01 (±2.47)	158.00 (±8.37)	0.16 (±0.05)	BDL	BDL	1620 (±178.9)	38.80 *Poor*
Piped Gro-und water	W. Fringe Behala	NA	NA	NA	NA	NA	NA	NA	NA
	E.Fringe Mukunda.	7.36 (±0.03)	173.30 (±5.32)	318.60 (±5.59)	0.88 (±0.08)	BDL	BDL	0	60.80 *Marg.*
Tube well	W. Fringe Behala	7.53 (±0.10)	48.21 (±5.91)	173.20 (±5.07)	0.70 (±0.27)	BDL	BDL	0	70.90 *Fair*
	E.Fringe Mukunda.	7.52 (±0.08)	295.54 (±8.99)	413.20 (±7.79)	1.26 (±0.05)	BDL	BDL	0	48.60 *Marg.*
Standards	Acceptable Limit	7.0–8.5	200	200	0.10	0.01	1.00	0	

Note 1: The acceptable limits prescibed in Indian standards (CPHEEO-2009) are presented in the table.
Note 2: Standard deviations where applicable are presented under the mean in parenthesis.

water in the study area suffers from other disadvantages like hardness and iron. Expensive control methods that are not economically feasible in large scale in Kolkata are necessary to contain hardness and iron. Users naturally have the preference for treated surface water.

The water quality as available to the users has been estimated from sample survey. Several types of potable water that are consumed in the peri-urban areas of Kolkata are collected. The sources and areas are detailed in Table 3. For each location five numbers of samples were tested for each of the potable water types. The acceptable limits under the Indian standard (CPHEOO 2010) are also shown in Table 3 for comparison.

Water samples collected from the study area were tested using standard laboratory procedures as per provisions of Standard Methods for the Examination of Water and Wastewater (APHA, 1995). A digital pH meter was used for measuring the pH. The Chloride content (as Cl^-) was estimated by Mohr's method. Hardness (as $CaCO_3$) was determined through EDTA method. Iron and Fluoride concentrations were determined by colourimetric procedure. Arsenic was measured by a spectrophotometer. Total coliform was determined by fermentation technique and results were reported as MPN (Most Probable Number) of organisms present. The results from the tests were presented in Table 3.

3.3 Water quality index

Water quality index (WQI) is a tool for simplifying the reporting of water quality data. Traditional reports similar to that of Table 3 consist of variable-by variable report which is of value to experts. It supplies a wealth of data but could be overwhelming to non-chemists. For example, the Indian code for drinking water (BIS10500-1991) has more than 30 variables. The managers and decision makers may still wish to have a simpler measurement for overall representation of water quality. Indexing is a tool for such representations. There is no single measure that can describe overall water quality for a body of water. Although there is no globally accepted composite index of water quality, some authorities are using, aggregated water quality data in the development of water quality indices (UNEP, 2007). Most water quality indices rely on standardizing and weighted average of parameters according to their perceived importance to overall water quality (Sargaonkar and Deshpande, 2003).

The water quality index proposed by the Canadian Council of Ministers of the Environment (CCME) is a standardized and flexible index used in Canada and in other countries (CCME 2001). It was developed by the environment department of Canada after unifying a number of provincial indexes. The index was standardized against the provisions of the Indian drinking water standards (BIS10500-1991) for this study. The software (CCME 2001) developed by the Canadian government was used. Similar indexing system has been used in Nagpur (Ramakrishnaiah et al., 2009) and Karanataka (Rajankar et al., 2009), India.

The water quality index is based on the CCME method (CCME, 2001). There are three factors namely scope, frequency and amplitude in the index that are scaled to a range between 0 and 100. The estimated WQI values ranges from 0 to 100 that are classified into five water quality categories namely excellent (95–100), good (80–94), fair (65–79), marginal (45–64) and poor (0–44). The WQI of the samples was estimated by the recommended software (CCME, 2001) and reported in Table 3. The acceptable limits of the variables in the sample are based on the Indian standard (CPEEHO, 1999) and is also reported in Table 3.

4   Costing

Life cycle costing (LCC) is usually adopted for CBA. It is the total cost of ownership of an asset including its cost of acquisition, operation, maintenance, conversion, and/or decommission (SAE, 1999). Since outflows occur over multiple time periods it is reduced to a present value using the standard principle of time value of money for standardized LCC methodology such as by US Government (Fuller and Petersen, 2002). The risk and uncertainty of cash flows are also considered in LCC (Emblemsvåg, 2003). A flow diagram of the life cycle cost of a typical water supply installation is presented in Fig. 3. The components of acquisition and annual sustenance costs are presented. In order to compress the stream of cash flows the economic adjustment components are considered. The components required for managing the variability of outcomes are also presented in Fig. 3.

The present value estimate that has been used in relation to this study is presented in Eq. (1).

$$Cp = \frac{\sum_{t=1}^{L} \frac{C_t(1+r)^t}{(1+i)^t}}{\sum_{t=1}^{Lc} Qt} + \sum_{t=1}^{L} \frac{\frac{O_t(1+r)^t}{Qt}}{(1+i)^t}. \tag{1}$$

Where, Cp = Present value of cost in INR/Kl, i = Discount rate at year t, Ct = Acquisition costs in INR incurred at year t, Ot = Sustenance costs in INR incurred at year t, Qt = Quantity of water in Kl generated at year t, L = Expected life of

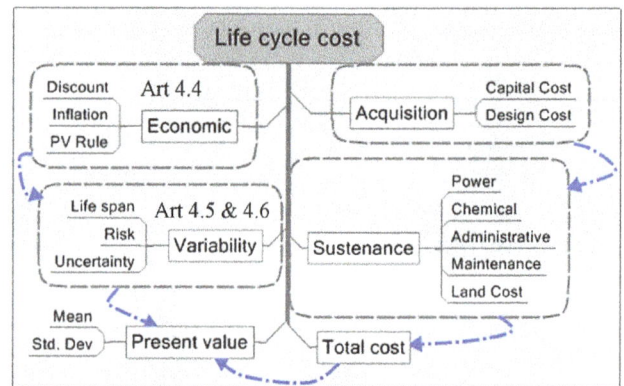

Figure 3. Block diagram for cost estimate.

the facilities, i = Discount rate at year t and r = inflation rate at year t.

The inputs in Eq. (1) would require complete technical design, drawing and specifications for a specific installation and estimation of all operations of the facility which might be exhaustive, time consuming and sometimes infeasible for new projects. In order to circumvent such issues data driven models like artificial neural network (ANN) has been proposed for cost estimate (Adeli, 2001; Boussabaine, 1996). The ANN is a mathematical model of theorized brain activity for processing an existing dataset (Zurada, 1992). Similar ANN techniques have been used in this case for estimating the cost of water supply.

4.1   Cost estimate

The costs of the facilities have been collected from the records of the concerned departments of GRWW in an earlier study. We (Dutta Roy et al., 2010) trained an ANN model among the existing cost data of the GRWW network with the help of NeuroSolutions software (NeuroDimensions, 2009) and derived a cost model. The ANN cost model was based on three input factors namely land rent, serving population and the distance from the treatment plant. The model is applied to the DRWW network and costs for the booster stations are estimated. The inputs as well as the ANN findings are presented in Table 4.

5   Benefit

5.1   Benefit analysis

The benefit of the project is first estimated based on a willingness to pay (WTP) survey which is fairly common even in Kolkata (World Bank, 2001; Roy et al., 2003; Guha, 2007; Majumdar and Gupta, 2009; Dutta Roy et al., 2010). The consumer's spending capacity and experience and awareness about a product have an impact on the WTP. As a result of urbanizations the population in the study area is increasing

Table 4. ANN Results of the DRWW Network.

Sl	Station	Tributary pop. (000)	Land rent (INR)	Supply (Kl/day)	Distance (KM)	ANN Cost (INR/Kl)
10	Kamalgazi	341.803	608986	51270.45	57.92	5.2990 ± 0.2161
11	Begampur	419.718	341032	62957.70	42.48	4.8360 ± 0.2166
12	Krishnamohanpur	474.514	243594	71177.10	41.15	4.7970 ± 0.2056

Table 5. Correlation matrix.

	WQI BFW	WQI DC	Fam Size	Avg Age	High Age	Low Age	Avg Edu	High Edu	Low Edu	Total Incm	PC Incm	Exp BFW	Dfclt BFW	Exp DC	Dfclt DC	Buy Amnt	Buy Mony	Awr Path	Awr Arsn	Awr Iron	Awr Total	WTP
WQI BFW	1																					
WQI DC	0.17	1																				
Fam Size	−0.01	0.16	1																			
Avg Age	−0.17	0.40	−0.12	1																		
High Age	−0.09	0.40	0.48	0.64	1																	
Low Age	−0.15	0.11	−0.53	0.66	0.04	1																
Avg Edu	−0.23	−0.27	−0.23	0.04	−0.07	0.10	1															
High Edu	−0.19	−0.10	0.10	0.10	0.24	−0.10	0.71	1														
Low Edu	−0.11	−0.32	−0.41	−0.13	−0.36	0.10	0.78	0.36	1													
Total Incm	−0.08	0.25	−0.04	0.16	0.16	0.05	0.42	0.39	0.33	1												
PC Incm	−0.04	0.16	−0.37	0.22	−0.07	0.33	0.48	0.30	0.48	0.73	1											
Exp BFW	−0.12	−0.40	0.16	−0.05	0.05	−0.12	0.08	0.16	0.11	−0.10	−0.11	1										
Dfclt BFW	−0.07	−0.04	0.22	−0.24	−0.04	−0.23	−0.14	−0.09	−0.12	−0.01	−0.10	0.13	1									
Exp DC	−0.10	−0.52	0.09	−0.11	−0.03	−0.12	0.14	0.13	0.20	−0.09	−0.09	0.85	0.05	1								
Dfclt DC	−0.15	−0.85	−0.11	−0.35	−0.38	−0.13	0.26	0.13	0.31	−0.10	−0.03	0.41	0.22	0.44	1							
Buy Amnt	0.09	0.78	0.36	0.30	0.44	−0.04	−0.18	−0.01	−0.31	0.13	−0.05	−0.30	−0.01	−0.40	−0.67	1						
Buy Mony	−0.02	0.61	0.31	0.25	0.35	−0.04	−0.09	0.09	−0.24	0.10	−0.05	−0.27	0.01	−0.34	−0.51	0.88	1					
Awr Path	0.07	0.48	−0.07	0.40	0.25	0.29	0.02	0.02	0.01	0.11	0.14	−0.20	−0.10	−0.20	−0.48	0.31	0.22	1				
Awr Arsn	−0.05	0.36	−0.11	0.29	0.22	0.16	0.09	0.06	0.09	0.06	0.11	−0.10	−0.01	−0.16	−0.42	0.25	0.14	0.57	1			
Awr Iron	0.05	0.57	−0.01	0.31	0.26	0.11	−0.05	0.03	−0.06	0.14	0.13	−0.21	−0.01	−0.21	−0.55	0.37	0.22	0.70	0.78	1		
Awr Total	−0.00	0.53	−0.05	0.37	0.29	0.18	0.01	0.04	0.00	0.12	0.13	−0.18	−0.03	−0.13	−0.55	0.35	0.23	0.82	0.90	0.94	1	
WTP	−0.09	0.01	−0.21	0.17	0.06	0.17	0.56	0.43	0.45	0.64	0.75	−0.02	−0.07	0.03	0.03	−0.04	−0.01	0.04	0.12	0.03	0.07	1

rapidly. The migrating residents are experiencing the ground water piped supply. The WTP for ground water piped supply in the peri-urban area is subjected to change because of actual experience and changing income profile among the population in the study area with the passage of time. In addition, the presence of harmful chemicals like arsenic in the study area has been published in technical literature and media. The spending capacity, experience and awareness among the residents are increasing that is expected to change the WTP for ground water piped supply with the passage of time in the study area.

Turpie (2003), in a study about valuation of biodiversity in South Africa, found that the WTP for conservation have changed when the respondents were informed about the risks. Brouwer (2006) found that WTP for bathing water after a draught experience is not the same as before. In this case, the WTP for ground water in the peri-urban area is expected to change with time because of factors like increase in the average income profile for urbanization in the area, negative experience for hardness and iron content and additional awareness about harmful chemicals like arsenic in ground water. The expected change of WTP is modeled by dynamic system analysis.

5.2 WTP survey

A WTP survey was carried out for estimating the consumer willingness to pay for water among 201 samples in the study area. In addition to WTP, data was collected for 21 other variables likely to influence the consumer WTP. A correlation matrix for these variables (Pearson) was calculated assessing the underlying correlations between these variables and is presented in Table 5. It is observed that two variables viz. WQI bathing/washing and difficulty for bathing/washing are poorly correlated (Pearson correlation coefficient < 0.30) with other variables. Since the variables are poorly correlated these are omitted from further analysis.

5.2.1 Factor analysis

A principal component analysis followed by a varimax rotation carried out on remaining 18 variables using SPSS (2009) for determining the underlying factors of WTP. The analysis was run following the Kaiser's criterion (Kaiser, 1958) of retaining the components with eigenvalues greater than one. The Kaiser's criterion seemed to be applicable here as the number of variables is less than 30 (SPSS 2009). The Kaiser-Meyer-Olkin measure of sampling adequacy is found to be 0.645 (> 0.5) suggesting the samples are adequate. The Bartlett's Test of Sphericity is found to be highly significant

Table 6. Varimax rotated factor loadings for five-factor solutions.

Factors	Variable Symbol	Variables	Selection	Factor Load
Factor 1	Awr_Total	Total Awareness	Yes	0.972
Variance = 19.61	Awr_Iron	Awareness about Iron Contamination		0.906
Eigenvalue = 5.62	Awr_Arsn	Awareness about Arsenic Contamination		0.904
	Awr_Path	Awareness about Pathogenic Contamination		0.778
Factor 2	Avg_Edu	Average Education Level of the Family	Yes	0.830
Variance = 19.25	WTP	Willingness to Pay		0.828
Eigenvalue = 3.98	PC_Incm	Monthly Per Capita Income of the Family	Yes	0.796
	Total_Incm	Monthly Total Income of the Family		0.780
	Low_Edu	Lowest Education Level in the Family		0.695
	High_Edu	Highest Education Level in the Family		0.691
Factor 3	WQI_DC	Water Quality Index for Drinking and Cooking	Yes	0.844
Variance = 18.79	Buy_Amt	Amount of Water Bought Per Month	Yes	0.822
Eigenvalue = 2.25	Buy_Mony	Money Spent for Buying Water Per Month	Yes	0.743
	Dfclt_DC	Degree of Difficulty with Drinking / Cooking	Yes	−0.720
	Exp_DC	Years of Using the Drinking and Cooking water	Yes	−0.718
Factor 4	High_Age	Highest Age in the Family		0.855
Variance = 11.03	Fam_Size	Number of Members in the family	Yes	0.684
Eigenvalue = 1.76				
Factor 5	Low_Age	Lowest Age in the Family		0.908
Variance = 10.70	Avg_Age	Average Age of the Family	Yes	0.708
Eigenvalue = 1.46				

($p < 0.001$) suggesting Factor Analysis is appropriate (SPSS 2009) in this case. The analysis extracted five factors with eigenvalues greater than one which together accounted for 79.38% of the explained variance. The average communality after extraction has found to be 0.84 (> 0.6) conforming to Kaiser's criterion. The details of factor loadings, explained variance and eigenvalues of the five extracted factors are given in Table 6. The items with loadings greater than 0.6, are used to define five extracted factors. The variables for the regression of WTP are decided based on the extracted factors and are indicated in Table 6.

5.2.2 Multiple regression

A multiple regression analysis has been conducted for WTP against ten explanatory variables selected from the factor analysis presented in Table 6. The mean and standard deviation for the selected variables as obtained from the survey is presented in Table 7. The R^2 for the regression is found to be 0.63. The analysis of variance (ANOVA) is presented in Table 8 and seems to be acceptable for such studies. The regression equation for WTP is presented in Eq. (2).

$$\text{WTP} = 99.28 - 2.06(\text{WQI\_DC}) + 4.98(\text{Fam\_Size}) + 0.41(\text{Avg\_Age}) +$$
$$11.29(\text{Avg\_Edu}) + 0.02(\text{PC\_Incm}) + 0.22(\text{Exp\_DC}) - 9.16(\text{Dfclt\_DC}) +$$
$$0.02(\text{Buy\_Amt}) + 0.03(\text{Buy\_Mony}) - 25(\text{Awr\_Total}) \qquad (2)$$

Table 7. Variables statictics for multiple regression.

Variables	Mean	Standard Deviation
WQI_DC	82.90	9.97
Fam_Size	4.55	1.60
Avg_Age	36.42	7.86
Avg_Edu	4.87	1.64
PC_Incm	2145.35	2883.28
Exp_DC	9.87	8.91
Dfclt_DC	1.93	1.24
Buy_Amt	306.24	274.54
Buy_Mony	125.22	110.42
Awr_Total	7.63	2.54
WTP	58.51	84.27

The explanatory variables as shown in Table 7 are employed in Eq. (2) for conducting a Monte Carlo's simulation. The monthly per family WTP is found to be 58.51 ± 67.43. A sensitivity analysis has been carried out and the tornado diagrams for regression coefficients and Spearman rank correlation coefficients are presented in Fig. 4.

The WTP is found to be most sensitive on per capita income (PC_Incm) and the least on awareness (Awr_Total). It has been hypothesized that the income would considerably

Table 8. ANOVA for multiple regression.

	Df	SS	MS	F	Significance F
Regression	10	450256.1	45025.61	15.59652	8.10×10^{-16}
Residual	90	259821.1	2886.901		
Total	100	710077.2			

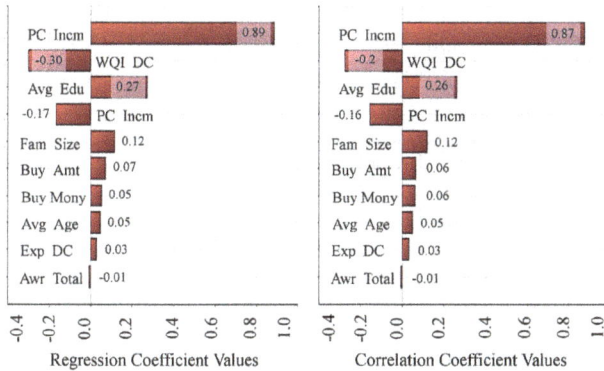

Figure 4. Tornado diagrams for sensitivity analysis of WTP

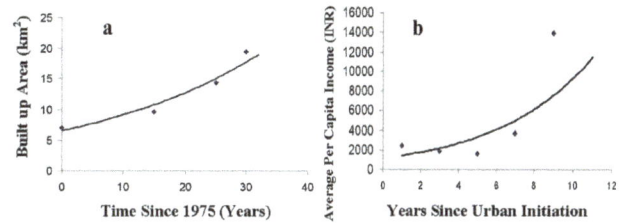

Figure 5. Trends of income and urban growths.

increase in the peri-urban areas with the growth of urbanizations. Such increase should be considered in CBA particularly for the peri-urban areas where the effect would be pronounced to reflect the effect of the expansion of the city with time. Dynamic analysis technique is used for estimating the increase of such income.

5.3 Dynamic system analysis

The observed WTP is dependant upon variables like income which changes over time. As a result WTP is likely to change over time as well. System dynamics analysis (SD) has been used to estimate the change of WTP with time. SD analysis was first proposed by Forrester (1991). It is a methodology for studying complex feedback systems over time. He applied SD to urban planning (Forrester, 1969) for its ability for long term effectiveness. Ho et al. (2007) chose SD after evaluating other optimizing models in connection with the conjunctive use of surface and subsurface water. They found that the SD model is a suitable methodology for constructing complex water resources models.

There are number of commercial software that has been developed for SD model. In this case, iThink (iSee 2009) has been used. It is a common SD application that has been used in several water quality books (Carlseen, 2004; Werick, 1994; Wurbs, 1994) and used by institutes like (UNESCO-IHE, 2010). In the present study, iThink (iSee 2009) software is used for analyzing system dynamics of income growth with time.

5.3.1 Urbanization and income growth

The growth of urbanization results in increase of built up areas. Such growth for peri-urban Kolkata has been estimate by Bhatta (September 2009) from satellite data and is presented in Fig. 2. The data has been combined with the geographical data for development of the regression based relationship presented in Eq. (3) and in Fig. 5a. The correlation (R^2) has been found to be 0.96.

$$B = 6.5295e^{0.0333t_a} \tag{3}$$

where B = built up area in km^2 and t_a = time in years since 1975.

Using Eq. (3) present built-up area (at year 2010) is calculated as $20.9435\ km^2$. The income and years of urbanization have been studied from the WTP survey data conducted in the eastern fringe of Kolkata viz. Mukundapur. A relationship is derived from regression as shown in Eq. (4) and in Fig. 5b. It has been observed that the average income of the population migrating to this area would increase as presented in Fig. 5b. The correlation (R^2) has been found to be 0.58. It is so expected because only progressively richer people could afford the price increase caused for rapid urbanization in this area.

$$S = 1172e^{0.2074t_i} \tag{4}$$

where S = Average per capita income and t_i = time in years of urbanization.

5.3.2 Dynamic simulation

The increase in built up area and per capita income in the fringes of Kolkata with the years of urbanization is evaluated over a period of 20 years in an iThink (2009) dynamic simulation model as shown in Fig. 6.

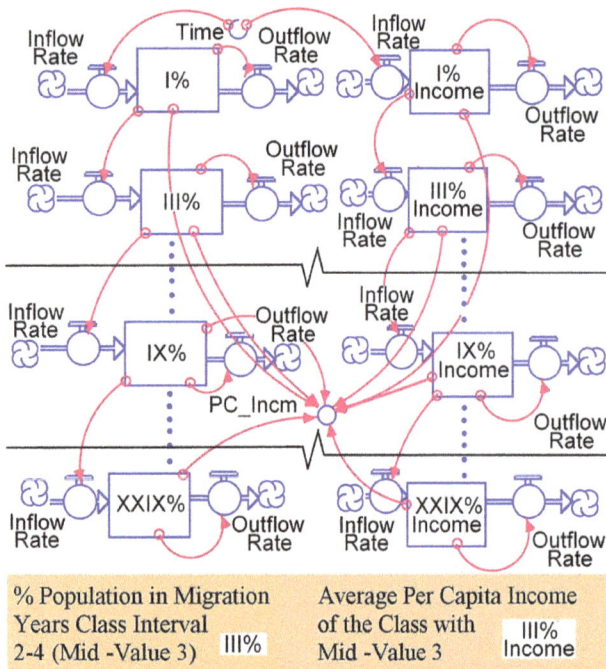

Figure 6. Dynamic simulation model in iThink.

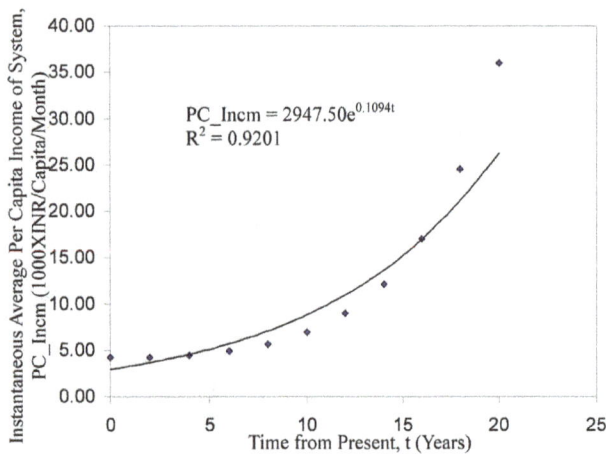

Figure 7. iThink model results.

The iThink software automatically simulates a time dependent dynamic system, comprising different input variables (shown in Fig. 6) based on assigned relationships. The model component relationships are presented in Table 9.

The model is simulated for ten time steps i.e. for 20 years and the average per capita incomes are presented in Table 10 and increase of income with time is presented in Fig. 7.

The regression relationship between incomes with time is presented in Eq. (13). The correlation (R^2) is found to be 0.92.

$$PC\_Incm = 2947.5e^{0.1094t} \qquad (13)$$

where, PC_Incm=Instantaneous Average Per Capita Income of the System (INR/Capita/Month), t = Year of Calculation (time in year from recent).

5.4 WTP estimate

The questionnaire survey conducted in the eastern fringes of Kolkata viz. Mukundapur provided the opportunity to estimate the WTP by regressing against the income as presented in Eq. (14) and based on Eq. (13) it is rewritten as in Eq. (15). The daily per capita WTP is presented in Eq. (16). Since the daily per capita water demand is 150 lpcd (CPHEOO, 1999) the WTP per cubic meter of water may be written as in Eq. (17).

$$WTP_m = 0.0211 PC\_Incm + 6.1137 \qquad (14)$$

$$WTP_m = 0.0211\left(2947.5e^{0.1094t}\right) + 6.1137 \qquad (15)$$

$$WTP_d = \frac{WTP_m \times 12}{365 \times Fam\_Size} \qquad (16)$$

$$WTP_0 = \frac{(WTP_d) \times 1000}{150} \qquad (17)$$

where, WTP_m = Monthly per Family WTP (INR/Family/Month) and PC_Incm = Monthly per family income (INR/Family/Month), WTP_0 = WTP per cubic meter of water (INR/CuM).

6 Benefit cost comparison

6.1 Net present value

The present value of WTP after correction for inflation for each cubic meter of water is estimated from Eq. (17) and presented in Eq. (18). The net present value of WTP for each cubic meter is presented in Eq. (19).

$$WTP_p = \frac{\sum_{T=1}^{75} \frac{WTP_0(1+r)^T}{(1+i)^T}}{T} \qquad (18)$$

$$NPV = WTP_p - Cp \qquad (19)$$

where, WTP_p = Present value of WTP for each CuM of water, r = average inflation rate, T = Expected life of the water treatment facility for which CBA is conducted, Cp = present value of cost of the facility for which CBA is conducted.

The WTPp (Eq. 18) is simulated over time as presented in Fig. 8a. The graphs in Fig. 8b–d show the variation of NPV of WTP for each CuM of water (Eq. 19) over time. The range between ± three standard deviation has been also presented as bands in Fig. 8. The costs of the facilities for booster stations in Kamalgazi, Begampur and Krishnamohanpur are obtained from the ANN model presented in Table 4.

Table 9. Relationships in iThink model.

Model Component Description	Relationships	Eq. No.
Built up Area in km^2 (B)	$B = 6.5295e^{0.0333(2t_i+35)}$	(5)
The rate of increase in built up area	$\frac{dB}{dt_i} = 0.4349e^{0.0333(2t_i+35)}$	(6)
Percentage of city population migrating to fringe areas (P)	$P = B \times \frac{100}{20.9435}$	(7)
Migrants increase rate	$\frac{dP}{dt_i} = \frac{dB}{dt_i} \times \frac{100}{20.9435}$	(8)
By Eqs. (6) and (8)	$\frac{dP}{dt_i} = 2.0764e^{0.0333(2t_i+35)}$	(9)
Per capita income of any first group from today (S)	$S = 1172e^{0.2074(2t_i+9)}$	(10)
Average per capita income increase rate of first group with respect to iThink time step	$\frac{dS}{dt_i} = 486.1456e^{0.2074(2t_i+9)}$	(11)
Instantaneous Average per Capita Income of the System (INR/Capita/Month)	$\frac{\sum_{\forall i} i\% \times i\% \text{INCOME}}{\sum_{\forall i} i\%}$, $\quad i \in [I, III, V, ...XXIX]$	(12)

Note: All relations are with respect to iThink time scale (1 time step = 2 years).
Where, P = population migrating to the fringe area in percentage of the present population, B = Built up area in km^2, S = Average per capita income of the migrants.

Table 10. iThink model results.

iThink Time Steps	Time Years	Average per Capita Income (INR/Capita/Month)
0	0	4207.73
1	2	4244.40
2	4	4440.84
3	6	4869.12
4	8	5638.99
5	10	6917.04
6	12	8955.88
7	14	12 138.16
8	16	17 043.27
9	18	24 548.19
10	20	35 980.16

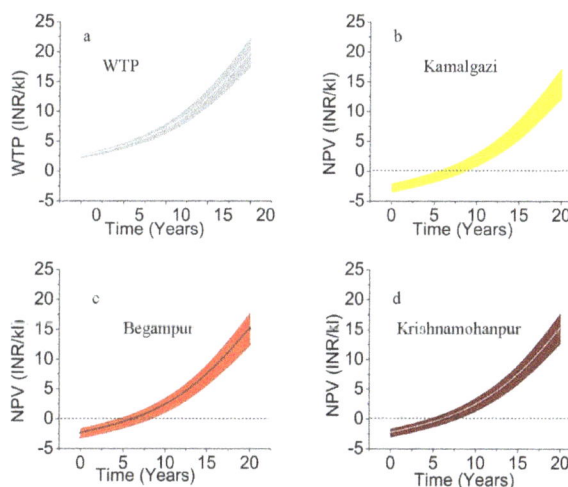

Figure 8. Variation of WTP and NPV with time.

6.2 Risk

The inputs for estimation of WTP and NPV are stochastic in nature. The common procedure for investigating such issues in CBA is by means of Monte Carlo's simulations (MCS). For example, Whittington et al. (2004) used MCS in Kathmandu, Nepal and Zhang (2009) used it in Singapore for water supply projects. MCS has been recommended by the regulatory authorities in counties like Canada (2007) and EU (Jasper, 2008). In MCS analysis, possible inputs depending upon its probability distributions are provided to the model and the resultant outcomes are noted. After numerous operations, the mean and standard deviations of the outcomes are assumed as the distributions of the possible outcome values.

The risks in inputs like the family size (Fam_Size), rate of inflation (r), rate of discount (i) are presented as triangular distribution in Table 11. The present value of ANN generated cost of supply (Cp) as shown in Table 5 are also presented. Monte Carlo's simulations consisting of 10 000 iterations have been conducted by Palisade @Risk (DTS, 2009) software for estimating the standard deviation of NPV as shown in Fig. 9. The results of the Eq. (19) for typical three cases have been presented in Fig. 9 as examples.

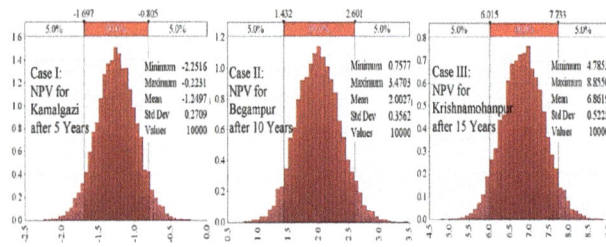

Figure 9. Distributions observed in typical simulation runs for NPV.

Table 11. Parameters for WTP and NPV simulation.

Symbol	Unit	Low	Average	High
Fam_Size	Number	2.00	4.55	8.00
R	%	5.00	6.00	7.00
I	%	6.00	7.00	8.00
Cp	INR/kl	Mean	Standard Deviation	
	Kamalgazi	5.2990	±0.2161	
	Begampur	4.8360	±0.2166	
	Krishnamohanpur	4.7970	±0.2056	

6.3 Uncertainty

The common procedure for investigating uncertainty in a CBA is done by means of sensitivity analysis (ADB, 1999; Baffoe-Bonnie, 2006; Canada, 2007; Florio, 2006). It identifies the critical parameters for the solution and determines systematically the influence of parameter variations on the solution (Fellin et al., 2005). The guidelines of OECD (2006) also specified sensitivity study for CBA. Sensitivity analysis is an iterative process. The variables in Eq. (19) are changed systematically and the resultant NPV is recorded. After numerous iterations, the data set would provide an idea about the input which has the largest effect on the NPV. The 'Toprank' software of the Palisade system (DTS, 2009) has been used to conduct the sensitivity analysis. The results of the sensitivity analysis are presented in the tornado diagram in Fig. 10. The inflation and discount rate is found to be much more significant than other factors like project cost (Cp). The power series relationship of inflation (r) and discount (i) in the NTP and the long time periods of 75 years have made it much more sensitive than other factors like the project cost.

7 Result and discussion

A previous CBA study (Dutta Roy et al., 2010) conducted for the city of Kolkata has been extended for the peri-urban areas in this article. The block diagram presented in Fig. 1 has been followed for a CBA methodology suitable to peri-urban areas. The urbanization is taking place rapidly in the

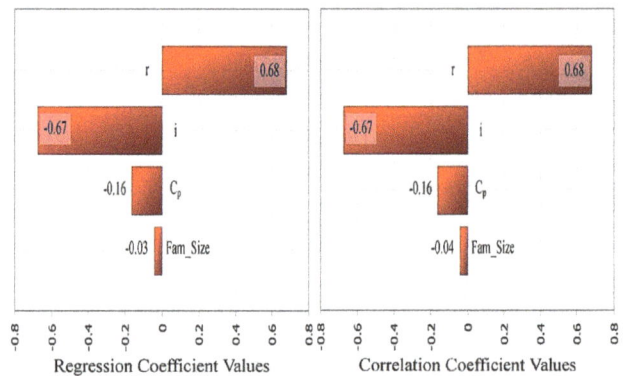

Figure 10. Tornado diagrams for sensitivity analysis of NPV.

suburbs of Kolkata as evidenced from Fig. 2 derived from the satellite data (Bhatta, September 2009). In the peri-urban areas, people mostly depend on ground water from individually owned bore wells or surface water from ponds. Only pockets of suburbs have small groundwater based piped water systems installed mainly by external grants (IBRD, 1973). The water samples from these sources have been tested according to standard methods (APHA, 1995), compared with the national standards (CEEPHO, 1999) and presented in Table 3.

It may be noted from Table 3 that water quality of the fringe areas are inadequate in a number criteria prescribed in the Indian standards (CEEPHO, 1999). WQI has therefore been used for simplifying the quality issues about variable by variable comparisons. Canadian WQI method (CCME, 2001) as presented in Table 4 and calibrated to Indian standards has been used in the present case. It is observed in Table 3 that the WQI of water from pond or well in the fringe areas are of "poor" quality. The WQI has been found to be improved to about "fair" quality for ground water. The piped surface water obtained from the treatment plant on the other hand has the WQI of above 90 indicating "good" quality. The users in the suburbs would naturally prefer piped surface water. In order to explore the consumers' preferences a water quality survey has been conducted in the peri-urban areas. It has been found from the survey that a large percentage of people actually buy water and most have a positive WTP for treated surface water.

In our previous study (Dutta Roy et al., 2010) a neural network model has been developed for rapid estimation of water supply network costs. The model has been employed for estimating costs of the booster pumping stations in the DRWW network as shown in Table 4.

The benefit of the water supply has been initially estimated from a WTP survey. Varieties of queries about the water supply have been added to the survey. A factor analysis presented in Table 6 indicates that five factors would account for about 80% of the variance. The multiple regression study with ten explanatory variables as selected in Table 6 has been conducted and the regression relationship is presented in Eq. (2). The mean and standard deviation for the selected explanatory variables as obtained from the survey is presented in Table 7. These statistics have been used for conducting a Monte Carlo's simulation of the Eq. (2). A sensitivity analysis has been carried out and a tornado diagram has been presented in Fig. 4. The WTP is found to be most sensitive on per capita income of the surveyed population.

The high dependence of WTP on income has created a special problem for the peri-urban areas. Since urbanizations are taking place rapidly in peri-urban areas as indicated in Fig. 2 the per capita income are also increasing faster in these areas. The CBA of a water supply project based on the present per capita income of the surveyed population may not be appropriate only after a few years because of increased per capita income caused by urbanizations. Since the water supply projects are usually for a long period of time the increase of per capita income with urbanizations should also be included for a realistic CBA.

A SD analysis has been conducted for ascertaining the increase of per capita income with time and the results have been plotted in Fig. 7. The regression relationship between income and time as derived from SD analysis is presented in Eq. (13). It may be noted that this per capita income rise in the peri-urban area is from the result of urbanization because progressively richer people could afford housing in these areas. The income rise as a result of inflation is a different issue and has been addressed in the present value estimate analysis in Eq. (18). The cost derived from Table 4 and the benefits derived from Eq. (13) has been used in Eq. (19) for estimating the net present value and the results are presented in Fig. 8.

It may be noted in Fig. 8 that NPV for each of the booster stations in the DRWW network namely Kamalgazi, Begampur and Krishnamohanpur are negative at the start. These projects should normally be rejected in a standard CBA. However, the NPV for each of these projects become positive after about 5–10 years because of increase of per capita income in the area as a result of urbanization as shown in Fig. 8. The impact of urbanizations should therefore be included in the CBA conducted in the peri-urban areas.

A Monte Carlos simulation of Eq. (19) has been conducted for estimating the risks associated with the NPV. Triangular distribution of risks for input variables like inflation and interest rates have been considered. The mean and standard deviation of the NPV for the three numbers booster pumping stations in the DRWW network are presented in Table 11. The distributions of the NPV for three typical cases have been presented in Fig. 9. The uncertainty associated with the NPV is investigated in the Tornado diagram in Fig. 10. The inflation and discount rate is found to be most significant. The power relationship of inflation (r) and discount (i) with NPV have made these more sensitive than other factors.

8 Conclusions

The development of the infrastructures like water supply is a challenge for mega cities in the developing world where urbanizations are taking place rapidly. Continuous expansion of the city area makes the project implementations more complex. A standard CBA analysis may reject a number of proposals in the peri-urban areas in a developing country mainly because of lesser WTP at the time of CBA. However, the urbanization of the area might change the financial characteristics and the resultant WTP may be sufficient to affirm a project otherwise unviable. The factors of such urbanizations are not included in a CBA methodology recommended in developed countries like US (Kopp et al., 1997) or EU countries (Baffoe-Bonnie et al., 2006), (Florio, 2006) probably because the developed world does not face such acute urbanization pressure in the peri-urban areas. However, the experience in Kolkata as presented in this article seems to show that these factors should be included in the peri-urban areas of the developing countries.

Acknowledgements. The authors thank Anupam Debsarkar, Nakibul Hossain Mondal, Nasim Mondal, Prasanta Mondal and Gitali Sarkar for extending their support and cooperation in carrying out the sample survey for the study.

References

Asian Development Bank (ADB): Handbook for the Economic Analysis of Water Supply Projects, http://www.adb.org/documents/handbooks/water_supply_projects/default.asp (last access: 30 March 2010), 1999.

Adeli, H.: Neural networks in civil engineering: 1989–2000, Comput.-Aided Civ. Inf., 16, 126–142, 2001.

Argent, R. and Houghton, B.: Land and water resources model integration: software engineering and beyond, Adv. Environ. Res., 5, 4, 351–359, 2001.

Antony, J., Kumar, V., Pathan, S., and Bhanderi, R. J.: Spatio-temporal analysis for monitoring urban growth – a case study of Indore city, J. Indian Soc. Remote Sens., 35, 1, 11–20, 2007.

American Public Health Association (APHA): Standard Methods for the Examination of Water and Wastewater, 19th Edn., USA, 1995.

Baffoe-Bonnie, B., Harle, T., Glennie, E., Dillon, G., and Sjovold, F.: Framework for operational cost benefit analysis in water supply, TECHNEAU, contract number 018320, 2006.

Banerjee, A. and Das, S.: Population environment and development: some issues in sustainability of Indian mega-cities of New Delhi and Kolkata (Calcutta), http://epc2006.princeton.edu/download.aspx?submissionId=60139 (last access: 30 July 2010), 2006.

Basu, S. R. and Main, H. A. C.: Calcutta's water supply: demand, governance and environmental change, Appl. Geogr., 21, 23–44, 2001.

Bhattacharya, P., Chatterjee, D., and Jacks, G.: Occurrence of arsenic-contaminated groundwater in alluvial aquifers from delta plains, eastern India: options for safe drinking water supply, J. Water Resour. Dev., 3, 79–82, 1997.

Bhatta, B., Saraswati, S., and Bandyopadhyay, D.: Quantifying the degree-of-freedom, degree-of-sprawl, and degree-of-goodness of urban growth from remote sensing data, Appl. Geogr., 30, 1, 96–111, 2010.

Bhatta, B.: Modelling of urban growth boundary using geoinformatics, Int. J. Digital Earth, 2, 4, 359–381, 2009.

Bhatta, B.: Analysis of urban growth pattern using remote sensing and GIS: a case study of Kolkata, India, Int. J. Remote Sens., 30, 8, 4733–4746, 2009.

Bureau of Indian Standard (BIS): BIS10500-1991, Indian Standard Drinking Water – Specification, 2003.

Boussabaine, H.: The use of artificial neural networks in construction, Constr. Manage. Econ., 14, 427–436, 1996.

Brouwer, R.: Do stated preference methods stand the test of time? A test of the stability of contingent values and models for health risks when facing an extreme event, Ecol. Econ., 60, 399–406, 2006.

Carlsen, W.: Watershed Dynamics. Cornell Scientific Enquiry Series, NSTA Press, USA, 2004.

President of the Treasury Board (Canada): Canadian cost-benefit analysis guide regulatory proposals, http://sciencepolicy. colorado.edu/students/envs_5120/CanadaCBA.pdf (last access: 21 September 2010), 2007.

Canadian Council of Ministers of the Environment (CCME): CCME water quality index 1.0 User's Manual, http://www.ccme. ca/ourwork/water.html?category_id=102 (last access: 30 June 2010), 2001.

Central Ground Water Board (CGWB): Ground Water Information Booklet, Kolkata Municipal Corporation, West Bengal, http://cgwb.gov.in/District_Profile/WestBangal/Kolkata% 20Municipal%20Corporation.pdf (last access: 30 July 2010), 2007.

Chatterjee, A., Das, D., Mandal, B. K., Roy Chowdhury, T., Samanta, G., and Chakraborti, D.: Arsenic in ground water in six districts of West Bengal, India: the biggest arsenic calamity in the world. Part I. Arsenic species in drinking water and urine of the affected people, Analyst, 120, 643–650, 1995.

Chung, G., Lansey, K., Blowers, P., Brooks, P., Ela, W., Stewart, S., and Wilson, P.: A general water supply planning model: evaluation of decentralized treatment, Environ. Model. Softw., 23, 893–905, 2008.

Cohen, B.: Urbanization in developing countries: current trends, future projections, and key challenges for sustainability, Technol. Soc., 28, 63–80, 2006.

Costanza, R. and Ruth, M.: Using dynamic modeling to scope environmental problems and build consensus, J. Environ. Manage., 22, 2, 183–195, 1998.

Central Public Health and Environmental Engineering Organization (CPHEEO): Manual on Water Supply and Treatment. Third Edition, Ministry of Urban Development, The Government of India, New Delhi, 1999.

Dietrich, B. and Henderson, J.: Urban Water Supply Conditions and Needs in Seventy-five Developing Countries, World Health Organization, Geneva, 1963.

Decision Tools Suite (DTS): Palisade Corporation, Ithaka, New York, USA, http://www.palisade.com/ (last access: 13 January 2010), 2009.

Dutta Roy, K. and Chakrabarty, S.: Assessment of production cost of municipal water: some case study, J. Inst. Public Health Eng., 07, 4, 16–19, 2006.

Dutta Roy, K., Thakur, B., Konar, T. S., and Chakrabarty, S. N.: Rapid evaluation of water supply project feasibility in Kolkata, India, Drink. Water Eng. Sci., 3, 29–42, doi:10.5194/dwes-3-29-2010, 2010.

Emblemsvåg, J.: Life-Cycle Costing Using Activity-Based Costing and Monte Carlo Methods to Manage Future Costs and Risks, John Wiley & Sons, Inc., 2003.

Fellin, W., Lessmann, H., Oberguggenberger, M., and Vieider, R.: Analyzing Uncertainty in Civil Engineering, Springer, Berlin, 2005.

Florio, M., Finzi, U., Genco, M., Levarlet, F., Maffii, S., Tracogna, A., and Vignetti, S.: Guide to Cost-Benefit Analysis of Investment Projects, Evaluation Unit, DG Regional Policy, European Commission, 2006.

Forrester, J.: System Dynamics and the Lessons of 35 Years, Sloan School of Management, Massachusetts Institute of Technology, 1991.

Forrester, J.: Urban Dynamics, Pegasus Communications, 1969.

Fuller, S. and Petersen, S.: Energy Price Indices and Discount Factors for Life-Cycle Cost Analysis – April 2002, US Govt., NST Handbook 135 http://fire.nist.gov/bfrlpubs/build02/PDF/ b02017.pdf (last access: 30 July 2010), 2002.

Gessler, M. and Brighu, U.: Monitoring Public Providers for the Poor, Case Study 1: Jaipur, India, Regulating Water and Sanitation for the Poor: Economic Regulation for Public and Private Partnerships, edited by: Franceys, R. and Gerlach, E., Earthscan, London, 2008.

Guha, S.: Valuation of clean water supply by willingness to pay method in a developing nation: a case study in calcutta, India, J. Young Invest., 17, 4, http://www.jyi.org/research/re.php?id= 1295 (last access: 25 June 2010), 2007.

Ho, C., Yang, C., Chang, L., and Yeh, M.: System Dynamics Modeling of the Conjunctive-Use of Surface and Subsurface Water, World Environmental and Water Resources Congress, 2007, Restoring Our Natural Habitat, ASCE, Tampa bay, Florida, USA, 2007.

IBRD: Appraisal of the Calcutta Urban Development Project, Washington DC, USA, 27 July, 1973.

isee system: iThink and Stella, systems thinking for education and research, http://www.iseesystems.com/softwares/Education/ StellaSoftware.aspx (last access: 30 July 2010), 2009.

Jaspers: Guidelines for Cost Benefit Analysis of Water and Wastewater Projects to be Supported By The Cohesion Fund And The European Regional Development Fund in 2007–2013, http://discutii.mfinante.ro/static/10/Mfp/evaluare/ Water_WasteWater_GuideFINAL.pdf (last access: 21 September 2010), 2008.

Kaiser, H.: The Varimax Criterion for Analytic Rotation in Factor Analysis, Psychometrika, 23, 187–200, 1958.

Kolkata Municipal Corporation for the Asian Development Bank (KMC): Environmental Assessment Report: Kolkata Environmental Improvement Project (Supplementary Financing), Project Number: 29466, August 2006.

Kolkata Municipal Corporation for the Asian Development Bank (KMC): Project Administration Memorandum for the Supplementary Loan: Kolkata Environmental Improvement Project,

June 2007.

Kopp, R. J., Krupnick, A. J., and Toman, M.: Cost-Benefit Analysis and Regulatory Reform: An Assessment of the Science and the Art, Discussion Paper 97-19, Resources for the Future, Washington, DC, USA, 1997.

Lata, K. M., Prasad, V. K., Badarinath, K. V. S., and Raghavaswamy, V.: Measuring urban sprawl: a case study of Hyderabad, http://www.gisdevelopment.net/application/urban/sprawl/urbans0004.htm, last access: 30 July 2010.

Llorente, M. and Zerah, M.: The urban water sector: formal versus informal suppliers in India, Urban India, Vol. XXIII, No. 1, National Institute Of Urban Affairs, January–June 2003.

Majumdar, C. and Gupta, G.: Willingness to pay and municipal water pricing in transition: a case study, J. Integr. Environ. Sci., 6, 4, 1–12, 2009.

NeuroDimensions Inc.: Neural network software, http://www.nd.com/ (last access: 31 July 2010), 2009.

Organization for Economic Co-operation and Development (OECD): Cost-Benefit Analysis and the Environment: Recent Developments, ISBN 92-64-01004-1, 2006.

Rajankar, P. N., Gulhane, S. R., Tambekar, D. H., Ramteke, D. S., and Wate, S. R.: Water quality assessment of groundwater resources in Nagpur region (India) based on WQI, E J. Chem., 6, 3, 905–908, 2009.

Ramakrshinaiah, C., Sadashivalah, C., and Ranganna, G.: Assessment of water quality index for the groundwater in Tumkur Taluk, Karnataka State, India, E J. Chem., 6, 2, 523–530, 2009.

Roy, J., Chattopadhyay, S., Mukherjee, S., Kanjilal, M., Samajpati, S., and Roy, S.: An economic analysis of demand for water quality: case of Kolkata, Econ. Polit. Weekly, January, 186–192, 2003.

Ruet, J.: Water supply and sanitation as "Urban Commons" in Indian metropolis: how redefining the State/Municipalities relationships should combine global and local de facto 'commoners', Centre de Sciences Humaines, New Delhi, India, 2002.

Society of Automotive Engineers (SAE): Reliability and Maintainability Guideline for Manufacturing Machinery and Equipment, M-110.2, Warrendale, PA, 1999.

Sahu, P. and Sikdar, P.: Assessment of aquifer vulnerability to groundwater pollution by multi-criteria analysis in and around East Calcutta Wetlands, West Bengal, India, Asian J. Water Environ. Pollut., 6, 2, 31–42, 2009.

Sargaonkar, A. and Deshpande, V.: Development of an overall index of pollution for surface water based on a general classification scheme in Indian context, Environ. Monit. Assess., 89, 43–67, 2003.

Sikdar, P., Sarkar, S., and Palchoudhury, S.: Geochemical evolution of groundwater in the quaternary aquifer of Calcutta and Howrah, India, J. Asian Earth Sci., 19, 5, 579–594, 2001.

Shekhar, S.: Urban sprawl assessment Entropy approach, Pune, http://www.archidev.org/article.php3?id_article=610, last access: 30 July 2010.

Statistical Package for the Social Sciences (SPSS): IBM 2009, Armonk, New York, USA, http://www.spss.com/, last access: 30 July 2010.

Sudhira, H., Ramachandra, T., and Jagadish, K.: Urban sprawl: metrics, dynamics and modelling using GIS, Int. J. Appl. Earth Obs. Geoinf., 5, 1, 29–39, 2004.

Turpie, J.: The existence value of biodiversity in South Africa: how interest, experience, knowledge, income and perceived level of threat influence local willingness to pay, Ecol. Econ., 46, 199–216, 2003.

United Nations Environment Programme (UNEP): Global Drinking Water Quality Index Development and Sensitivity Analysis Report, United Nations Environment Programme, Global Environment, Monitoring System/Water Programme, 2007.

UNESCO-IHE Institute for Water Education: http://www.unesco-ihe.org/courses/module/5696F5400DF3040BC12575F60029C10E (last access: 30 July 2010), 2010.

US EPA: http://www.epa.gov/waterscience/standards/about/rev.htm (last access: 30 July 2010), 2010.

World Bank: Final Report: An Assessment of Social Aspect and Willingness to Pay, Project Preparation Studies for Calcutta Water Supply, Sewerage and Drainage Projects, World Bank Project No. P.O. 50648, Kolkata, 2001.

Werick, W.: Managing Water for Drought: National Study of Water Management During Drought, US Army Corps of Engineers, 1994.

Whittington, D., Lauria, D. T., Prabhu, V., and Cook, J..: An economic reappraisal of the Melamchi water supply project – Kathmandu, Nepal, Portuguese Econ. J., 3, 157–178, 2004.

Wurbs, R.: Computer Models for Water-Resources Planning and Management: National Study of Water Management During Drought, US Army Corps of Engineers, 1994.

Zhang, S.: ArchitectingWater Supply System – A Perspective from Value of Flexibility, Second International Symposium on Engineering Systems, MIT, Cambridge, USA, 15–17 June 2009.

Zurada, M.: An Introduction to Artificial Neural Systems, PWS Publishing Company, Mumbai, India, 1992.

12

Removal and transformation of pharmaceuticals in wastewater treatment plants and constructed wetlands

E. Lee[1], S. Lee[2], J. Park[3], Y. Kim[4], and J. Cho[3]

[1]Department of Civil, Environmental and Architectural Engineering, University of Colorado Boulder, Boulder, CO 80309, USA

[2]Woongjin Chemical Co., Ltd, KANC 906-10, lui-dong, Yeongtong-gu, Suwon-si, Gyeonggi-do 443-270, Korea

[3]School of Civil and Environmental Engineering, Yonsei University, Yonsei-ro 50, Seodaemun-gu, Seoul 120-749, Korea

[4]Department of Environmental Science and Engineering, Gwangju Institute of Science and Technology (GIST), 1 Oryong-dong, Buk-gu, Gwangju, 500-712, Korea

Correspondence to: E. Lee (eunkyung.lee83@gmail.com)

Abstract. Since trace organic compounds such as pharmaceuticals in surface water have been a relevant threat to drinking water supplies, in this study removal of pharmaceuticals and transformation of pharmaceuticals into metabolites were investigated in the main source of micropollutants such as WWTPs and engineered constructed wetlands. Pharmaceuticals were effectively removed by different WWTP processes and wetlands. Pharmaceutical metabolites with relatively low log D value were resulted in the low removal efficiencies compared to parent compounds with relatively high log D value, indicating the stability of metabolites. And the constructed wetlands fed with wastewater effluent were encouraged to prevent direct release of micropollutants into surface waters. Among various pharmaceuticals, different transformation pattern of ibuprofen was observed with significant formation of 1-hydroxy-ibuprofen during biological treatment in WWTP, indicating preferential biotransformation of ibuprofen. Lastly, transformation of pharmaceuticals depending on their structural position was investigated in terms of electron density, and, the electron rich $C_1 = C_2$ bond of carbamazepine was revealed as an initial transformation position.

1 Introduction

For several decades, pharmaceuticals and personal care products (PPCPs) have been noticed as emerging problematic compounds (Ternes et al., 1998; Snyder et al., 2003). In order to reduce residual concentrations of PPCPs, various advance treatments have been studied by many research groups (Lee and Gunten, 2010; Rosal et al., 2010 and many others). However, high levels of PPCPs are still detected in wastewater effluent, surface waters, and drinking waters (Kim et al., 2007; Lee et al., 2012; Yoon et al., 2010; Benotti et al., 2009). Since those trace organic compounds have been detected even in treated drinking waters by Benotti et al. (2009), control of micropollutants has been important especially in the wastewater treatment plant, the main source of the micropollutants in the aquatic environments.

Meanwhile, constructed wetlands have been introduced as an alternative to wastewater treatment for micropollutants removal (Matamoros and Bayona, 2006). And few reports have been focused on the relationship between log D and removal of PPCPs in wetland systems. In previous study (Lee et al., 2011), removal efficiency in constructed wetlands has been investigated using corresponding octanol-water partitioning coefficient of pharmaceuticals.

And few studies have been reported with regarding to behaviors of pharmaceutical metabolites in various environments (Stumpf et al., 1998; Quintana et al., 2005). The potential ecotoxicological effect of pharmaceutical metabolites is still not known, thus, there is necessity to study further about the fate of pharmaceutical metabolites in various wastewater treatment processes.

In this study, removal and transformation of pharmaceuticals has been investigated with regarding to physicochemical and structural properties of pharmaceuticals and their metabolites in various environments such as WWTPs and constructed wetlands receiving wastewater effluent.

2 Materials and methods

2.1 Target compounds

Various micropollutant compounds, including 9 pharmaceuticals, 11 selected metabolites, and 1 personal care product, were selected in this study. Acetaminophen (ACT), atenolol (ATN), carbamazepine (CBZ), diclofenac (DCF), glimepiride (GMP), ibuprofen (IBU), naproxen (NPX), O-desmethyl-naproxen (O-desmethyl-NPX), sulfamethoxazole (SMX), and tri(2-chloroethyl) phosphate (TCEP) were obtained from Sigma-Aldrich (St. Louis, MO). Caffeine (CAF) was purchased via Fluka Chemie GmbH (Buchs, Switzerland). 1-hydroxy ibuprofen (IBU-1OH), 2-hydroxy ibuprofen (IBU-2OH), ibuprofen carboxylic acid (IBU-CA), iopromide (IOP), and N-acetyl-sulfamethoxazole (N-acetyl-SMX), paraxanthine, paraxanthine-1-methyl-d3, 4-hydroxy diclofenac were purchased via Tronto Research Inc (Tronto, Canada). Authentic carbamazepine metabolites 10, 11-dihydro-10, 11-epoxycarbamazepine (CBZ-EP), 2-hydroxycarbamazepine (CBZ-2OH), 3-hydroxycarbamazepine (CBZ-3OH), and 10, 11-dihydro-10-hydroxycarbamazepine (CBZ-10OH) were provided from Norvatis Pharma AG (Basel, Switzerland). Carbamazepine-d_{10} and $^{13}C_1$-naproxen-d_3 were purchased from Cambridge Isotope Laboratories (Andover, MA, US). Atenolol-d_7 and diclofenac-d_4 were obtained from C/D/N Isotopes (Pointe-Claire, Canada). Ibuprofen-d_3, N^4-acetylsulfamethoxazole-d_4, and sulfamethoxazole-d_4 were obtained from Tronto Research Inc (Tronto, Canada). HPLC grade methanol was obtained from J.T. Baker (Philipsburg, NJ, US). Methyl *tert*-butyl ether (MTBE) and formic acid were obtained from Sigma-Aldrich (St. Louis, MO, US). Information of target compounds and their metabolites is listed in Table 1. The log K_{ow}, log D_{ow}, and pK_a for investigated compounds was calculated based on the molecular structures using ChemAxon Marvin Calculator Plugin. The log D_{ow} is a pH dependent log K_{ow} value and in this study log D at pH 7 was used.

2.2 Sample collection

In June 2010, water samples were collected from each process (influent and effluents of the various unit operations) in three different wastewater treatment plant. One

is Gwangju primary municipal wastewater treatment plant, which is operated with the nitrogen and phosphorus removal treatment system ($600\,000\,m^3\,day^{-1}$). Second sampling sites was Gwangju secondary municipal wastewater treatment plant, which had two different treatment trains (Conventional Activated Sludge and Modified Ludzack-Ettinger (MLE) process), having total capacity up to $120\,000\,m^3\,day^{-1}$. In those two different Gwangju municipal WWTPs, sodium hypochlorite was added as a disinfectant and final effluent samples were collected after disinfection process. Additionally, water samples from both Damyang wastewater treatment plant and Damyang constructed wetlands were collected and studied. There was no disinfection system in Damyang WWTP and final effluent was collected after secondary treatment. Damyang constructed wetlands connected to Damyang wastewater treatment plant is free surface flow constructed wetlands, which have two different ponds, containing *Aporus* ponds followed by *Typha* ponds. The hydraulic retention time of wetlands is approximately 6 h and flow rate is $1800\,m^3\,day^{-1}$. Every sample was spiked with a biocide sodium azide and ascorbic acids to quench any residual oxidant in the field.

2.3 SPE and LC-MS/MS analysis

After filtration using glass fiber membrane filter, all analytes were extracted by using AutoTrace automated solid phase extraction (SPE) system (Caliper Corporation, Hopkington, MA), as depicted by Vanderford and Snyder (2006). Briefly, the 6 mL, 500 mg hydrophilic-lipophilic balance (HLB) glass cartridges (Waters Corporation, Milford, MA) were preconditioned in the following order: 5 mL of MTBE, 5 mL of methanol, and 5 mL of deionized water. 500 mL of samples, spiked with the addition of standards for internal calibration, were loaded onto the cartridges at 15 mL min^{-1} in duplicate, after which the cartridges were rinsed with 5 mL of deionized pure water, and then dried with a steam of air for 50 min. The cartridges were eluted with 5 mL of methanol, followed by 5 mL of 1/9 (v/v) methanol/MTBE. The eluted solution was concentrated in a water bath at 40 °C with a gentle stream of air to a final volume of 500 μL, which was a concentration factor of 1000. The levels of pharmaceuticals and their metabolites were then measured using a Water 2695 Separations Module (Waters, Milford, MA) coupled with a Micromass Quattro Micro triple quadrupole tandem mass spectrometer (Micromass, Manchester, UK) in electrospray ionization mode (ESI). A 20 μL sample loop and 150 × 2.1 mm SunFire C18 column with a particle size of 3.5 μm (Waters, Milford, MA) was employed for analyte separation. A binary gradient, consisting of 0.1 % formic acid (eluent A) and 100 % acetonitrile (eluent B), was used at a flow rate of 0.2 mL min^{-1}. Selected PPCPs and their metabolites was analyzed using two different gradients. The gradient used for the PPCPs and most of the metabolites was: gradient with 15 % of B was held for 4 min, increased linearly to 80 % for

Table 1. Tested parent compounds and metabolites.

Analytes	Uses	Structure	pK_a^a	$\log K_{ow}^a$	$\log D^a$ at pH7
Acetaminophen	Analgesic		9.46	0.91	0.91
Glimepiride	Anticholesterol		4.32	3.12	2.18
tri(2-chloroethyl) phosphate (TCEP)	Flame retardant		N.E	2.11	2.11
Carbamazepine (CBZ)	Anticonvulsant		N.E	2.77	2.77
CBZ-EP	Carbamazepine metabolite		N.E.	1.97	1.97
CBZ-2OH	Carbamazepine metabolite		9.3	2.66	2.66
CBZ-3OH	Carbamazepine metabolite		9.46	2.66	2.66
CBZ-10OH	Carbamazepine metabolite		14.1	1.73	1.73
Sulfamethoxazole	Antibiotic		6.16	0.79	0.14
N[4]-acetylsulfamethoxazole	Sulfamethoxazole metabolite		5.88	0.86	0.1
Diclofenac	Analgesic		4.00	4.26	0.96

Table 1. Continued.

4-hydroxyl-diclofenac	Diclofenac metabolite		3.76, 8.61	3.96	0.89
Naproxen	Analgesic		4.19	2.99	0.25
O-desmethyl-naproxen	Naproxen metabolite		4.34, 9.78	3.9	0.23
Ibuprofen	Analgesic		4.85	3.84	1.71
1-hydroxyl-ibuprofen	Ibuprofen metabolite		4.90	2.85	0.77
2-hydroxyl-ibuprofen	Ibuprofen metabolite		4.63	2.08	0.03
Ibuprofen carboxylic acid	Ibuprofen metabolite		3.97, 4.77	2.78	-2.36
Caffeine	Stimulant		N.E	-0.55	-0.55
Paraxanthine	Caffeine metabolite		10.76	0.24	0.24

[a] log P, log D, and pK_a value were calculated from the Software Calculator Plugins.
[b] N.E: nonexistent at pH range 1–14.

6 min, held for 3 min with 80 % of B and then linearly increased to 100 % for 7 min. The gradient used for the carbamazepine metabolites was: gradient with 10 % of B increased linearly to 40 % for 15 min, increased linearly to 90 % for 10 min, and held until 30 min (Kang et al., 2008). A 5 min equilibration step with gradient of 10 % B was used at the beginning of each run. Detail LC-MS/MS analysis condition and analytical parameters of target compounds are shown in Table 2.

2.4 Molecular orbital calculations

Molecular orbital were calculated single determinant (Hartree-Fock) for optimization bearing the minimum energy obtained at the AM1 level. All semi-empirical calculations to obtain the point charge and electron density for pharmaceuticals and metabolites were performed in MO-G with a SCI-GRESS package version 7.7 (Fujitsu Co. Ltd.) (Watanabe et al., 2003).

Table 2. Analytical parameters of selected compounds (MDL: method detection limit; RL: reporting limit).

Compound	Retention time (min)	Cone voltage (V)	Collision energy (eV)	Parent ion (m/z)	Daughter ion (m/z)	MDL (ng L^{-1})	RL (ng L^{-1})
ESI negative							
Diclofenac	15.53	15	10	294	249	1.3	3.9
Diclofenac-d$_4$	11.23	15	12	298	254		
4-OH-DCF	13.70	20	12	310	266	5.1	15.3
Ibuprofen	15.70	15	8	205	161	1.2	3.6
Ibuprofen-d$_3$	11.63	15	8	208	164		
IBU-1OH	12.87	15	7	221	177	5.8	17.5
IBU-2OH	12.23	17	7	221	177	0.9	2.7
IBU-CA	12.27	12	5	235	191	2.3	6.8
Naproxen	9.84	10	8	229	185	3.3	10.0
Naproxen-d$_3$	9.83	10	6	233	189		
O-desmethyl-NPX	8.10	15	12	215	171	1.2	3.5
N-acetyl-SMZ	7.71	30	13	294	198	0.9	2.7
TCEP	9.12	30	16	285	161	7.9	23.8
ESI positive							
Acetaminophen	3.98	28	17	152	110	0.5	1.5
Acetaminophen-d$_4$	3.98	28	17	156	114		
Glimepiride	11.34	28	13	491	352	1.5	4.6
Sulfamethoxazole	7.83	30	18	254	156	0.6	1.7
Sulfamethoxazole-d$_4$	7.80	25	15	258	160		
Caffeine	6.11	35	20	195	138	1.5	4.6
Paraxanthine	2.72	30	22	181	124	9.3	27.9
Paraxanthine-d$_3$	2.72	30	22	184	124		
Carbamazepine	19.31	35	18	237	194	0.7	2.1
Carbamazepine-d$_{10}$	8.74	35	18	247	204		
CBZ-EP	15.64	28	24	253	180	1.2	3.6
CBZ-2OH	14.34	35	20	253	210	13.0	3.9
CBZ-3OH	15.78	35	20	253	210	1.5	4.5
CBZ-10OH	13.63	30	20	255	194	0.2	0.5

Table 3. Concentrations of selected PPCPs in WWTPs and wetlands.

PPCPs	Gwangju Primary WWTP		Gwangju Secondary WWTP		Damyang WWTP and Constructed wetlands				
	WWTP influent (ng L^{-1})	WWTP effluent (ng L^{-1})	WWTP influent (ng L^{-1})	WWTP effluent (ng L^{-1})	WWTP influent (ng L^{-1})	WWTP effluent (ng L^{-1})	Acorus wetland (ng L^{-1})	Typha wetland (ng L^{-1})	Wetland effluent (ng L^{-1})
Caffeine	72 471.2	< 4.6	45 457.0	1857.8	36 880.6	< 4.6	< 4.6	47.0	< 4.6
Carbamazepine	2085.4	108.3	1668.8	362.8	844.5	417.2	387.1	161.4	268.3
Sulfamethoxazole	6048.6	88.8	8092.9	70.7	409.7	27.4	17.9	7.4	17.4
Acetaminophen	162 159.9	1705.8	227 705.4	180.5	194 586.8	189.7	354.8	549.5	349.6
Ibuprofen	2626.0	< 3.6	4576.5	166.6	1645.3	17.9	43.2	52.9	47.2
Naproxen	1140.5	< 10.0	1778.5	10.7	203.5	< 10.0	11.4	< 10.0	< 10.0
Diclofenac	2673.5	111.7	2702.8	< 3.8	150.5	19.8	20.15	< 3.8	8.9
Glimepiride	18 895.3	197.4	40 338.6	86.6	36.8	12.3	11.3	< 4.6	14.4
TCEP	6011.8	320.3	4774.7	< 23.8	2608	340.6	277.85	186.35	268.4

Table 4. Concentrations of parent pharmaceuticals and their metabolites in WWTPs and wetlands.

Parent compounds and their metabolites	Gwangju Primary WWTP		Gwangju Secondary WWTP		Damyang WWTP and Constructed wetlands				
	WWTP influent (ng L^{-1})	WWTP effluent (ng L^{-1})	WWTP influent (ng L^{-1})	WWTP effluent (ng L^{-1})	WWTP influent (ng L^{-1})	WWTP effluent (ng L^{-1})	Acorus wetland (ng L^{-1})	Typha wetland (ng L^{-1})	Wetland effluent (ng L^{-1})
Caffeine	72 471.2	<4.6	45 457.0	<4.6	36 880.6	<4.6	<4.6	47.0	<4.6
Paraxanthine	8215.3	33.9	10 442.1	<27.9	3222.3	<27.9	<27.9	<27.9	<27.9
Sulfamethoxazole	6048.6	88.8	8092.9	166.6	409.7	27.4	17.9	7.4	17.4
N-acetyl-SMZ	5224.8	64.5	6224.2	72.4	152.6	4.9	<2.7	<2.7	5.3
Naproxen	1140.5	<10.0	1778.5	10.7	203.5	<10.0	11.4	<10.0	<10.0
O-desmethyl-NPX	191.5	31.7	125.0	31.1	245.0	9.3	<3.5	<3.5	7.75
Diclofenac	2673.5	111.7	2702.8	86.6	150.5	19.8	20.15	<3.8	8.9
4-OH-DCF	530.6	582.1	396.0	504.6	212.0	40.2	21.15	<15.3	<15.3

3 Results and discussion

3.1 Removal of pharmaceuticals and personal care products in WWTPs and wetlands

Removal of PPCPs in municipal wastewater treatment plants and constructed wetlands were investigated and summarized by comparing the concentrations of PPCPs in the influent and final effluent of each WWTP in Table 3. Unfortunately, the contribution of different transport mechanisms such as biodegradation, adsorption to sludge and sediments, and oxidation during disinfection was not considered here. Gwangju WWTP was found as a major source, releasing tons of micropollutants to the Yeongsan River water, and, the concentrations of PPCPs in Gwangju WWTPs ($\sim \mu g L^{-1}$) were generally much higher than in Damyang WWTP ($\sim ng L^{-1}$), presumably resulting from the dense populations in Gwangju area.

Particularly, high concentrations of caffeine and acetaminophen in WWTP influents reflect frequent use and ingestion of those compounds in this urban area. Most of PPCPs exhibited high removal efficiency (>90 %) during the WWTPs processes, except for carbamazepine removal in Gwangju secondary WWTP (removal efficiency at 74 %). After discharge of WWTP effluent into connected constructed wetlands, levels of some PPCPs (carbamazepine, sulfamethoxaolze, diclofenac, and TCEP) slightly decreased. This result indicates that operation of wastewater stabilization ponds or wetlands would be helpful to prevent release of micropollutants into surface water. Our research group has previously been studied the Damyang constructed wetlands with respect to the control of organic micropollutants, and, reported the efficiency of wetland treatments depending on the wetland characteristics and properties of micropollutants (Park et al., 2009). Many other studies have also been supported the necessity of additional wetland treatments for mi-

cropollutants control (Matamoros and Bayona, 2006; Conkle et al., 2008).

3.2 Transformation of pharmaceuticals and formation of metabolites in WWTPs

Table 4 shows occurrence of parent pharmaceuticals and their metabolites in WWTPs and wetlands system. In the WWTPs influent, most metabolites were detected at high level of concentrations with range of $100 \sim 10\,000\,ng\,L^{-1}$ compared to WWTP effluents, indicating dominant transformation pathway of metabolite resulting from human body rather than microbial transformation. However, after passing the WWTPs, concentrations of most metabolites were observed higher than that of parent compounds, indicating the structural stability of the metabolites during the WWTP process relative to their parent compounds. The stability of metabolites may be related to the difference of the log D value between parent pharmaceutical compounds and pharmaceutical metabolites. As pharmaceuticals transformed into their metabolites, their log D value at pH 7 slightly decreased (Table 1). Considering the dominant removal mechanism of micropollutants in WWTP process (i.e., sorption and biodegradation), decrease of log D may result in the reduction of sorption of pharmaceutical metabolites on the sludge surface. Consequently, pharmaceutical metabolites would be detected at higher levels than parent compounds in WWTP effluents.

In case of sulfamethoxazole metabolite, even though the concentrations of N-acetyl-sulfamethoxazole were low enough in wastewater effluents, by considering retransformation of acetylated metabolite back into its parent compound, N-acetyl-sulfamethoxazole should be assumed as a pharmaceutically active parent compound (Göbel et al., 2005). And based on this reason, monitoring of N-acetyl-sulfamethoxazole is important and necessary

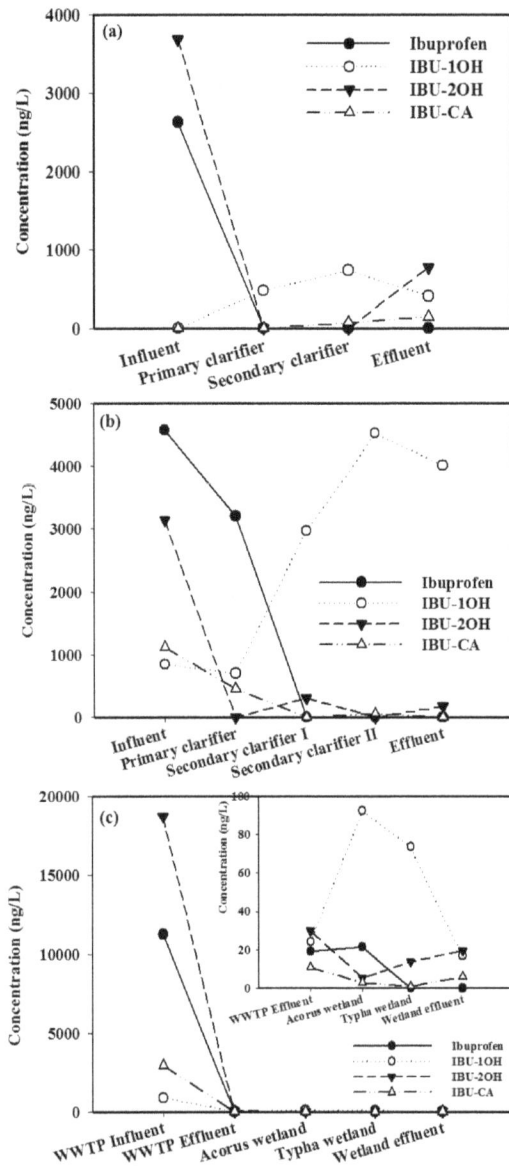

Figure 1. Transformation of ibuprofen in different WWTPs system: (**a**) Gwangju primary WWTP, (**b**) Gwangju secondary WWTP, (**c**) Damyang WWTP and constructed wetlands.

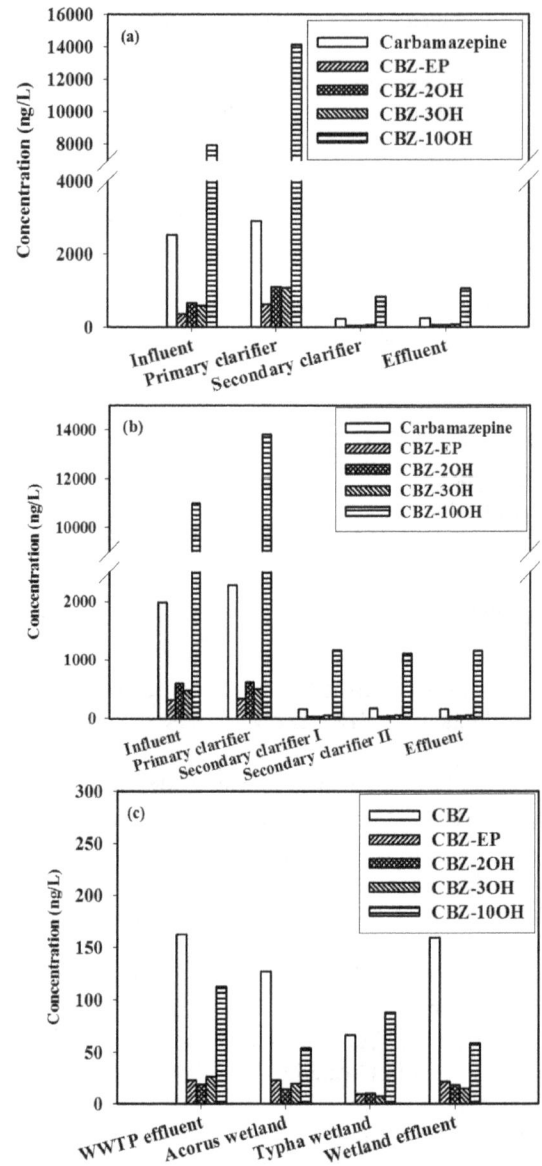

Figure 2. Distribution of carbamazepine metabolites in different WWTPs system: (**a**) Gwangju primary WWTP, (**b**) Gwangju secondary WWTP, (**c**) Damyang WWTP and constructed wetlands.

for the effective control of micropollutants in aquatic environment.

During the engineered constructed wetland treatments, there was no specific change of pharmaceutical metabolites in exception with ibuprofen metabolites. As can be seen in Fig. 1, ibuprofen was differently transformed depending on the treatment process. In WWTPs influent, 2-hydroxyibuprofen, the well-known human metabolite of ibuprofen was dominant, whereas, after the activated sludge treatment, concentrations of 1-hydroxyibuprofen indicate the highest level in all WWTPs and wetlands. Significant formation of 1-hydroxyibuprofen in WWTPs might be possibly explained by preferential microbial metabolism in ac-

tivated sludge treatment process. In the additional experimental results conducted in river waters, completely different composition of ibuprofen metabolites was observed. 2-hydroxy-ibuprofen was found to be dominant in both river waters, Yeongsan River and Seomjin River (data are not shown). This might infer that 2-hydroxyibuprofen is much more persistent and stable than 1-hydroxyibuprofen and even than parent compound, ibuprofen, as previously mentioned by Weigel et al. (2004).

In contrast, carbamazepine did not exhibit any change in transformation behavior during WWTPs and wetland treatments (Fig. 2). All the selected carbamazepine metabolites, 10,11-epoxy-carbamazepine,   2-hydroxycarbamazepine,

Table 5. Calculations of frontier electron density for carbamazepine.

Atom List	HOMO Density	Nucleophilic Frontier Density	Electrophilic Frontier Density	Radical Frontier Density	LUMO Density
C_1	0.148	0.202	0.215	0.208	0.131
C_2	0.119	0.252	0.175	0.214	0.169
C_3	0.064	0.222	0.108	0.165	0.136
C_4	0.118	0.113	0.184	0.148	0.057
C_5	0.046	0.201	0.088	0.145	0.090
C_6	0.067	0.113	0.136	0.124	0.049
N_7	0.074	0.003	0.211	0.107	0.000
C_8	0.039	0.141	0.080	0.111	0.054
C_9	0.010	0.117	0.036	0.076	0.030
C_{10}	0.068	0.214	0.111	0.163	0.133
C_{11}	0.012	0.088	0.044	0.066	0.013
C_{12}	0.046	0.094	0.117	0.105	0.047
C_{13}	0.024	0.048	0.076	0.062	0.007
C_{14}	0.111	0.129	0.173	0.151	0.072
C_{15}	0.016	0.043	0.078	0.060	0.010
C_{16}	0.012	0.010	0.021	0.016	0.000
N_{17}	0.007	0.002	0.072	0.037	0.000
O_{18}	0.018	0.002	0.064	0.033	0.000
H_{19}	0.000	0.001	0.001	0.001	0.000
H_{20}	0.000	0.000	0.001	0.001	0.000
H_{21}	0.001	0.002	0.002	0.002	0.001
H_{22}	0.002	0.001	0.003	0.002	0.000

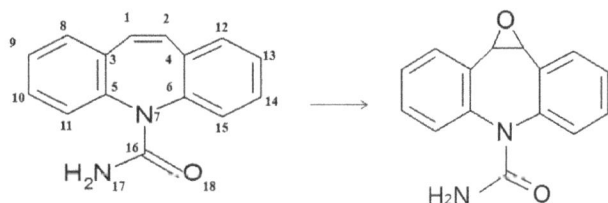

Figure 3. Structure of carbamazepine.

3-hydroxycarbamazepine and 10-hydroxycarbamazepine was detected in influents, effluents of WWTPs and wetlands. Even though the levels of 10,11-epoxy-carbamazepine was low, the eco-toxicological effect of CBZ-EP on the environment is worthy of further examination due to its pharmaceutically active property like its parent compound, carbamazepine. The most dominant carbamazepine metabolite was 10-hydroxy-carbamazepine. The predominant formation of CBZ-10OH might be due to the different distribution of electron density of atoms in carbamazepine.

3.3 Effect of electron density on transformation of carbamazepine into metabolites

In previous studies, electron density distribution of chemicals has been used to find the initial positions of OH radical attack in oxidation (Watanabe et al., 2003; Jung et al., 2010; Heimstad et al., 2009). A higher electron density indicates more

electrons in the bonds, resulting in electrophilic reaction (Horikoshi et al., 2004; Kaneco et al., 2006). In this study, electron density of pharmaceutical was examined to find the initial transformation position in various transformation mechanisms including biological process and photochemical oxidations. In previous study, Park et al. (2009) suggested that hydrolysis reaction at the amide and urea functional groups may lead to the biological transformation of carbamazepine. Frontier electron density of carbamazepine calculated by MOPAC, Scigress software is shown in Table 5. Radical frontier density, the averaged value of nucleophilic frontier density and electrophilic frontier density, are used in this study. Based on the electron density of carbamazepine, the carbon bonds between C_1 and C_2 showed the largest frontier electron density (0.208 and 0.214, respectively). This electron rich carbon bond of the olefin structure may provide the initial position of oxidation. Prevalent occurrence of specific carbamazepine metabolites such as CBZ-EP, CBZ-10OH, and CBZ-DiOH in environment are presumably related with this oxidation pathway. However, other metabolites were difficult to find a relationship between transformation and electron density. The reason for this low relationship is thought that there might be some other different preferred transformation pathways depending on the intrinsic property of chemicals rather than electron density derived transformation. However, electron density of chemicals is still believed as a key parameter to elucidate unknown pathway in various processes, especially in oxidation process.

4 Conclusions

Removal and transformation of pharmaceuticals in WWTPs and constructed wetlands were extensively investigated in this study. Pharmaceuticals were effectively removed by different WWTP processes and wetlands. From this study, the additional operation of wastewater management wetlands was encouraged to prevent direct discharge of micropollutants into surface waters. And additionally, pharmaceutical metabolites were found to be more stable than the parent compounds during WWTP processes due to the lower log D value of metabolites than parent compounds. Different transformation pattern of pharmaceuticals was also observed, especially in transformation of ibuprofen. 1-hydroxyibuprofen was dominantly formed during biological treatment in WWTP, indicating preferential biotransformation of ibuprofen. At last, electron density of carbamazepine was examined to elucidate the transformation pathway. The electron rich $C_1 = C_2$ bond in olefin structure of carbamazepine was revealed as an initial transformation position.

Acknowledgements. This research was supported by the National Research Foundation of Korea (NRF) grant funded by the Korea government (MEST) (No. 2012047029).

References

Benotti, M. J., Trenholm, R. A., Vanderford, B. J., Holady, J. C., Stanford, B. D., and Snyder, S. A.: Pharmaceuticals and endocrine disrupting compounds in U.S. drinking water, Environ. Sci. Technol., 43, 597–603, 2009.

Conkle, J. L., White, J. R., and Metcalf, C. D.: Reduction of pharmaceutically active compounds by a lagoon wetland wastewater treatment system in Southeast Louisiana, Chemosphere, 73, 1741–1748, 2008.

Göbel, A., Thomsen, A., McArdell, C. A., Joss, A., and Giger, W.: Occurrence and sorption behavior of sulfonamides, macrolides, and trimethoprim in activated sludge treatment, Environ. Sci. Technol., 39, 3981–3989, 2005.

Heimstad, E. S., Bastos, P. M., Eriksson, J., Bergman, A., and Harju, M.: Quantitative structure – Photodegradation relationships of polybrominated diphenyl ethers, phenoxyphenols and selected organochlorines, Chemosphere, 77, 914–921, 2009.

Horikoshi, S., Tokunaga, A., Hidaka, H., and Serpone, N.: Environmental remediation by an integrated microwave/UV illumination method: VII. Thermal/non-thermal effects in the microwave-assisted photocatalyzed mineralization of bisphenol-A, J. Photoch. Photobio. A, 162, 33–40, 2004.

Jung, Y. J., Oh, B. S., Kim, K. S., Koga, M., Shinohara, R., and Kang, J. W.: The degradation of diethyl phthalate (DEP) during ozonation: oxidation by-products study, J. Water Health., 8, 290–298, 2010.

Kaneco, S., Katsumata, H., Suzuki, T., and Ohta, K.: Titanium dioxide mediated photocatalytic degradation of dibutyl phthalate in aqueous solution – kinetics, mineralization and reaction mechanism, Chem. Eng. J., 125, 59–66, 2006.

Kang, S.-I., Kang, S.-Y., and Hur, H.-G.: Identification of fungal metabolites of anticonvulsant drug carbamazepine, Appl. Microbiol. Biotechnol., 79, 663–669, 2008.

Kim, S. D., Cho, J., Kim, I. S., Vanderford, B. J., and Snyder, S. A.: Occurrence and removal of pharmaceuticals and endocrine disruptors in South Korean surface, drinking, and waste waters, Water Res., 41, 1013–1021, 2007.

Lee, E., Lee, S., Kim, Y., Huh, Y.-J., Kim, K.-S., Lim, B.-J., and Cho, J.: Wastewater Treatment Plant: Anthropogenic Micropollutant Indicators for Sustainable River Management, in: Encyclopedia of Sustainability Science and Technology, edited by: Meyers, R. A., Springer, 17, 11911–11932, 2012.

Lee, S., Kang, S. I., Lim, J. L., Huh, Y. J., and Cho, J.: Evaluating controllability of pharmaceuticals and metabolites, in biologically-engineered processes, using corresponding octanol-water partitioning coefficient with consideration of ionizable functional groups, Ecol. Eng., 37, 1595–1600, 2011.

Lee, Y. and von Gunten, U.: Oxidative transformation of micropollutants during municipal wastewater treatment: Comparison of kinetic aspects of selective (chlorine, chlorine dioxide, ferrate VI, and ozone) and non-selective oxidants (hydroxyl radical), Water Res., 44, 555–566, 2010.

Matamoros, V. and Bayona, J. M.: elimination of pharmaceuticals and personal care products in subsurface flow constructed wetlands, Environ. Sci. Technol., 40, 5811–5816, 2006.

Park, N., Vanderford, B. J., Snyder, S. A., Sarp, S., Kim, S. D., and Cho, J.: Effective controls of micropollutants included in wastewater effluent using constructed wetlands under anoxic condition, Ecol. Eng., 35, 418–423, 2009.

Quintana, J. B., Weiss, S., and Reemtsma, T.: Pathways and metabolites of microbial degradation of selected acidic pharmaceutical and their occurrence in municipal wastewater treated by a membrane bioreactor, Water Res., 39, 2654–2664, 2005.

Rosal, R., Rodríguez, A., Perdigón-Melón, J. A., Petre, A., García-Calvo, E., Gómez, M. J., Agüera, A., and Fernández-Alba, A. R.: Occurrence of emerging pollutants in urban wastewater and their removal through biological treatment followed by ozonation, Water Res., 44, 578–588, 2010.

Snyder, S. A., Westerhoff, P., Yoon, Y., and Sedlak, D. L.: pharmaceuticals, personal care products, and endocrine disruptors in water: implications for the water industry, Environ. Eng. Sci., 20, 449–469, 2003.

Stumpf, M., Ternes, T. A., Haberer, K., and Baumann, W.: Isolation of ibuprofen-metabolites and their importance as pollutants of the aquatic environment, Vom Wasser, 91, 291–303, 1998.

Ternes, T. A.: Occurrence of drugs in German sewage treatment plants and rivers, Water. Res., 32, 3245–3260, 1998.

Vanderford, B. J. and Snyder, S. A.: Analysis of pharmaceuticals in water by isotope dilution liquid chromatography/tandem mass spectrometry, Environ. Sci. Technol., 40, 7312–7320, 2006.

Watanabe, N., Horikoshi, S., Kawabe, H., Sugie, Y., Zhao, J., and Hidaka, H.: Photodegradation mechanism for bisphenol A at the TiO_2/H_2O interfaces, Chemosphere, 52, 851–859, 2003.

Weigel, S., Berger, U., Jensen, E., Kallenborn, R., Thoresen, H., and Hühnerfuss, H.: Determination of selected pharmaceuticals and caffeine in sewage and seawater from Tromsø/Norway with emphasis on ibuprofen and its metabolites, Chemosphere, 56, 583–592, 2004.

Yoon, Y., Ryu, J., Oh, J., Choi, B. G., and Snyder, S. A.: Occurrence of endocrine disrupting compounds, pharmaceuticals, and personal care products in the Han River (Seoul, South Korea), Sci. Total Environ., 408, 636–643, 2010.

13

Effect of biostimulation on biodegradation of dissolved organic carbon in biological filters

K. Tihomirova, A. Briedis, J. Rubulis, and T. Juhna

Department of Water Engineering and Technology, Riga Technical University, Azenes street 16/20-263,
LV-1048, Riga, Latvia

Correspondence to: K. Tihomirova (kristina.tihomirova@rtu.lv)

Abstract. The addition of labile organic carbon (LOC) to enhance the biodegradation rate of dissolved organic carbon (DOC) in biological columns was studied. Acetate standard solution (NaAc) and Luria Bertrani (LB) medium were used as LOC as biostimulants in glass column system used for measurements of biodegradable dissolved organic carbon (BDOC). The addition of LOC related with the increase of total DOC in sample. The concentration of BDOC increased up to 7 and 5 times and was utilized after 24 min. contact time. The biodegradation rate constant was increased at least 26 times during adaptation-biostimulation period. There was a strong positive correlation between the biodegradation rate constant and the concentration of BDOC. Biostimulation period ranged from 24 to 53 h for NaAc biostimulant and from 20 to 168 h for LB. The study has shown that LOC could be used as stimulator to enhance the biodegradation rate of DOC during biofiltration.

1 Introduction

During ozonation dissolved organic carbon (DOC) is transformed to biodegradable organic carbon (BDOC), which then is metabolized by bacteria in the biofilter (Fahmi et al., 2003; Hammes and Vital, 2008; Volk et al., 1993). The amount of BDOC fraction depends on the type of natural organic matter (NOM) present in the water (Huck, 1990; Kaplan et al., 1994). The NOM compounds have different biodegradation kinetics: (i) fast biodegradable, (ii) more resistant to biodegradation or slow biodegradable and (iii) not biodegradable (Carlson and Amy, 2001; Klevens et al., 1996; Yavich et al., 2004). About 30 % of DOC is usually removed after biofiltration (Volk et al., 2002) which are usually designed for empty bed contact time (EBCT) of less than 30 min. In Boreal regions, where the surface waters contain high concentrations of NOM or organic matter substances having a low biodegradation rate and biofilters are operated at low temperatures, NOM removal in biofilter is not effective, and can reach only 15–19 % measured as DOC, or 75 % measured as neutral fraction or fast biodegradable part of NOM (Tihomirova, 2011). The slow biodegradable part of BDOC which is not removed in the biofilter will enter the distribution network and will be used as a substrate for bacteria (Eikebrokk et al., 2007; Tihomirova, 2011). To increase the biodegradation rate for removal of recalcitrant organic substances the addition of labile organic carbon (LOC) is a widely used practice for the remediation of contaminated soils, sediments and sewage (Brandt et al., 2003; Shimp and Pfaender, 1985; Spain et al., 1980; Wiggins and Alexander, 1988), however applicability of this approach for the removal of compounds resistant to biodegradation from drinking water during biofiltration has not yet been studied.

The aim of this paper was to evaluate the effect of addition of LOC in water to enhance the biodegradation rate of DOC during biofiltration.

Biostimulation effect was determined using bioreactor set-up used to measure BDOC (Eikebrokk et al., 2007) and developed within the EU project TECHNEAU, which yields information not only on final concentrations of BDOC but also on the degradation kinetics. The study was carried out in a laboratory scale using treated humic rich water after biofilters.

2 Materials and methods

2.1 Glassware

All glassware used in these experiments were cleaned thoroughly with a 10 % solution of potassium dichromate in concentrated sulphuric acid and rinsed with ultra pure water (Elga PureLab Ultra, Veolia Water Ltd., UK), dried and covered with aluminium septum heated for 6 h at +250 °C in order to avoid organic carbon release (Van der Kooij et al., 1982).

2.2 Reagents

Acetate stock solution (NaAc) with concentration γ (DOC) = 1 g l^{-1}, was made in a 1000 ml volumetric flask where 5.6648 g of sodium acetate trihidrate (CH$_3$COONa · 3H$_2$O, Ultra, $\geq$ 99.5 %, Fluka, Germany) was dissolved and made up to volume with water. The solution is stable at 4 °C for about 6 months.

Luria Bertani (LB) medium stock solution with concentration γ (DOC) = 7.5 g l^{-1}, was made in a 1000 ml volumetric flask where LB medium contains 10 g Peptone, 5 g Yeast extract, 10 g NaCl in 1000 ml sterile ultra pure water. Both mentioned solutions were used as LOC for the biostimulation experiments.

To avoid inorganic nutrient limitation all samples (100 ml) were supplemented with 100 µl solution of inorganic nutrients. The solution was prepared by dissolving 4.55 g (NH$_4$)$_2$SO$_4$, 0.2 g KH$_2$PO$_4$, 0.1 g MgSO$_4$· 7H$_2$O, 0.1 g CaCl$_2$· 2H$_2$O and 0.2 g NaCl in sterile ultra pure water (1000 ml) (Miettinen et al., 1999).

2.3 DOC determination

The concentrations of DOC were measured with a TOC-5000A Analyzer (Shimadzu Corporation, Kyoto, Japan) according to European Standard EN 1484:1997. The 0.45 µm pore size membrane filters (Millipore Corporation, USA or Sartorius AG, Germany) used for DOC measurements were carefully rinsed, first with ultra pure water and then with the water sample. The blank and control solution were analyzed with each series of DOC sample in order to verify the accuracy of the results obtained by the method. Every DOC sample was tested in duplicate and the mean values were calculated (instrumental relative standard deviation (CV) $\leq$ 2 %).

2.4 Study site

The experiments were done using water samples taken after passing through biologically activated carbon (BAC) filters from the surface water treatment plant (WTP) in Riga, Latvia. The raw water in River Daugava contains 15.34 ± 3.84 mg l^{-1} of DOC and is treated conventionally by coagulation-sedimentation and filtration in rapid sand filters.

Then the water is ozonated and filtrated through the BAC filters. Water after biofilter contains 5.33 ± 1.45 mg l^{-1} of DOC (Tihomirova et al., 2010).

The BAC samples were taken without strict frequency over the period of 1 yr. In total 30 experiments were carried out in this study.

2.5 Experimental design and sampling procedure

All water samples (2000 ml) from BAC filtration were collected in glass bottles completely filled with the sample and subsamples (50 ml) from glass column system were collected in sterile glass bottles and stored in a refrigerator at temperature in the range of 2 to 5 °C, before the analyses were done.

To evaluate the effect of addition of LOC on biodegradation the BDOC experimental set-up contains several chromatography glass columns with different height (H, from 5 to 25 cm) (Chromaflex, USA) coupled in-series system (Fig. 1), which was filled with glass carrier beads ($\varnothing$ = 6 mm, specific surface area = 3.76 cm^2 g^{-1}) as a support media for bacteria (Eikebrokk et al., 2007) was used.

The EBCTs in system was 272 min. The samples were continuously pumped upward the columns using a peristaltic pump (Masterflex L/S, Cole-Parmer, USA). An optimal flow rate of 2–5 ml min^{-1} was used. The biodegradation kinetics of the sample was measured using the intermediate samples (i.e. EBCTs with 8 min intervals).

The biomass in all the columns was kept constant by the homogenization after each experiment. The glass beads were removed from the columns and homogenized by shaking for 24 h and reused after. The biomass concentration in all columns was 5.23 × 10^{11} cell cm^{-2} recalculated from adenosine 5-triphosphate (ATP) measurements, according to Magic-Knezev and Van der Kooij (2004).

2.6 Determination of biomass concentration

The concentration of biomass in the column system was measured as the concentration of ATP. The total ATP concentration was determined as described by Berney et al. (2006) using the Promega Bac Titer-Glo Microbial Cell Viability Assay (Promega Corporation, USA) and the calculations were based on the standard curve made with known ATP standard dilutions (Promega Corporation, USA) in sterile ultra pure water. The measurements of ATP were made in the solution obtained from 5 g of glass beads from each column collected in a sterile plastic tube filled with 25 µl ultra pure water and treated by sonification for 2 min with 40 % amplitude or 115 J of energy (Ultra Sonic processor, Cole Parmer, USA) and in the effluent water after each column during the sampling process for DOC samples. The bacterial ATP was calculated by subtracting the extracellular ATP from the total ATP (Hammes and Vital, 2008). The luminescence was measured as an integral over 10 s in relative light units (RLU)

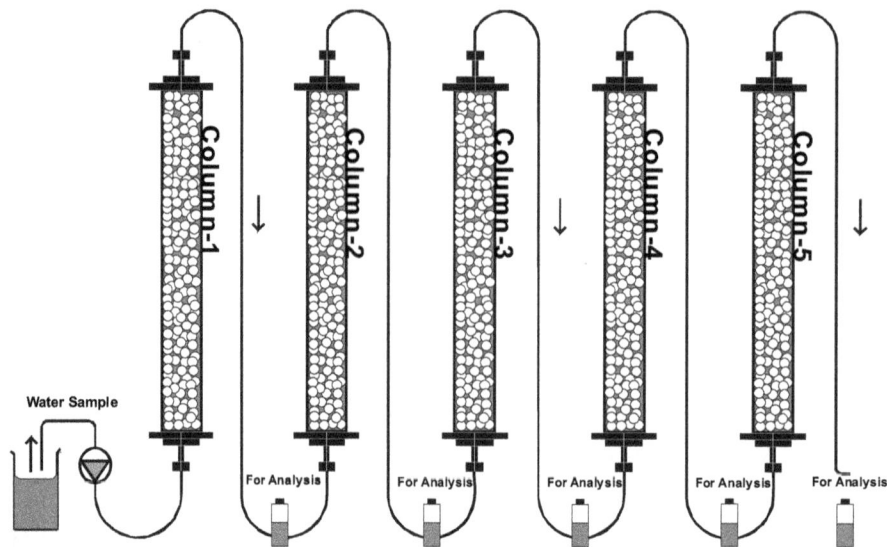

Figure 1. BDOC experimental set-up (adapted from Tihomirova, 2011).

using a luminometer (Hygiene International, Pi-102, Germany). All the samples were measured in triplicate.

2.7 Calculations

The k degradation rate was obtained by fitting the experimental data to the exponential function and expressed as first order kinetic constant (min^{-1}). The equation was fitted separately to data for period in which the minimum concentration of DOC or maximum concentration of BDOC was reached. The regression coefficients (R^2) for exponential curve and Pearson criteria (P) were used (Microsoft Office Excel, 2003). To compare the degradation rate values of statistically significant assays the Moment correlation coefficient (r) was used (Fower et al., 1998).

3   Results

To stimulate biodegradation rate of slowly degradable part of DOC in drinking water treatment the effect of addition of LOC (NaAc or LB) in water was tested in the glass column system.

The average concentration of BDOC of water sample from BAC without addition of LOC was only 0.49 ± 0.29 mg l^{-1} ($n = 14$) or 7 % of DOC$_\text{BAC}$ in this study. The biodegradation rates constant (k) of BDOC in BAC water samples with NaAc and LB as biostimulants were higher (about 4.16×10^{-2} and 2.57×10^{-2} min^{-1}, respectively), whereas the biodegradation rate in the sample without biostimulant was one order of magnitude lower $(0.10 \times 10^{-2}$ min$^{-1})$, see Fig. 2. The biodegradation rates in the samples with NaAc and LB as biostimulant were at least 26 times higher compared with the BAC sample without biostimulant.

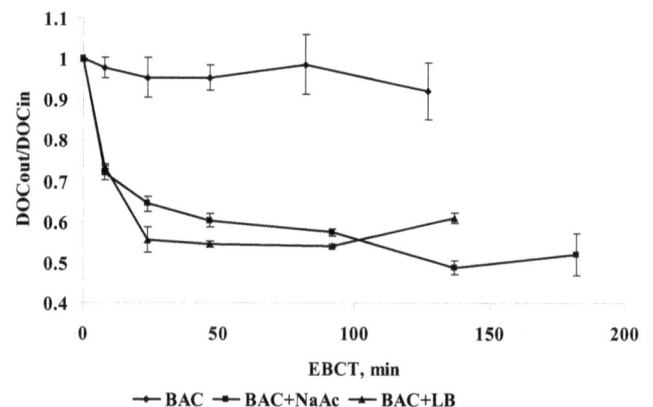

Figure 2. Average DOC changes versus EBCT (adapted from Tihomirova, 2011). Legends: water sample after biofilters from Daugava WTP (BAC, $n = 14$); water sample after biofilters from Daugava WTP with biostimulant sodium acetate after 30 h feeding (BAC+NaAc; $n = 3$) and with Luria Bertrani broth after 48 h feeding (BAC+LB; $n = 3$), respectively.

Traditionally the concentrations of BDOC analysed with dynamic method in glass column system with attached biofilm were calculated as the difference in concentration of DOC between the inlet water sample in column system and the effluent water sample with lowest concentration of DOC, namely DOC$_\text{BAC}$ and DOC$_\text{min}$ (Fig. 3) based on the definitions developed by other researchers (Fahmi et al., 2003; Hammes and Vital, 2008; Volk et al., 1993; Ribas et al., 1991 and Yavich et al., 2004) and in this study called BDOC$_\text{BAC}$. The total concentration of BDOC (namely BDOC$_\text{total}$) which accounts for both parts – quantity of biodegradable part in sample (DOC$_\text{BAC}$) and LOC (DOC$_\text{LOC}$) was not considered in this paper since it was not comparable with measurements

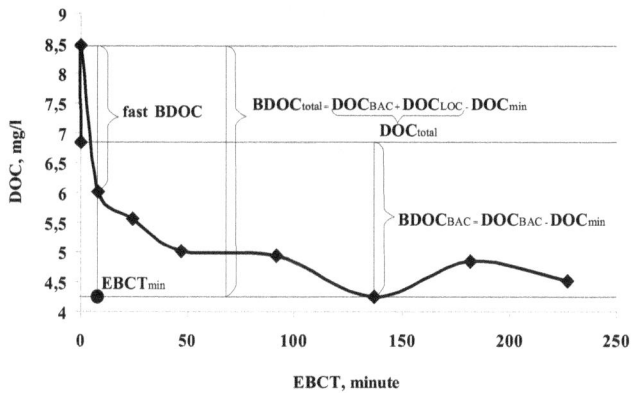

Figure 3. Principle of BDOC quantification shown on an example of 30 h feeding of BAC water sample supplemented with NaAc as a biostimulant.

of BDOC in WTP. The initial concentration of substrate in the BAC sample (DOC_{BAC}) for the series of experiments with NaAc and LB was 5.87 ± 0.96 ($n = 9$) and 4.73 ± 0.19 ($n = 7$) mg l^{-1}, accordingly. For given DOC measurement this is variation of DOC measurements in inflow to the column during the all experiments due to natural variation of DOC in the BAC samples. The concentration of dose of biostimulant (DOC_{LOC}) of NaAc and LB was 1.81 ± 0.36 ($n = 9$) and 1.25 ± 0.25 ($n = 7$) mg l^{-1}, respectively.

The period of biodegradation time when $BDOC_{BAC}$ was $< 15\%$ in each water sample supplemented with LOC can be named an adaptation period of biomass (Tihomirova et al., 2012). During the adaptation period a decrease of minimal EBCT was observed (Table 1). The results showed that during the experiment the biodegradation rate constant using biostimulants NaAc and LB increased up to 41 and 26 times, respectively, and this time interval can be called the biostimulation period. The maximum concentration of BDOC and the maximum biodegradation rate was reached after 51 and 48 h feeding with NaAc and LB, respectively (Table 1). The biostimulation period with NaAc was accomplished after 53 h, after which both $BDOC_{BAC}$ and biodegradation rate decreased. In water samples supplemented with LB the biostimulation period was accomplished after 168 h. The removal efficiency of $BDOC_{BAC}$ reached up to 49 % and 37 % at maximal biodegradation rate with both NaAc and LB, respectively which is significantly higher result compared with sample after BAC filters (Fig. 2). There was a strong positive correlation between biodegradation rate constant and $BDOC_{BAC}$ ($r = 0.63$ for NaAc; $r = 0.65$ for LB; $P = 0.6$ for both biostimulants). The biodegradation process can be divided in two periods – adaptation or coadaptation (20–24 h with biostimulants in this study) and biostimulation period, which was limited from 24 to 53 h for NaAc and from 20 to 168 h for LB biostimulant.

The experiments with BAC samples and NaAc and LB biostimulants showed that after the adaptation phase the fast

degradable part increased to 28.9 and 29.3 % of DOC, the slowly degradable increased to 20.9 and 10.7 % and the non-degradable part decreased to 50.2 and 60 %, respectively (data not shown). The BDOC was 50 and 40 % of DOC_{total}, respectively.

4 Discussions

Biostimulation approach was based on the hypothesis that the biodegradability in water samples is higher than the average concentration of BDOC measured at the effluent from BAC filters in WTP. This was concluded by Tihomirova (2011) over regular three year monitoring period of DOC and BDOC in WTP. The concentration of BDOC decreased from 1.47 ± 1.05 mg l^{-1} in the raw water to 0.59 ± 0.51 mg l^{-1} (10 % and 9 % of the DOC, respectively) after rapid filter and increased to 0.82 ± 0.38 mg l^{-1} (15 %) after BAC filtration. High error range might be attributed to the fluctuation of raw water and treated water quality. Consequently, biological stability of water samples increased during coagulation-sedimentation and rapid sand filtration and ozonation, but it decreased again after BAC filters. DOC was higher in the river water than in the drinking water, but the ratio of BDOC/DOC (%) was higher in the drinking water (15 %). This indicate that after ozonation concentration of biodegradable organic compounds increased, but removal of biodegradable organic matter in biofilters or the production of biologically stable water is not efficient. Results of this study showed that average BDOC in BAC samples ($n = 14$) was 7 % only.

The addition of LOC was related to time of adaptation of biomass in glass column system which for NaAc and LB biostimulants was 24 and 20 h, respectively (Table 1) (Tihomirova et al., 2012), the time interval during which the BDOC value was negligible or biodegradation is not detected (Wiggins and Alexander, 1988). As shown previously (Yavich et al., 2004), the addition of a small amount of biostimulant resulted in an increase in BDOC concentration and a sharper decrease in "fast" BDOC.

As shown in this study for samples taken at Daugava WTP the biodegradation rate constant k of samples containing NaAc and LB as biostimulants increased up to 26 times compared with the BAC sample without biostimulant. Thus the biodegradation rates can be enhanced by using biostimulants.

According to obtained measurements in this study and recalculation biomass concentration was 806 ng ATP per cm^3 in experimental system. As shown previously (Magic-Knezev and Van der Kooij, 2004), ATP concentration can ranged 25 to 5000 ng ATP per cm^3 in biofilters at different WTP. The results (Table 1) showed that biomass concentration detached from biofilter when biodegradation process was completed and decrease of biodegradation rate constant. Recalculated total biomass amount in system

Table 1. Biodegradation of substrate in glass column system feed using water samples supplemented with labile organic carbon depending on the adaptation time of the experimental system.

Adaptation time, h	BAC+NaAc				BAC+LB			
	$EBCT_{min}$, min	$k \times 10^{-2}$, min^{-1}	$BDOC_{BAC}$, %	ATP, cells $\times 10^8$ ml^{-1}	$EBCT_{min}$, min	$k \times 10^{-2}$, min^{-1}	$BDOC_{BAC}$, %	ATP, cells $\times 10^8$ ml^{-1}
0	> 272	0	0	60	> 272	0.31	0.4	394
4	> 272	0	0	26	> 272	0.41	2	412
20	–	–	–	–	137	0.33	20	333
24	272	0.18	27	52	–	–	–	–
30	137	0.19	38	65	–	–	–	–
44	–	–	–	–	47	1.30	27	291
48	24	3.22	34	65	24	2.57	37	419
51	24	4.16	49	49	–	–	–	–
53	8	3.70	20	65	–	–	–	–
168	–	–	–	–	272	0.18	30	552
192	47	0.53	6	105	227	0.11	10	656
240	92	0.25	1	1602	–	–	–	–

Legends: the minimal EBCT ($EBCT_{min}$), biodegradation rate constant (k), concentration of BDOC (%) and biomass concentration (determined as ATP, cell per ml) in effluent are shown.

was 405 170 ng ATP, and max detached biomass amount −33 640 ng ATP, what is 8 % of total biomass concentration in system. There was not significant correlation between cell concentration in the effluent and the ATP concentrations in the biofilter. These results showed that bacterial release into the effluent sample may be due to substrate concentration limitation after biodegradation.

The rate of biodegradation is highly variable and depends on many parameters including chemical structure and concentration of the organic substance in water samples, applied oxidant (ozone, chlorine) dose, temperature, salinity, pH, concentration of oxygen, and inorganic nutrients and concentration of active microorganisms (Becker et al., 2006; Brandt et al., 2003, 2004; Kulp et al., 2007; Sanchez et al., 2007; Spain et al., 1980; Spain and Veld, 1983; Steiner and Sauer, 2001; Swindoll et al., 1988). Some organic compounds, especially aromatic compounds are rather resistant to natural biodegradation and biodegradation of a compound of the mixture can be strongly influenced by the presence of other components in the mixture (Tsai and Juang, 2006). It has been shown previously that the addition of LOC results in an increase of the concentration of easy degradable part of mixture and the stimulated biodegradation results in decrease in minimum EBCT compared to that without biostimulation (Spain et al., 1980; Yavich et al., 2004). The biodegradable DOC was utilized after short contact time or at the top of the biofilter (Moll et al., 1998). Direct addition of a nutrient can stimulate the activity of microbial communities (Macbeth et al., 2004) and the presence of easy degradable carbon sources can enhance the biodegradation of more persistent compounds (Brandt et al., 2003; Shimp and Pfaender, 1987).

Method of co-adaptation and its impact on the biodegradation of different chemicals has been used to degrade resistant chemical pollutions in different environmental settings (Schmidt and Alexander, 1985; Shimp and Pfaender, 1987; Wiggins and Alexander, 1988). Biostimulation method described in this study can also be compared with the other (e.g. phosphorus dosing into biofilters to stimulate the activity of the biomass, Rubulis, 2006).

Addition of biostimulant increased $BDOC_{total}$ in influent water; however increase in the total consumption of BDOC allowed to reduce both BDOC added as biostimulant and native BDOC in the sample. In this study biostimulants (NaAc or LB) are easy convertible substrate for bacterial biomass. In these experiments NaAc was used as single carbon source and LB medium as multicarbon source. Results obtained in this study showed that BDOC concentration increased significantly and was utilized after 24 min contact time due to biostimulation comparing with sample without biostimulant.

This study showed that the biofiltration systems of the WTPs in Boreal regions treating humic rich water can be improved and with biostimulation approach biodegradation of humic rich water can reach up to 29 %, which is very close to the theoretical −30 % as DOC (Volk et al., 2002). LOC addition stimulated biodegradation rate of slowly biodegradable organic matter. The biomass consumed easy degradable substrate and the adaptation period contributed to that slowly degradable organic matter was used faster as a secondary substrate. From the results it can be concluded that LOC can be used as a stimulator of biodegradation which would allow to effective remove of slowly degradable organic matter within EBCT which is usually about 30 min. The biostimulation period is the period limited in time when

BDOC and the biodegradation rate constant substantially increase.

This study demonstrates principle of biostimulation applicability for enhanced DOC removal in drinking water technology. Before implementation in real plant the technology should be engineered to avoid both problems with release of biomass and other problems, which can related to optimal condition for filter operations such as application of automatic control of biomass and substrate dosing. In practice the optimal concentration of LOC would be depended from the DOC concentration in influent water. LOC work as stimulator, and to avoid the increase the BDOC in system after biofilter, LOC concentration should be verified for every individual plant.

5 Conclusions

From the obtained results it can be concluded that:

- The addition of LOC was related to the increase of the DOC_{total}. At the same time $BDOC_{total}$ increased significantly (up to 7 and 5 times higher with NaAc and LB as biostimulants, respectively) and it was utilized after 24 min contact time due to biostimulation.

- The study demonstrates that biodegradation rates can be increased by using a biostimulant. The biodegradation rates of BDOC in BAC water samples with NaAc and LB as biostimulants were 4.16×10^{-2} and 2.57×10^{-2} min^{-1} whereas the degradation rate in the sample without biostimulant was 0.10×10^{-2} min^{-1}.

- There was a strong positive correlation between the constant of biodegradation rate and $BDOC_{BAC}$ ($r = 0.63$ for NaAc; $r = 0.65$ for LB; $P = 0.6$ for both biostimulants).

- The adaptation time for the mixture of sample and biostimulant was 20 and 24 h for LB and NaAc, respectively.

- The biostimulation period for NaAc and LB biostimulants was from 24 to 53 h, and from 20 to 168 h, respectively.

- The biostimulation period was accomplished with increasing EBCT, decreasing BDOC and biomass detachment.

Acknowledgements. This work has been partly supported by the European Social Fund within the National Programme "Support for the carrying out doctoral study program's and post-doctoral researches" project "Support for the development of doctoral studies at Riga Technical University". This work has been undertaken as a part of the research project "Technology enabled universal access to safe water – TECHNEAU" (Nr. 018320) which is supported by the European Union within the 6th Framework Programme. There hereby follows a disclaimer stating that the authors are solely responsible for the work. It does not represent the opinion of the Community and the Community is not responsible for any use that might be made of data appearing herein.

References

Becker, J. G., Berardesco, G., Rittmann, B. E., and Stahl, D. A.: Effects of endogenous substrate on adaptation of anaerobic microbial communities to 3-Chlorobenzoate, Appl. Environ. Microbiol., 72, 449–456, 2006.

Berney, M., Weilenmann, H. U., and Egli, T.: Flow-cytometric study of vital cellular functions in Escherichia coli during solar disinfection (SODIS), Microbiology, 152, 1719–1729, 2006.

Brandt, B. W., Van Leeuwen, I. M. M., and Kooijman, S. A. L. M.: A general model for multiple substrate biodegradation. Application to co-metabolism of structurally non-analogous compounds, Water Res., 37, 4843–4854, 2003.

Brandt, B. W., Kelpin, F. D. L., Van Leeuwen, I. M. M., and Kooijman, S. A. L. M.: Modeling microbial adaptation to changing availability of substrates, Water Res., 38, 1003–1013, 2004.

Carlson, K. H. and Amy, G. L. Ozone and biofiltration for multiple objectives, J. Am. Water Works Assoc., 93, 88–98, 2001.

Eikebrokk, B., Juhna, T., Melin, E., and Osterhus, S. W.: Deliverable 5.3.2a. Water Treatment by Enhanced Coagulation and Ozonation-Biofiltration. Intermediate report on operation optimization procedures and trials. Report of Integrated Project TECHNEAU Funded by the European Commission under the Sustainable Development, Global Change and Ecosystems Thematic Priority Area, http://www.techneau.org/index.php?id=120, last access: 12 August 2009, 2007.

European Standard EN 1484:1997: Water analysis – Guidelines for the determination of total organic carbon (TOC) and dissolved organic carbon (DOC), 1997.

Fahmi, Nishijima, W., and Okada, O.: Improvement of DOC removal by multi-stage AOP-biologycal treatment, Chemosphere, 50, 1043–1048, 2003.

Fower, J., Cohen, L., and Jarvis, P.: Practical Statistics for field biology, 1st Edn., John Wiley & Sons Inc, USA, 1998.

Hammes, F. and Vital M.: Deliverable 3.3.9. A Report on the Growth of Pathogenic Bacteria on Natural Assimilable Organic Carbon. Report of Integrated Project TECHNEAU Funded by the European Commission under the Sustainable Development, Global Change and Ecosystems Thematic Priority Area, http://www.techneau.org/index.php?id=120, last access: 17 August 2009, 2008.

Huck, P. M.: Measurement of biodegradable organic matter and bacterial growth in drinking water, J. Am. Water Works Assoc., 82, 121–132, 1990.

Kaplan, L. A., Reasoner, D. J., and Rice, E. W.: A survey of BOM in United-Satates drinking waters, J. Am. Water Works Assoc., 86, 78–86, 1994.

Klevens, C. M., Collins, M. R., Negm, R., and Farrar, M. F.: Characterization of NOM removal by biological activated carbon, in: Advances in Slow Sand and Alternative Biological Filtration, edited by: Graham, N. and Collins, R., John Wiley & Sons, 79–87, 1996.

Kulp, T. R., Han, S., Saltikow, C. W., Lanoil, B. D., Zargar, K., and Oremland, R. S.: Effect of imposed salinity gradients on dissimilatory arsenate reduction, sulphate reduction, and other microbial processes in sediments from two California soda lakes, Appl. Environ. Microbiol., 73, 5130–5137, 2007.

Macbeth, T. W., Cummings, D. E., Spring, S., Petzke, L. M., and Sorenson, K. S.: Molecular characterization of a dechlorinating community resulting from in situ biostimulation in trichlororthene-contaminated deep, fractured basalt aquifer and comparison to a derivative laboratory culture, Appl. Environ. Microbiol., 70, 7329–7341, 2004.

Magic-Knezev, A. and Van der Kooij, D.: Optimization and significance of ATP analysis for active biomass in granular activated carbon filters used in water treatment, Water Res., 38, 3971–3979, 2004.

Miettinen, I. T., Vartiainen, T. K., and Martikainen, P. J.: Determination of assimilable organic carbon (AOC) in humus-rich waters, Water Res., 33, 2277–2282, 1999.

Moll, D. M., Summers, R. S., and Breen, N.: Microbial characterization of biological filters used for drinking water treatment, Appl. Environ. Microbiol., 64, 2755–2759, 1998.

Ribas, F., Frias, J., and Lucena, F.: A new dynamic method for the rapid determination of the biodegradable dissolved organic carbon in drinking water, J. Appl. Microbiol., 71, 371–378, 1991.

Rubulis, J.: The influence of phosphorus on the formation of biofilm in drinking water, Ph.D. thesis, Riga Technical University, Riga, Latvia, 2006.

Sanchez, B., Champomier-Verges, M.-C., Collado, C., Anglade, P., Baraige, F., Sanz, Y., De Los Reyes-Gavilan, C. G., Margolles, A., and Zagorec, M.: Low-pH adaptation and the acid tolerance response of Bifidobacterium longum biotype longum, Appl. Environ. Microbiol., 73, 6450–6459, 2007.

Schmidt, S. K. and Alexander, M.: Effect of dissolved organic carbon and second substrates on the biodegradation of organic compounds at low concentrations, Appl. Environ. Microbiol., 49, 822–827, 1985.

Shimp, R. J. and Pfaender, F. K.: Effect of adaptation to phenol on biodegradation of monosubstituted phenols by aquatic microbial communities, Appl. Environ. Microbiol., 53, 1496–1499, 1987.

Spain, J. C. and Veld, P. A. V.: Adaptation of natural microbial communities to degradation of xenobiotic compounds: effects of concentration, exposure time, inoculum, and chemical structure, Appl. Environ. Microbiol., 45, 428–435, 1983.

Spain, J. C., Pritchard, P. H., and Bourguin, A. W.: Effect of adaptation on biodegradation rates in sediments/water cores from Estuarine and freshwater environments, Appl. Environ. Microbiol., 40, 726–734, 1980.

Steiner, P. and Sauer, U.: Proteins induced during adaptation of Acetobacter aceti to high acetate concentrations, Appl. Environ. Microbiol., 67, 5474–5481, 2001.

Swindoll, C. M., Aelion, C. M., and Pfaender, F. K.: Influence of inorganic and organic nutrients on aerobic biodegradation and adaptation response of subsurface microbial communities, Appl. Environ. Microbiol., 54, 212–217, 1988.

Tihomirova, K.: NOM Removal from Water and its Influence on the Drinking Water Quality: Natural organic matter removal from water and its influence on the water quality in distribution network, LAP LAMBERT Academic Publishing, ISBN: 978-3-8465-2340-7, 172 pp., 2011.

Tihomirova, K., Rubulis, J., and Juhna, T.: Changes of NOM Fractions during conventional drinking water treatment process in Riga, Latvia, Wa. Sci. Technol., 10, 157–163, doi:10.2166/ws.2010.652, 2010.

Tihomirova, K., Briedis, A., Rubulis, J., and Juhna, T.: Effect of biomass adaptation to biodegradation of dissolved organic carbon in water, Biodegradation, 23, 319–324, doi:10.1007/s10532-011-9511-z, 2012.

Tsai, S.-Y. and Juang, R.-S.: Biodegradation of phenol and sodium salicylate mixtures by suspended Pseudomona Putilda CCRC 14365, J. Hazard. Mater., 138, 125–132, 2006.

Van der Kooij, D., Visser, A., and Hijnen, W. A. M.: Determination the concentration of easily assimilable organic carbon in drinking water, J. Am. Water Works Assoc., 74, 540–545, 1982.

Volk, C., Wood, L., Johnson, B., Robinson, J., and Kaplan, L.: Monitoring dissolved organic carbon in surface and drinking water, J. Environ. Monit., 4, 43–47, 2002.

Volk, L., Renner, C., Roche, P., Paillard, H., and Joret, J. C.: Effects of ozone on the production of biodegradable dissolved organic carbon (BDOC) during water treatment, Ozone: Sci. Eng., 15, 389–404, 1993.

Wiggins, A. B. and Alexander, M.: Role of chemical concentration and second carbon sources in acclimation of microbial communities for biodegradation, Appl. Environ. Microbiol., 54, 2803–2807, 1988.

Yavich, A. A., Lee, K.-H., Chen, K.-C., Pape, L., and Masten, S. J.: Evaluation of biodegradability of NOM after ozonation, Water Res., 38, 2839–2846, 2004.

Application of optical tomography in the study of discolouration in drinking water distribution systems

P. van Thienen, R. Floris, and S. Meijering

KWR Watercycle Research Institute, P.O. Box 1072, 3430 BB Nieuwegein, The Netherlands

Abstract. Theories describing the turbulent deposition of particles from aerosols have recently been applied to drinking water distribution. In order to allow the study of these processes in a quantitative way and internally observe a cloud of suspended particles in a pipe, we have developed an optical tomography technique and measuring device using low cost electronic components specifically for this application. The mathematical methodology and the electronic device are described in this paper, and tests of both the mathematical approach and the actual device are presented. We conclude that the mathematical framework presented is suitable and that the technical implementation works in a test setting. The described methodology may provide a valuable tool for the study of processes related to drinking water discolouration in the lab.

1 Introduction

Water discolouration continues to be one of the prime reasons for customer complaints relating to water quality to be received by water companies (Husband and Boxall, 2011). The resuspension of particles present in the drinking water distribution system by a hydraulic disturbance is usually invoked to explain its occurrence (Vreeburg and Boxall, 2007). A chain of events or processes can be identified in the generation of a discolouration event, i.e. origin, evolution, deposition, cohesion/adhesion and resuspension of particles.

The suggestion that currently applied models do not encompass the complete range of physical processes responsible for the occurrence of discolouration (Blokker et al., 2010) has led van Thienen and Vreeburg (2010) and van Thienen et al. (2011b) to theoretically investigate the possibility of processes other than gravitational settling affecting particle deposition. Their findings predict the occurrence of the turbulent process of turbophoresis under specific conditions which may be expected in transport mains. In order to experimentally verify these predictions, a method of internally observing a pipe in which this process is supposed to take place is required. In this paper, the required methodology, of which a preview has been presented by van Thienen et al. (2011a), is developed using the principles of optical tomography. The method allows its user to study particle transport, deposition and resuspension behaviour in a transparent pipe both qualitatively (in terms of different structures in the particle concentration field relating to different mechanisms) and quantitatively (in terms of radial transport velocities). The technique uses a sophisticated inversion scheme to include and deal with the effects of measurement uncertainties and noise.

Tomography is a well established technology in the field of medicine (see e.g. Bibb and Winder, 2010), allowing doctors a non-destructive view inside a patient. It is also used e.g. in geophysics to view inside the Earth using seismic waves (e.g. Bijwaard and Spakman, 1999), and has been applied to industrial gravity chutes (Zheng et al., 2006). However, its application in drinking water research is new.

2 Methodology

2.1 Principle

In the present application, we use visible light to internally observe a moving volume of water containing suspended particles. Light is shone through a perspex pipe perpendicular to the pipe direction from several points and at several angles of incidence. At the same time, the transfer of this light through the perspex pipe and the particle bearing water it is filled with is measured. Scattering and absorption reduce the intensity of the light falling on the sensors. By mathematically

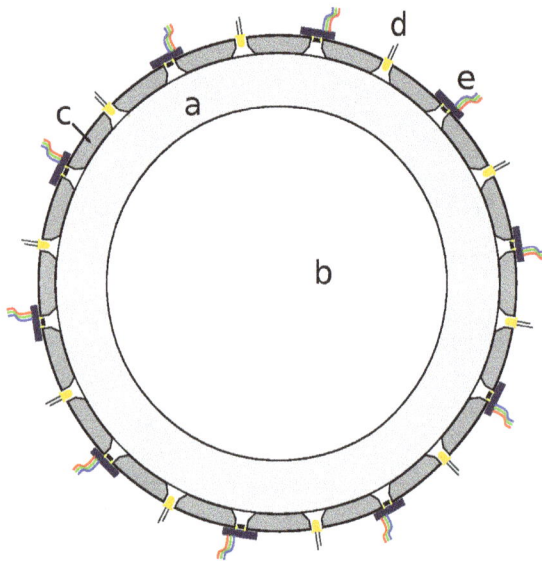

Figure 1. Schematic set-up of the measurement device. The image shows a cross section of the perspex pipe (**a**) filled with water (**b**). A tightly fitting PVC ring (**c**) is fitted around the pipe. Tapered holes have been drilled in this ring, into which alternating LEDs (**d**) and light sensors (**e**) have been placed.

combining these measurements, a clear image of the cross-sectional particle concentration field can be obtained.

2.2 Measuring device set-up

A PVC ring containing ten evenly spaced LEDs with ten evenly spaced sensors in-between is fixed around the transparent tube (see Fig. 1). The LEDs are high-power white LEDs; the sensors are Phidget light sensors (Phidgets, 2010). An Arduino Mega micro-controller (Arduino, 2010) with custom software is used to control the LEDs (using switching electronics) and also contains the AD-converters which are used to digitally read the light sensors (10 bit resolution). The Arduino board is connected to a computer using a USB cable. At regular intervals (typically a few seconds), the LEDs are briefly switched on successively. While a single LED is illuminated, all light sensors are read. This entire measurement cycle takes only a fraction of a second. When all sensor readings have been collected, they are sent to the computer for processing. We note that the measuring device was constructed exclusively from low cost components.

2.3 Refraction and reflection

Because of significantly different optical densities, light is refracted and reflected both at the air-perspex interfaces and at the perspex-water interfaces. The Fresnel equations (see e.g. Brass, 2010) allow the calculation of the relative amounts of light which are reflected and refracted. Snell's law subsequently provides the angle of refraction. Based on these

expressions, light beams can be traced from their emission from the LED to their absorption by a sensor. Figure 2 shows a set of light beams shone into the perspex pipe from the center of the light emitting surface of a single LED at different angles of incidence. Light can be seen to reach all sensors through combinations of reflections and/or refractions with different intensities (indicated by colour). In the case where LED and light sensor are adjacent (Fig. 2a), most light falling on the sensor comes from a reflection on the outside of the perspex pipe and as a result does not contain any information on the light absorption by material inside the pipe. The other cases (Fig. 2b–e), however, show that most light reaching the sensors follows the most direct route through the pipe, i.e. by subsequent refraction at the air-perspex, perspex-water, opposite water-perspex, and opposite perspex-air transitions, which is generally close to the imaginary line connecting LED and sensor.

2.4 Mathematical description

2.4.1 Physical assumptions

In the following, a number of assumptions will be made. These are:

- The fraction of light which is scattered by suspended particles increases in some way with the particle concentration. As a result, the relative amount of light which passes through the suspension decreases with increasing particle concentration.

- A simple optical model is used in which particles only cast shadows and their concentration is sufficiently low for particles not to be in each other's shadows. No scattering or reflection of light off particles is taken into account.

2.4.2 Model set-up

The cross-sectional domain is discretized into a number of triangles. On the corners of each of these triangles, an (initially unknown) light absorption coefficient is defined. For each light beam (path from LED to light sensor), an equation can be written which describes the decrease in intensity of the light beam as a function of the (unknown) light absorption coefficients. The relative light intensity decrease can be written as the ratio of the intensity decrease to the reference intensity:

$$\left(\frac{(I_E - I_R)}{I_E} \right) = \Sigma_{j=1}^{n} \int_{\text{path}} c(l) dl \tag{1}$$

In this expression, I_E is the unperturbed or reference light intensity at the receiver and I_R is the measured light intensity at the receiver in the perturbed experimental situation. A summation is made over all n elements in the discretization, and the unknown absorption or "shadow" coefficients

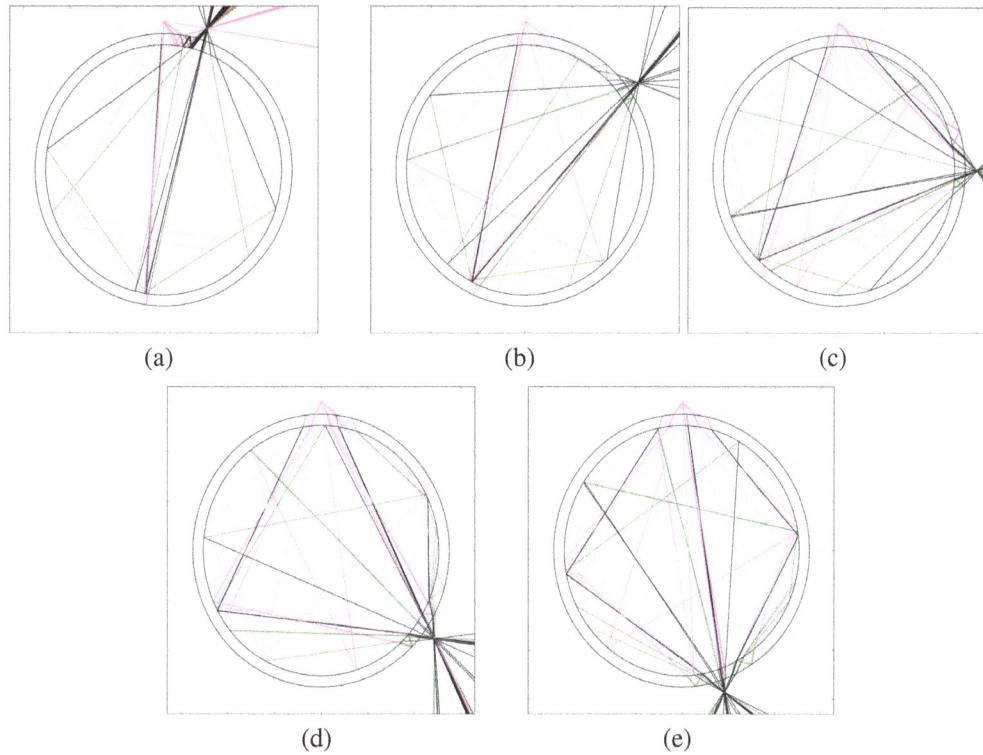

Figure 2. Light paths from a single LED (top of graphs) to five sensors (**a–e**; the other five sensors are their mirror images), including refraction and reflection (up to 7 times). Colours indicate beam intensity relative to emission, ranging from green (0.001) to pink (0.1), with black segments indicating a relative intensity of less than 0.001.

c are integrated over the path of the light beam inside each of the elements it passes through. Combining all measurements $((I_F - I_R)/I_F)_i$ in a vector r, all unknown absorption coefficients c in a vector s, and all integration coefficients in a matrix $\mathbf{A}$, we get a system of equations:

$$\mathbf{A}s = r \qquad (2)$$

Solving this system of equations is an inverse problem.

2.5 Light beam paths

As indicated in Sect. 2.3, light reaches the sensors through a complex set of paths which contribute in varying amounts to the total signal. However, in most cases, 98 % or more of the light which reaches a sensor arrives through a direct path without reflections. Therefore, a simple approximation of this set of paths, a straight line connecting a LED to a beam, can be applied instead. In the following, we apply both this simple approximation (*single beam model*) and a full set of light paths (*multiple beam model*).

2.5.1 Inversion procedure

When the number of beams is equal to the number of unknown coefficients, in principle we have a determined system of equations, which can possibly be solved. This would

result in knowing the values of the light absorption coefficients in all nodes of the discretization, which are a proxy for the local particle concentration. However, measurement and discretization errors dominate the result if this procedure is applied. In addition to this, refraction and incidental low beam intensities reduce the number of beams which can actually be used, resulting in a underdetermined problem. A more refined inversion procedure is required to get meaningful results. Suitable methods have been developed and applied in fields in which inversion is done of measurements of "natural experiments" which are not controlled by the researcher, specifically geophysics (e.g. Bijwaard and Spakman, 1999; Fokker et al., 2010). These generally suffer both from being underdetermined and having significant measurement uncertainties. Applying these methods (Tarantola, 2005; Muntendam-Bos et al., 2008), the system (Eq. 2) is rewritten as:

$$s = \mathbf{C}_m \mathbf{A}^T (\mathbf{A}\mathbf{C}_m \mathbf{A}^T + \mathbf{C}_d)^{-1} r \qquad (3)$$

In this expression, $\mathbf{C}_m$ is the covariance matrix of the prior model matrix. It is produced by computing the covariance matrix for a large number of plausible models which span the parameter space of expected results, and contains all prior knowledge, uncertainties and variations of the model. $\mathbf{C}_d$ is the prior data covariance matrix, which has the variances σ^2

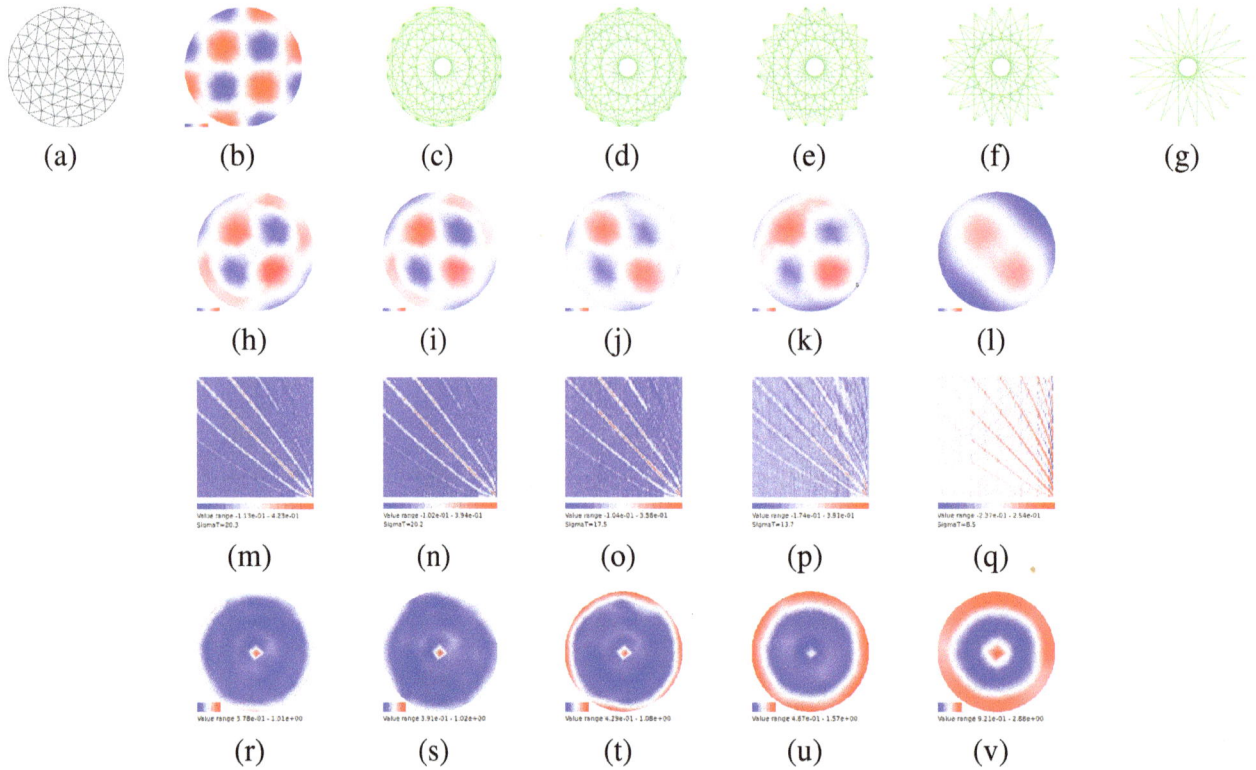

Figure 3. Resolution and resolvability as a function of ray coverage for the single beam model. **(a)** Domain discretization; **(b)** synthetic absorption coefficient field; **(c–g)** set of beam paths used; **(h–l)** simulated tomograms for these beam path sets; **(m–q)** corresponding resolution matrices (note that SigmaT indicates the sum of the main diagonal of the resolution matrix); **(r–v)** posterior model variance for all degrees of freedom. Note that frames **(b)** and **(h–l)** all use the same colour scale.

of the measurements on the main diagonal and zeros off the main diagonal. The total variances consist of the actual measurement variances $\sigma^2_{\mathrm{m}ij}$ which are obtained for each LED-sensor combination ij from multiple samples which are taken and a system variance σ^2_{s} which is assumed to be the same for all measurements and includes all errors which are not included in σ^2_{m}. A representative value of σ_{s} is chosen. The variances are summed following Bienaymé's formula:

$$\sigma^2_{ij} = \sigma^2_{\mathrm{m}ij} + \sigma^2_{\mathrm{s}} \tag{4}$$

2.5.2 Resolution and covariance

In order to understand what quality of results can be expected from the inversion procedure, we compute the resolution matrix and compare prior and posterior covariance matrices. The resolution matrix is defined as (Tarantola, 2005; Muntendam-Bos et al., 2008):

$$\mathbf{R} = \mathbf{C}_{\mathrm{m}} \mathbf{A}^T (\mathbf{A} \mathbf{C}_{\mathrm{m}} \mathbf{A}^T + \mathbf{C}_{\mathrm{d}})^{-1} \mathbf{A} \tag{5}$$

For a perfectly resolved system, the resolution matrix is equal to the identity matrix.

The posterior covariance matrix can be computed from the resolution matrix (Tarantola, 2005; Muntendam-Bos et al., 2008):

$$\mathbf{C} = (\mathbf{I} - \mathbf{R}) \mathbf{C}_{\mathrm{m}} \tag{6}$$

This expression shows that a well-resolved posterior model with low variances is obtained for a model which has a resolution matrix close to the identity matrix. Also, it is clear from Eq. (6) that uncertainties in the prior model directly propagate into the posterior model. The magnitude of this propagation depends on the resolution. For a resolved system, posterior variances should be smaller than prior variances.

We shall use matrices $\mathbf{R}$ and $\mathbf{C}$ to evaluate the mathematical quality of the results of the inversion procedure.

3 Tests

3.1 Mathematical approach

A series of simulations has been performed in order to assess the capabilities and limits of the mathematical model in the context of the measurement device setup. Figure 3 summarizes the results of these tests for a single beam model

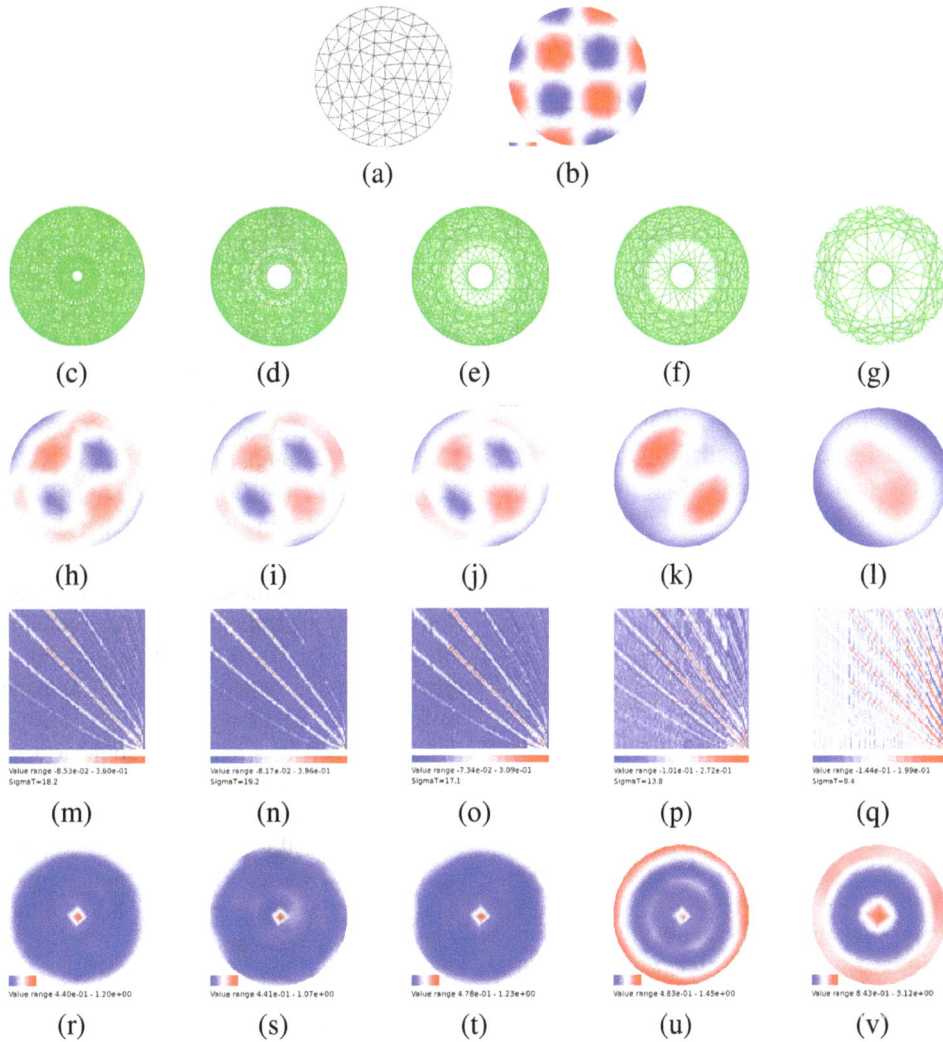

Figure 4. Resolution and resolvability as a function of ray coverage for multiple beam model. Frame descriptions are the same as in Fig. 3.

(see Sect. 2.5). Figure 4 does the same for a multiple beam model as described in Sect. 2.3. A synthetic light absorption coefficient field was generated exhibiting a checkerboard pattern (see Figs. 3b and 4b). Corresponding synthetic measurements were computed for all beams based on this pattern (10 samples per beam), adding white noise to the measurements with a maximum amplitude of 10 %. These synthetic measurements were fed into the inversion procedure, assuming all uncertainty to originate from the measurement error (Eq. 4, $\sigma_s = 0$). For different beam sets displayed in Fig. 3c–g, this results in the corresponding tomograms of Fig. 3h–l for the single beam model and Fig. 4h–l for the multiple beam model. Traditionally, checkerboard tests have often been used to illustrate the quality (or lack thereof) of an inversion procedure (e.g. Morgan et al., 2002). In the tomograms of Figs. 3h–l and 4h–l, a decreasing correspondence with the original synthetic input field (Figs. 3b/4b) can be observed for decreasing beam coverage. However, in frames

h, i and j (Figs. 3 and 4), the reconstructed image is still acceptably close. An alternative, more quantitative way of determining the quality of the tomograms is by means of the resolution and posterior covariance matrices, as defined in Sect. 2.5.2. As has been indicated above, a resolution matrix is close to the identity matrix for a well resolved model. The sum of the main diagonal of this model indicates how many degrees of freedom are resolved. Figures 3m–q and 4m–q show the resolution matrix for our five different beam coverage scenarios for the single beam model and the multiple beam model, respectively. It is clear that in no case is the resolution matrix close to the identity matrix. This means that additional information is taken from the prior model covariance matrix in the inversion procedure. The resulting (posterior) uncertainty in the individual degrees of freedom is plotted in Figs. 3r–v and 4r–v. These images show that the central degree of freedom is not well resolved in all cases compared to the surrounding area. In addition to this, the degrees of freedom close to the outer wall become increasingly badly resolved as the beam coverage decreases, since it is in this region where the coverage is removed.

Table 1 shows how the maximum value of the prior model covariance matrix compares to the maximum value of the posterior model covariance matrix, both for the single beam and the multiple beam optical models. It is clear that in each case, a significant reduction of the variance is obtained, which means that we have gained information relative to the prior model. As the available number of observations is reduced by ignoring more measurements – moving from c to g (Table 1 and Figs. 3 and 4) – the variance reduction does decrease, however. The numbers shown in Table 1 do not show a monotonously decreasing series for the multiple beam model. This is due to the fact that all numbers are based on simulations with a significant amount of random noise in the input. As a result, all numbers presented are valid for a single simulation but will be somewhat different when the simulation is repeated. However, the presented set of results does illustrate a clear trend.

3.2   Inversion results

Now that the theoretical resolvability and resolution have been ascertained, lab tests are required, in which an object representative of a particle cloud is placed inside the device and located using the device. For this purpose, a curtain was made from wire mesh, which is expected to have optical properties similar to that of a cloud of dark, non-reflecting particles suspended in water and can be easily positioned inside the device (see Fig. 5a, c). This curtain was placed inside the device in different configurations and orientations (Fig. 5b, d). The resulting images are shown in Fig. 5e–g for the single beam model and in Fig. 5h–j for the multiple beam model, for curtain configurations/orientations indicated by a dashed line (note that the actual area occupied by the curtain is wider than the dashed line). The following can be said for all three cases e–g (Fig. 5) as well as for cases h–j (Fig. 5):

- The curtain is well resolved in the tomograms.

- The image of the curtain is somewhat wider than the actual curtain.

- The ends of the curtain are not so accurately resolved.

The width can be partially explained from the limited resolution of the tomogram (12 elements in its diameter). The ends of the curved curtain in Fig. 5f–g and h–j may be related to the fact that we are imaging an object consisting of lamellae of wire mesh, the visibility of which strongly depends on their orientation relative to the different light beams which are used in its reconstruction in the tomogram. The fact that the linear curtain in Fig. 5e does not appear to touch the walls is a resolution issue. As can be seen in Figs. 3t and 4t, the posterior model variance at the pipe wall is rather

Table 1. The maximum variance in prior and posterior models for the different beam coverage scenarios depicted in Figs. 3c–g and 4c–g.

	c	d	e	f	g
prior	7.58	7.58	7.58	7.58	7.58
posterior; single	1.01	1.02	1.08	1.57	2.88
posterior; multiple	1.20	1.07	1.23	1.45	3.12

large, meaning that this region is not well resolved. This is due to the lack of usable light beams shining more or less parallel to the wall in this region (Fig. 3e). In the case of the multiple beam model, Fig. 3e does show significant beam coverage close to the wall. However, the amount of light carried by these beams is so small that they contribute little to the final tomogram.

In addition to these dry tests, wet tests have been performed with the tomography device fitted to a vertical piece of perspex pipe filled with water. Small amounts of coffee granules were dropped in the water by hand in different patterns, forming a sinking column of coffee particle suspension when dropped in a single location or a wider cloud when dropped over a larger area. Working in this way, it was difficult to control the exact location and shape of the particle clouds sinking through the beam field of the measurement device. However, the resulting tomograms were consistently in agreement with the visually observed locations of the particle columns and clouds.

4   Discussion

4.1   Functioning

Both the mathematical consideration of the resolution of the inversion results and the dry tests which have been performed show that the device functions as intended. Also, it is capable of resolving particle concentration variations in the inside of a moving body of water which can not be determined by direct observation from the outside. The more complex and computation time consuming *multiple beam model* does not appear to result in significantly better images than the *single beam model*, but it does have a better formal resolution close to the wall when ray coverage is reduced.

4.2   Practical considerations

During the test phase of the device, several practical issues were come across and resolved. These include the following:

- Because the variations in light intensity which are the basis of the inversion procedure are relatively small, proper shielding of the device and the section of transparent pipe on which it is mounted is essential. In case the shielding is incomplete, recalibration is required whenever the lighting conditions in the lab change, re-setting I_E in expression (Eq. 1).

- The optical properties of the particles are important in the sense that highly reflective and/or scattering particles reduce the quality of the inversion results, since reflection and scattering off the particles are not included in the simple optical model which is applied. For this reason, a test material was selected which is dull and dark.

4.3 Particle concentrations

In the above, tomograms are obtained which show the distribution of a "light absorption coefficient" throughout a cross-section. This is very useful when one wants to study concentration *differences*, but the actual concentration values are not obtained. In order to convert absorption coefficients to concentrations, a calibration curve needs to be constructed from experiments with known particle concentrations. It is expected that different curves are obtained for different materials with different optical surface properties.

4.4 Applications

The intended operating environment for the method which has been described in this paper is in the lab. Under normal conditions, the particle concentration in drinking water is much too low to be detectable with the present optical set-up, except when hydraulic disturbances cause discolouration. In the lab, it is possible and often desirable to work with particle concentrations which are much higher than in drinking water distribution systems to facilitate observation and reduce time scales of e.g. accumulation. The first investigation in which this technique has been applied is the experimental verification of a theoretical mechanism map for turbulent particle deposition in drinking water distribution systems (Floris et al., 2011). This presents results describing the conditions of initiation of particle deposition by turbophoresis, building on theoretical predictions by van Thienen et al. (2011b). A number of additional research areas in drinking water distribution in which the method can be applied can be thought of, such as particle resuspension, saltation and bed transport.

4.5 Limitations and possible improvements

Some limitations of the current setup are:

- Since we are considering a moving body of water, we may expect the plug of water to advance by some distance in the course of the measurement cycle. The resulting image is therefore only meaningful if variations in the particle concentration field do not occur on this time and length scale. At typical flow velocities of

0.01 to $1 \, \text{m s}^{-1}$ and measurement cycle times (depending primarily on the number of samples taken) of 300 to 1500 ms, the length of the imaged plug is 3 mm to 1.5 m.

- The resolution of the tomograms is limited by the number of LEDs and sensors which have been installed.

Several improvements can be made to the device and procedure. We list the most obvious and effective:

- One of the main issues with the current setup is the reduced ray coverage close to the pipe wall due to the effects of refraction. By proper shaping of the outside of the transparent pipe or by choosing a different material with a lower refractive index, this effect can be reduced. For the latter option, however, the choice of alternative materials is not obvious, since the refractive index of perspex is already relatively low.

- Because of the time required to take all measurements, which may add up to hundreds of milliseconds, the image obtained is to some degree an average of the situation over this time period. Using faster electronics would allow a reduction of the measurement time and thus a representation which is closer to a "real" snapshot.

- A more sophisticated optical model including light scattering (and possibly reflection) by particles should increase the accuracy of the tomograms and allow the use of a wider range of particle materials. This would alleviate the following limitations of the current optical model: (1) in the current model, when particles which do significantly scatter and/or reflect light are studied, more light from a single source reaches all light sensors in varying amounts, depending on the particle distribution. The method tries to explain this in terms of the applied model, i.e. a smaller amount of light being blocked by particles on the modeled beam paths, meaning fewer particles in these paths. (2) When high particle concentrations are used, particles start to be in each other's shadows, more so on longer beam paths than on shorter paths. This results in a smaller amount of light intensity reduction for long beams (through the centre of the domain) per particle than for shorter beams (along the wall). As the optical model assumes equal light blocking for each particle regardless of the concentration, a uniform actual particle concentration throughout the domain would result in a tomogram with a somewhat higher light absorption coefficient along the walls than in the center.

- Finally, the resolution of the obtained images can be increased by increasing the number of light paths through the number of LEDs and/or sensors.

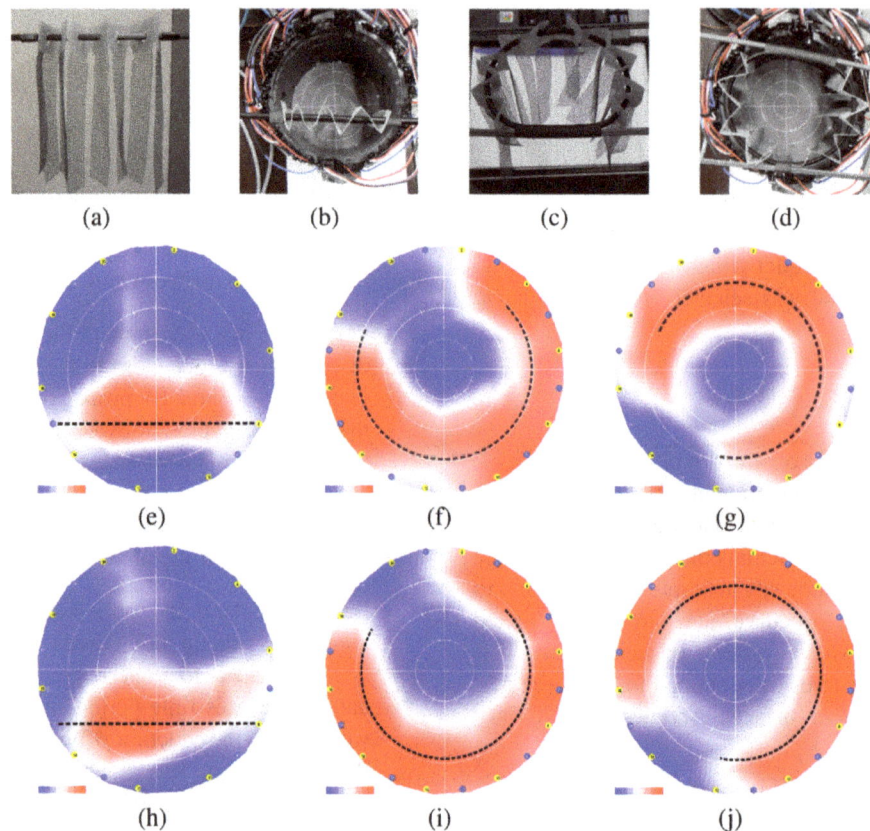

Figure 5. Dry test results of the optical tomography device. (**a**) Linear wire mesh curtain used to represent a particle cloud; (**b**) positioning of the linear curtain; (**c**) curved wire mesh curtain; (**d**) positioning of the curved curtain; (**e**) inversion result for linear curtain (single beam path); (**f, g**) inversion results for curved curtain (single beam path), (**h**) inversion result for linear curtain (multiple beam paths); (**i, j**) inversion results for curved curtain (multiple beam paths). A dashed line indicates the actual position of the curtain in (**e–g**).

5   Conclusions

A method for studying particle processes in situ has been presented and tested. The following can be concluded:

- The mathematical framework presented here is suitable for obtaining meaningful images from light measurements.

- The technical implementation is capable of resolving semi-transparent objects in a test setting.

The presented methodology provides a useful and promising technique for the study of several areas in drinking water discolouration research. The first application of the device and methodology is presented in Floris et al. (2011).

Acknowledgements. We thank Karin van Thienen-Visser for help with the inversion procedure, Melanie Tankerville for reviewing the English of this paper, and Mirjam Blokker and two anonymous referees for helpful comments on the manuscript.

References

Arduino: http://www.arduino.cc, last access: 6 December 2010.

Bibb, R. and Winder, J.: A review of the issues surrounding three-dimensional computed tomography for medical modelling using rapid prototyping techniques, Radiography, 16, 78–83, 2010.

Bijwaard, H. and Spakman, W.: Tomographic evidence for a narrow whole mantle plume below Iceland, Earth Planet. Sc. Lett., 266, 121–126, 1999.

Blokker, E., Vreeburg, J., Schaap, P., and van Dijk, J.: The self-cleaning velocity in practice, in: Water Distribution System Analysis conference, Tucson, Arizona, 2010.

Brass, M. E.: Handbook of Optics. Volume I: Geometrical and Physical Optics, Polarized Light, Components and Instruments, McGraw-Hill, third edn., 2010.

Floris, R. and van Thienen, P.: Experimental investigation of turbulent particle radial transport processes in DWDS using optical tomography, Drink. Water Eng. Sci. Discuss., 4, 61–83, doi:10.5194/dwesd-4-61-2011, 2011.

Fokker, P. A., Visser, K., Peters, E., Kunakbayeva, G., and Muntendam-Bos, A. G.: Inversion of Surface Subsidence Data to Quantify Reservoir Compartmentalization: A Field Study, in: Proceedings of the Society of Petroleum Engineers Annual Technical Conference and Exhibition, Florence, Italy, 2010.

Husband, P. S. and Boxall, J. B.: Asset deterioration and discolouration in water distribution systems, Water Res., 45, 113–124, 2011.

Morgan, J. V., Christeson, G. L., and Zelt, C. A.: Testing the resolution of a 3D velocity tomogram across the Chicxulub crater, Tectonophysics, 355, 215–226, 2002.

Muntendam-Bos, A. G., Kroon, I. C., and Fokker, P. A.: Time-dependent Inversion of Surface Subsidence due to Dynamic Reservoir Compaction, Math. Geosci., 40, 159–177, doi:10.1007/s11004-007-9135-3, 2008.

Phidgets: http://www.phidgets.com, last access: 6 December 2010.

Tarantola, A.: Inverse Problem Theory and Methods for Model Parameter Estimation, Society for Industrial and Applied Mathematics, Philadelphia, 2005.

van Thienen, P. and Vreeburg, J.: Turbulent processes in drinking water distribution, in: Water Distribution System Analysis conference, Tucson, Arizona, 2010.

van Thienen, P., Floris, R., Vreeburg, J. H. G., and Blokker, E. J. M.: Lab experiments on turbulent processes causing discolouration potential, in: Proceedings of the World Environmental and Water Resources Congress, Palm Springs, California, 2011a.

van Thienen, P., Vreeburg, J., and Blokker, E.: Radial transport processes as a precursor to particle deposition in drinking water distribution systems, Water Res., 45, 1807–1818, 2011b.

Vreeburg, J. and Boxall, J.: Discolouration in potable water distribution systems: A review, Water Res., 41, 519–529, 2007.

Zheng, Y., Liu, Q., Li, Y., and Gindy, N.: Investigation on concentration distribution and mass flow rate measurement for gravity chute conveyor by optical tomography system, Measurement, 39, 643–654, 2006.

15

Technical Note: Wet validation of optical tomography for drinking water discolouration studies

R. Floris, P. van Thienen, and H. Beverloo

KWR Watercycle Research Institute, P.O. Box 1072, 3430 BB Nieuwegein, the Netherlands

Correspondence to: R. Floris (roberto.floris@kwrwater.nl)

Abstract. This paper presents a set of validation experiments for the reconstruction of a cross-sectional particle concentration field in a transparent pipe filled with a fluid using optical tomography.

1 Introduction

Discoloured water still represents the second most common reason (after not getting any water) of customers' complaints to water companies (Vreeburg and Boxall, 2007; Husband and Boxall, 2011).

The origin of discoloured water is correlated with the presence of waterborne particles into the drinking water distribution system for which fate and behaviour are to a large degree controlled by the flow conditions (Husband et al., 2008). Changes in the flow velocity modify the magnitude of the forces acting upon the particles and as a consequence affect the radial transport mechanisms leading to deposition or resuspension. Theoretical investigations were conducted in order to understand the relation between flow conditions and particle transport mechanisms and how they affect the cross-sectional particle distribution (van Thienen et al., 2011b). An optical tomography system was developed and tested for the first time in drinking water research towards an experimental verification of these theoretical findings (van Thienen et al., 2011a). More specifically, it was aimed at the detection of the transition from turbulent diffusion dominated radial transport to turbophoresis dominated radial particle transport. The former mechanisms works to homogenize the particle concentration field, whereas the latter drives particles towards the wall. These particle configurations were recreated in the lab for the validation of the tomography system (see Fig. 1).

Tomography is a non-invasive technique of imaging that allows the study of the internal structures of an object or a system without influencing or conditioning it. The common basic principle of these techniques is to analyze the interaction between the system studied and a physical field and then to translate this information into images of the system under study.

The optical tomography technique was used here to reconstruct cross-sectional particle concentration fields in a transparent pipe using the interaction between light and particles suspended in a fluid. van Thienen et al. (2011a) give a full description of the applied methodology, including synthetic tests of the mathematical procedure and dry tests of the actual device, using wire mesh curtains to represent particle clouds in the measurement domain. Following their recommendation, a more complex optical model was implemented in the inversion procedure, which is described below.

Although the previous results of van Thienen et al. (2011a) showed the validity of the mathematical framework and capability of the optical tomography system in resolving semi-transparent objects, test conditions which are more representative of the actual intended application of the device are necessary. Because cross-sectional particle distributions in flowing water in a pipe are difficult to control, wet quasi-static tests in a highly viscous fluid were performed in order to asses the value of the tomograms obtained from measurements. Here we present the validation of the system distinguishing different particle clouds patterns expected by the theoretical model. Also, a suitable value for the system variance as used in the computation of the tomograms was determined from the observations.

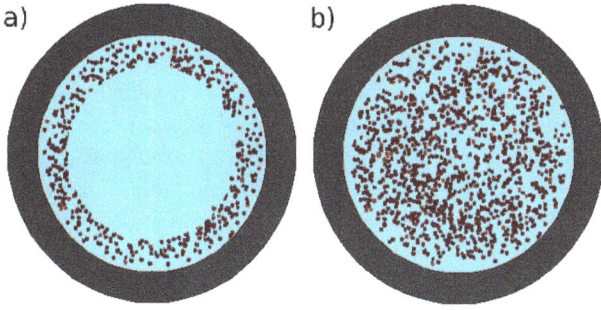

Figure 1. A schematic representation of the two model patterns created for the validation of the tomography system. (**a**) centered full ring pattern 1 cm thick; (**b**) homogeneous particle distribution.

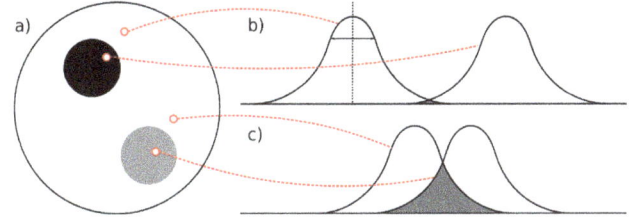

Figure 2. Tomogram showing two distinct circular features (**a**). The first of these is very different from the background, resulting in a small overlap (**b**) of the associated probability distributions for the two sampled points. The bottom feature is much closer to the background and has a larger overlap (**c**) of the associated probability distributions for the two sampled points.

2 Methods

2.1 Optical tomography

van Thienen et al. (2011a) give a complete description of the methodology and optical tomography measuring device (OTD), which is briefly summarized here. The OTD consists of a PVC ring containing, evenly spaced and alternatingly, 10 high power light emitting diodes (LED) and 10 light sensors (LS). The LEDs and the LSs are controlled by an Arduino board (Arduino, 2010) on which is installed custom made control software. The PVC ring is mounted on a transparent section of pipe which is part of a lab test rig, in which hydraulic conditions relevant for drinking water distribution can be simulated. The device generates light pulses using the LEDs and detects the result of the interaction between these pulses and the suspension using the light sensors. The LEDs are operated sequentially and for every light pulse data are collected by every LS. The data collected are transmitted to the workstation which computes a tomogram of the light absorption coefficient field with an inverse problem algorithm (van Thienen et al., 2011a). The tomogram is a two-dimensional image representing the variation of the light absorption coefficient representative of the local particle concentration through a cross-section of the suspension flow.

The mathematical framework of the tomography is based on work by Tarantola (2005) and Muntendam-Bos et al. (2008) and revolves around the following equation:

$$s = \mathbf{C}_\mathrm{m}\mathbf{A}^T(\mathbf{A}\mathbf{C}_\mathrm{m}\mathbf{A}^T + \mathbf{C}_\mathrm{d})^{-1}r \qquad (1)$$

In this expression, vector r contains all measurements, vector s has all unknown absorption coefficients and matrix $\mathbf{A}$ contains all integration coefficients. $\mathbf{C}_\mathrm{m}$ is the covariance matrix of the prior model matrix and $\mathbf{C}_\mathrm{d}$ is the prior data covariance matrix, which has the variances σ^2 of the measurements on the main diagonal and zeros off the main diagonal. For a more detailed description, the reader is referred to van Thienen et al. (2011a). The total variances consist of the actual measurement variances σ^2_{mij} which are obtained for each LED-sensor combination ij from multiple samples and a sys-

tem variance σ^2_s, which are summed following Bienaym's formula:

$$\sigma^2_{ij} = \sigma^2_{mij} + \sigma^2_\mathrm{s} \qquad (2)$$

Using this mathematical framework, a mathematical solution can always be found for the inversion problem (i.e. the reconstruction of the cross-sectional particle concentration field from the light measurements). The question, however, is to what extent the obtained image represents physical features instead of numerical artifacts and noise. The applied method not only computes a model for the physical configuration, but also the corresponding posterior variances for the model values at all nodes. When a feature is observed in a tomogram, the variances can answer the question whether the feature is statistically significant, i.e. whether it is really different from the background. A small overlap of the probability density functions (Fig. 2a, b) signifies a low likelyhood of the values actually being the same, whereas a large overlap (Fig. 2a, c) suggests a high probability. By choosing a confidence level, e.g. 95 %, a maximum amount of overlap for which values are considered different is obtained, in this case 0.05.

The main parameters affecting the width of the probability distributions are the different components of the prior data variance (van Thienen et al., 2011a). These are the measurement component σ_m and the system component σ_s. The former parameter is computed from a series of measurements and represents the noise in the measurements. All other errors and uncertainties in the system, i.e. in the discretization, systematic errors in measurement and analog-digital conversion, etc. are included in the latter. This parameter is difficult to determine exactly. Therefore, it should be estimated on the basis of validation experiments such as those presented in this paper. In these validation experiments, the physical structure which is reconstructed in the tomogram is known, so the appropriate value for σ_s can be selected which results in an accurate reconstruction without artefacts.

Figure 3. Mounting of the optical tomograohy device on a vertical end of transparent pipe.

2.1.1 Enhancement of the optical model

The optical model was improved relative to the approach described in van Thienen et al. (2011a). Originally, it was assumed that particles are never in the shadows of other particles, which means that the light intensity decreases linearly with the number of particles encountered. This is a reasonable approximation for low particle concentrations, but not for the higher concentrations which are applied in this work. Here, we apply an exponential extinction model based on the Lambert-Beer law:

$$\mathrm{d}I = -\frac{\sigma N}{\mathrm{d}x} \tag{3}$$

with I the light intensity, σ the absorption cross section and N the number density of absorbers, assuming a linear variation of the particle concentration between model nodes.

2.1.2 Experimental set-up

The wet quasi-static tests were performed mounting the Optical Tomography Device on a vertical transparent tube containing a glycerol solution (80 % in volume), see Fig. 3. Used coffee granules (i.e. the leftovers after making coffee) with a particle size of 500–600 µm were chosen as a testing material because of three factors: their density (900–1100 kg m^{-3}), which is close to the density of sediment samples collected from DWDSs (Boxall et al., 2001) their optical properties (high light absorption and low light reflection) and their physical and chemical stability after extraction (the soluble substances have already been extracted in the process of making coffee). Particles were left in the glycerol solution overnight before the experiments in order to obtain a stable dispersion at high concentration (400 g L^{-1}). The OTD was placed far from the free surface of the liquid (15 cm) in order to avoid artefacts originating from light reflections at this interface. Two model cross-sectional particle concentration distributions (a centered full ring pattern 1 cm thick and homogeneous pattern) were created injecting the high concentrated coffee dispersion into the clear glycerol solution using a syringe. A schematic representation of the two patterns is given in Fig. 1. Complete screening of the experimental setup from external light sources was ensured during the measurements.

3 Results

Tomograms for a single selected representative frame for each of the two experimental configurations described above are presented in Fig. 4. These show tomograms for four different values of the system error σ_s (0.08, 0.12, 0.15 and 0.20). Also included are probability density distributions for four selected nodes and their respective amounts of overlap.

4 Discussion

As can be seen in Fig. 4, both the ring structure and the homogeneous structure can be clearly resolved when an appropriate value of σ_s is chosen. For the ring structure to be statistically significant in the tomogram, i.e. for the outside part of the tomogram to be significantly different (95 % confidence interval) from the inside, we find the requirement that $\sigma_s \leq 0.15$. For the homogeneous structure to be statistically significant in the tomogram, i.e. for the outside part of the tomogram to be not significantly different (95 % confidence interval) from the inside, the requirement is found that $\sigma_s \geq 0.12$. Therefore, there is a range of σ_s between 0.12 and 0.15 where a statistically significant distinction can be made between ring structures and homogeneous structures for the applied particle concentration. The particle concentration which was used is 400 g L^{-1}. Some local mixing after injection resulted in a slightly lower effective particle concentration. When this value is also strived for in dynamic

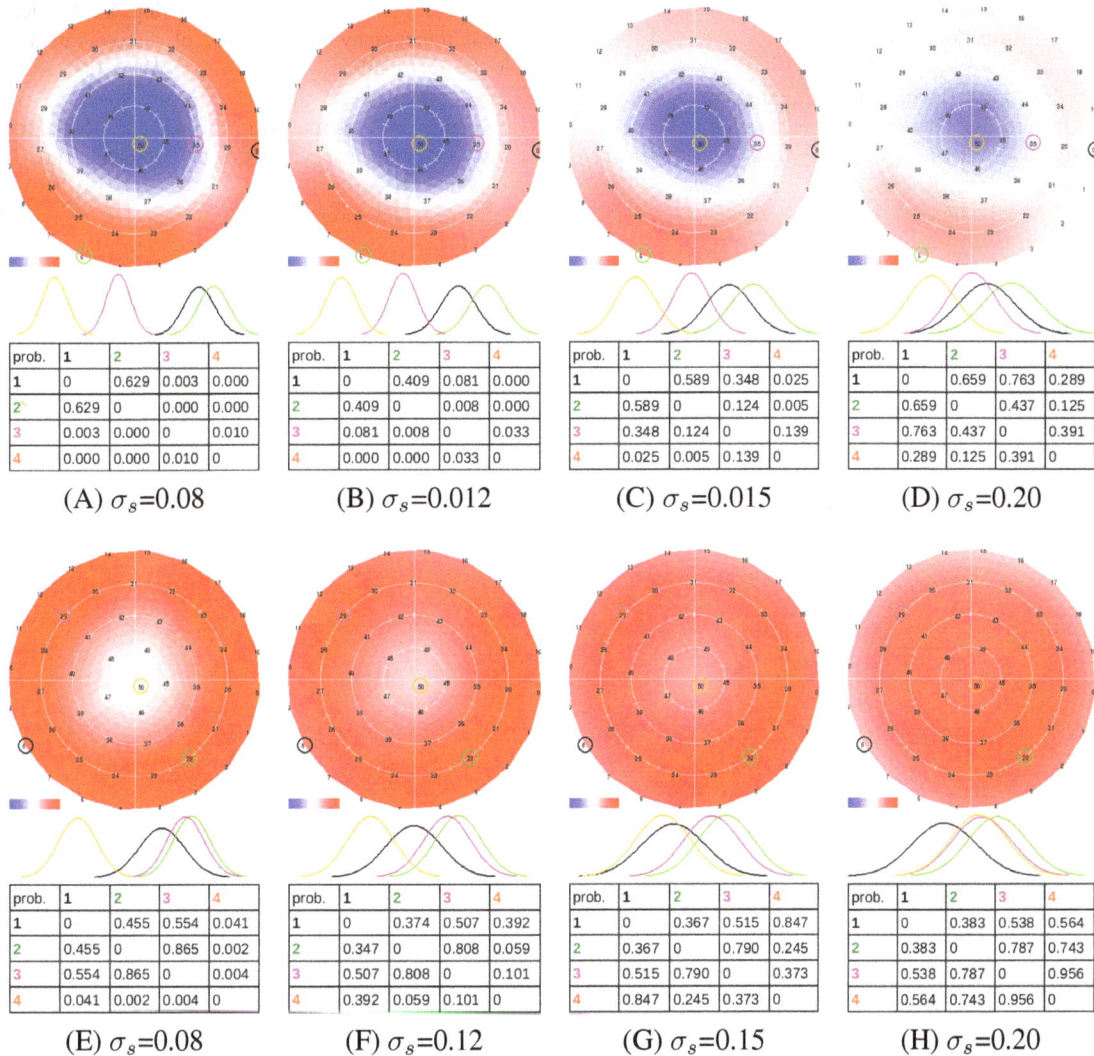

prob.	1	2	3	4
1	0	0.629	0.003	0.000
2	0.629	0	0.000	0.000
3	0.003	0.000	0	0.010
4	0.000	0.000	0.010	0

(A) σ_s=0.08

prob.	1	2	3	4
1	0	0.409	0.081	0.000
2	0.409	0	0.008	0.000
3	0.081	0.008	0	0.033
4	0.000	0.000	0.033	0

(B) σ_s=0.012

prob.	1	2	3	4
1	0	0.589	0.348	0.025
2	0.589	0	0.124	0.005
3	0.348	0.124	0	0.139
4	0.025	0.005	0.139	0

(C) σ_s=0.015

prob.	1	2	3	4
1	0	0.659	0.763	0.289
2	0.659	0	0.437	0.125
3	0.763	0.437	0	0.391
4	0.289	0.125	0.391	0

(D) σ_s=0.20

prob.	1	2	3	4
1	0	0.455	0.554	0.041
2	0.455	0	0.865	0.002
3	0.554	0.865	0	0.004
4	0.041	0.002	0.004	0

(E) σ_s=0.08

prob.	1	2	3	4
1	0	0.374	0.507	0.392
2	0.347	0	0.808	0.059
3	0.507	0.808	0	0.101
4	0.392	0.059	0.101	0

(F) σ_s=0.12

prob.	1	2	3	4
1	0	0.367	0.515	0.847
2	0.367	0	0.790	0.245
3	0.515	0.790	0	0.373
4	0.847	0.245	0.373	0

(G) σ_s=0.15

prob.	1	2	3	4
1	0	0.383	0.538	0.564
2	0.383	0	0.787	0.743
3	0.538	0.787	0	0.956
4	0.564	0.743	0.956	0

(H) σ_s=0.20

Figure 4. Tomograms, posterior probability density functions and associated amounts of overlap for selected nodes in the tomograms. Selected node markers and associated curves are marked in the same colour in each case. Both a ring structure: (**A–D**) and a homogeneous case (**E–H**) are displayed.

experiments (in contrast to the current quasi-static) in which the optical tomograph is applied, the same resolvability of the ring and homogeneous structures is expected. Finally, when applying the insights into statistical significance presented in the technical note to the results of the dry validation experiments of the optical tomograph in van Thienen et al. (2011a), we find that the observed linear and annular structures in that work are also statistically significant.

assuming a system standard deviation $0.12 \le \sigma_s \le 0.15$. Therefore, it may be applied to experimentally validate location of the transition between turbulent diffusion and turbophoresis dominated radial particle transport predicted from theory by van Thienen et al. (2011b).

5 Conclusions

The optical tomography setup described in van Thienen et al. (2011a) is capable of distinguishing (95 % confidence interval) between ring shaped and homogeneous particle concentration fields in transparent pipe sections sampled by the tomograph at particle concentrations of $400\,\mathrm{g\,L^{-1}}$ when

References

Arduino: http://www.arduino.cc, last access: 6 December 2010.

Boxall, J. B., Skipworth, P. J., and Saul, A. J.: A novel approach to describing sediment movement in distribution mains, based on measured particle characteristics, Proc. International CCWI conference, 3–5 September 2001, De Montfort University, UK, 2001.

Husband, P. S. and Boxall, J. B.: Asset deterioration and discolouration in water distribution systems, Water Res., 45, 113–124,
2011.

Husband, P. S., Boxall, J. B., and Saul, A.: Laboratory studies investigating the processes leading to discolouration in water distribution networks, Water Res., 42, 4309–4318, 2008.

Muntendam-Bos, A. G., Kroon, I. C., and Fokker, P. A.:
Time-dependent Inversion of Surface Subsidence due to Dynamic Reservoir Compaction, Math. Geosci., 40, 159–177,
doi:10.1007/s11004-007-9135-3, 2008.

Tarantola, A.: Inverse Problem Theory and Methods for Model Parameter Estimation, Society for Industrial and Applied Mathematics, Philadelphia, 2005.

van Thienen, P., Floris, R., and Meijering, S.: Application of optical tomography in the study of discolouration in drinking
water distribution systems, Drink. Water Eng. Sci., 4, 61–69,
doi:10.5194/dwes-4-61-2011, 2011a.

van Thienen, P., Vreeburg, J., and Blokker, E.: Radial transport processes as a precursor to particle deposition in drinking water distribution systems, Water Res., 45, 1807–1818, 2011b.

Vreeburg, J. and Boxall, J.: Discolouration in potable water distribution systems: A review, Water Res., 41, 519–529, 2007.

16

Method development for arsenic analysis by modification in spectrophotometric technique

M. A. Tahir[1], H. Rasheed[1], and A. Malana[2]

[1]National Water Quality Laboratory, Pakistan Council of Research in Water Resources, Kheyaban-e-Johar, H-8/1, Islamabad, Pakistan
[2]Department of Chemistry, Bahawuddin Zikriya University, Multan, Pakistan

Correspondence to: M. A. Tahir (matahir2k@hotmail.com)

Abstract. Arsenic is a non-metallic constituent, present naturally in groundwater due to some minerals and rocks. Arsenic is not geologically uncommon and occurs in natural water as arsenate and arsenite. Additionally, arsenic may occur from industrial discharges or insecticide application. World Health Organization (WHO) and Pakistan Standard Quality Control Authority have recommended a permissible limit of 10 ppb for arsenic in drinking water. Arsenic at lower concentrations can be determined in water by using high tech instruments like the Atomic Absorption Spectrometer (hydride generation). Because arsenic concentration at low limits of 1 ppb can not be determined easily with simple spectrophotometric technique, the spectrophotometric technique using silver diethyldithiocarbamate was modified to achieve better results, up to the extent of 1 ppb arsenic concentration.

1 Introduction

Arsenic is steel grey, very brittle, crystalline in nature and oxidizes on rapid heating to arsenous oxide with an odor of garlic. Arsenic exists as inorganic and organic compounds. In the environment, it combines with oxygen, chlorine and sulfur to form inorganic arsenic compounds. Arsenic also combines with carbon and hydrogen to form organo-arsenic compounds in animals and plants. Inorganic arsenic compounds are mainly used to preserve wood, and organic arsenic compounds are used as pesticides – primarily for cotton crop (Carapella, 1973; Calvert, 1975). Therefore, arsenic can be released into the environment from sources such as pesticides applications, wood preservatives, mining activities and petroleum refining. It is found exclusively as arsenite (Arsenic-III) or arsenate (Arsenic-V) in groundwater. Arsenite can be converted to arsenate under oxidizing conditions (e.g. well-aerated surface water). Likewise, arsenate can become arsenite under reducing conditions (e.g. anaerobic groundwater).

The groundwater pollution caused by arsenic in various countries of the world has led to major environmental issues. A number of networks supplying water in the United States, particularly in the Midwest and West regions, contain arsenic. In West Bengal of India, arsenic found in groundwater at several places is a hundred times above the permissible limits set for drinking water. The situation in Bangladesh is even worse than West Bengal. In many areas, arsenic contamination is found above $3000\,\mu g^{-1}$ compared to the recommended level of $10\,\mu g^{-1}$ (Guha Mazumder et al., 1988). Similarly, arsenic contamination is observed in Argentina, Canada, Chile, China, Greece, Japan, Mexico, Mongolia, New Zealand, South Africa, Philippines, Taiwan, Thailand and countries of the former USSR. Therefore, the arsenic poisoning is emerging as a global issue.

Arsenic monitoring is utmost important nowadays. Palmer (2001) reported atomic spectroscopy is the most widely-used method for the arsenic determination. Atomic spectroscopy involves use of the absorption characteristic of metals (Andreae, 1977; Christian and Feldman, 1970; Chu et al., 1972; Clement et al., 1973; Fishman and Spencer, 1977). USEPA (1999) reported that GFAAS (graphite furnace atomic absorption spectrometry) is an approved method by USEPA for measuring arsenic in drinking water. The

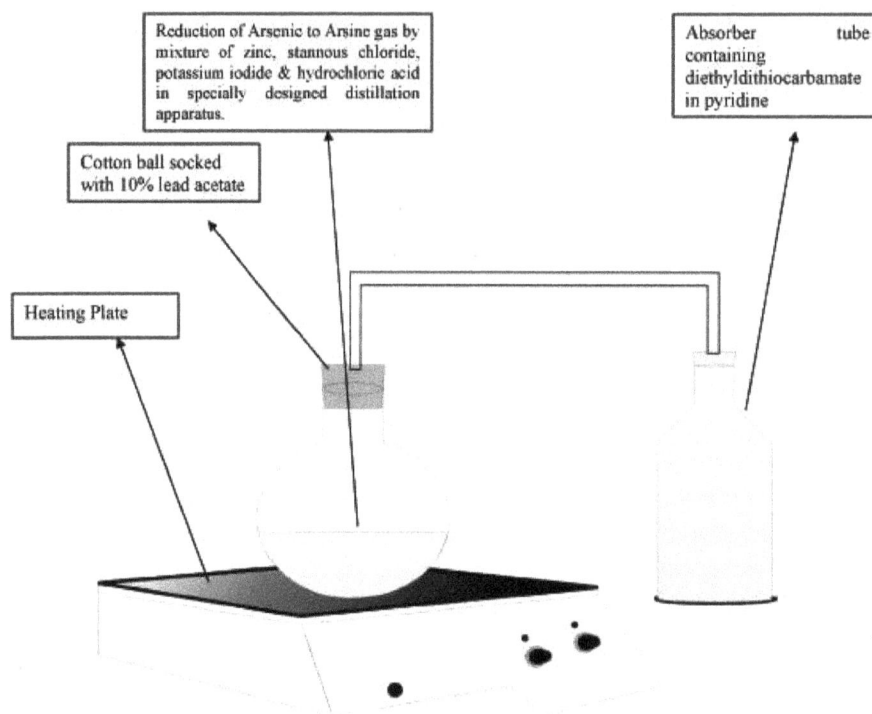

Figure 1. Arsine generation system.

detection limit for this method is 1–5 ppb. USEPA (2003) reported on the use of an ICP-AES (inductively coupled plasma atomic emission spectroscopy) instrument for the same purpose. It utilizes optical spectrometry to measure the characteristic atomic emission spectra of the analyte.

For arsenic analysis, detection limits are required to be very low (≤1 ppb). It can be achieved only by state-of-the-art, latest versions of equipment like the atomic absorption spectrometer or ICP. Such high-tech equipment facilities and the trained technical manpower to operate this equipment are lacking in most laboratories in the country. Spectrometers are available in most water quality laboratories; however, arsenic analysis in drinking water is quite difficult with such equipment due to the required low detection limits for arsenic. Therefore, a more practicable method for the analysis of arsenic at low detection limits on spectrophotometer was developed, which will be useful for all simple and modest laboratories.

2 Methodology

Analytical methods for inorganic arsenic are reported in the reference handbook (Michael, 1982) "Evaluation of Water for Pollution Control". The reference procedure is a photometric measurement using silver diethyldithiocarbamate. Arsenic analysis with the help of atomic absorption spectrometer is reported as a secondary method. We report on how the reference method was modified to some extent to get better

results owing to present research work. The principle of the modified method is based on reduction of arsenic to arsine gas by a mixture of zinc, stannous chloride, potassium iodide and hydrochloric acid in a specially-designed distillation apparatus. The arsine (AsH_3) is passed through a scrubber containing cotton saturated with lead acetate and then into an absorber tube containing silver diethyldithiocarbamate in pyridine. The arsenic reacts to form a red complex, which can be read on the spectrometer.

Apparatus for the experiment consists of mainly the arsine generator, the absorption tube and the Spectrophotometer (U-1100), Hitachi, Japan. Reagents consist of: arsenic standard solution BDH, UK, Hydrochloric Acid (ASC); lead acetate solution (10 %); potassium iodide solution (20 %); pyridine silver diethyldithiocarbamate (SDDC); stannous chloride solution (40 %); and zinc (0.3–1.5 mm or 14–50 mesh).

Reagent preparation for this experiment is done in the following way:

1. 1.25 g of silver diethyldithiocarbamate (SDDC) was transferred into a 250 ml volumetric flask and mixed well. The reagent was stored in an amber bottle. This reagent was used as an absorber for the arsenic.

2. Standards of 0, 5, 15, 25, 35 and 45 ppb or µg^{-1} arsenic were prepared from 1 ppm standard (BDH Cat. No. 455042K, Lot No. 105016109, UK).

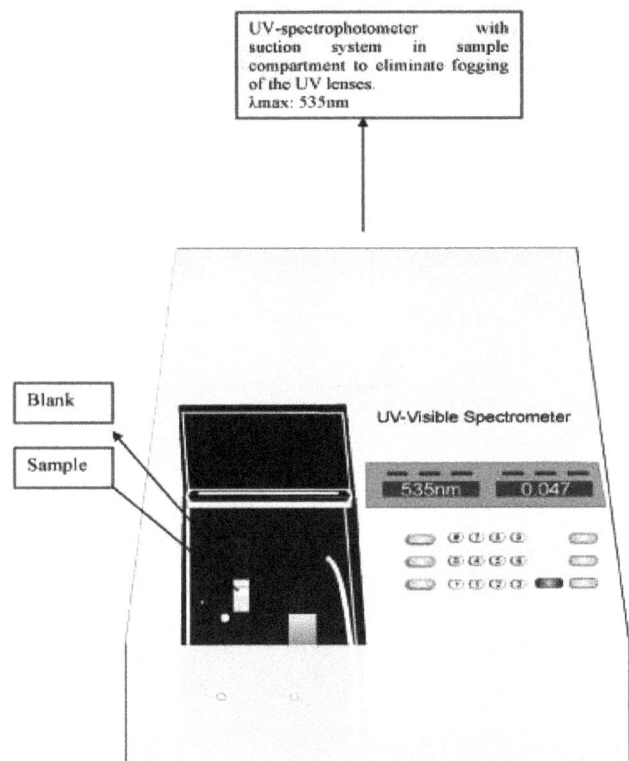

Figure 2. Suction system attached to spectrophotometer to rectify fogging problem.

3 Procedure (modified technique)

For evaluation of this modified technique, the standard addition technique was used to prepare arsenic-fortified solutions. For the addition of 20 ppb arsenic in each sample, five ml of 1-ppm arsenic standard was transferred to a 250 ml measuring flask; the flask was then filled to the mark with the water sample. After proper mixing, this sample was transferred into the arsenic generation or distillation apparatus (Fig. 1) under the fume hood to vent toxic fumes. A cotton ball soaked with 10 % lead acetate solution was placed in the gas scrubber. Prepared arsenic absorber solution (25 ml) was transferred into the gas bubbler assembly, which was then attached to the distillation apparatus. Then 25 ml HCl, 1 ml of stannous chloride solution and 3 ml potassium iodide solution were added to the flask, respectively. After 12 min, 6 g of 14–50 mesh-sized zinc was also added to flask. The temperature was maintained at 40 °C for 12 min in the beginning after adding potassium iodide solution, 60 °C for another 12 min in the middle after adding zinc, and then set at 40 °C for 16 min with continuous stirring at the rate of 100 to 150 rpm.

After the completion of reaction time (about 40 min), a dry sample cell was filled with un-reacted arsenic absorber solution (the blank) and the other one filled with reacted arsenic absorber. Finally, arsenic concentration was deter-

mined by using concentration mode on the pre-calibrated spectrophotometer at 535 nm (λmax.). The added concentration, i.e. 20 ppb, was deducted from final concentration calculations of each sample. Standard addition is a widely accepted technique for checking the validity of test results. It is known as "Spiking" and "Known addition". This technique is also used to check the performance of the reagents, instrument, apparatus, procedure and also to enhance the lower detection limit.

It was observed that fumes of arsenic absorber solution present in the 10 mm rectangular quartz cell with lid were responsible for fogging the lenses installed in the sample compartment, which were ultimately responsible for undesired results. Another source of unwanted results, especially in such a low concentration, is the electricity supplied to the spectrometer (variation in voltage). Making amendments in the equipment, as shown in Fig. 2, rectified the problem of fogging. Amendments included suction tube at upper side of the lenses (window) followed by a suction pump and a stabilizer. The problem of voltage variation was controlled with the help of the addition of a stabilizer as shown in Fig. 3.

4 Results and discussions

According to procedure, arsenic was reduced to arsine gas and formed a red complex with silver diethyldithiocarbamate. For the selection of suitable wavelength, absorbances and transmittances (%) were noted on spectrophotometer for known standard solutions, i.e. 0, 25, 35 and 45 ppb, after adopting the above procedure. Results are presented in Table 1.

From the results, it was concluded that maximum absorbances were noted at 535 nm in all cases. However, absorbances were found to be almost the same from 535 to 540 nm. The chances of interferences in this method are almost negligible except for antimony salts, which may interfere in color development. Therefore, 535 nm was selected as λmax as shown in Fig. 4.

Correlation between absorbance vs. known concentrations was developed with the help of the regression model in light of experimental data shown in Table 2. Graphical views (xy) are shown in Figs. 5 and 6. First and second views reflected relationship between absorbance and concentration with or without straight line. The degree of fitness or R square is near to 1, which indicated good relationship between x and y. The developed co-relation, degree of fitness or R square and value of constant, are given below:

Concentration of As = 983.4 X Absorbance − 0.8014
Degree of fitness or R_2 = 0.9958
Constant = −0.8014
X Coefficient = 983.4

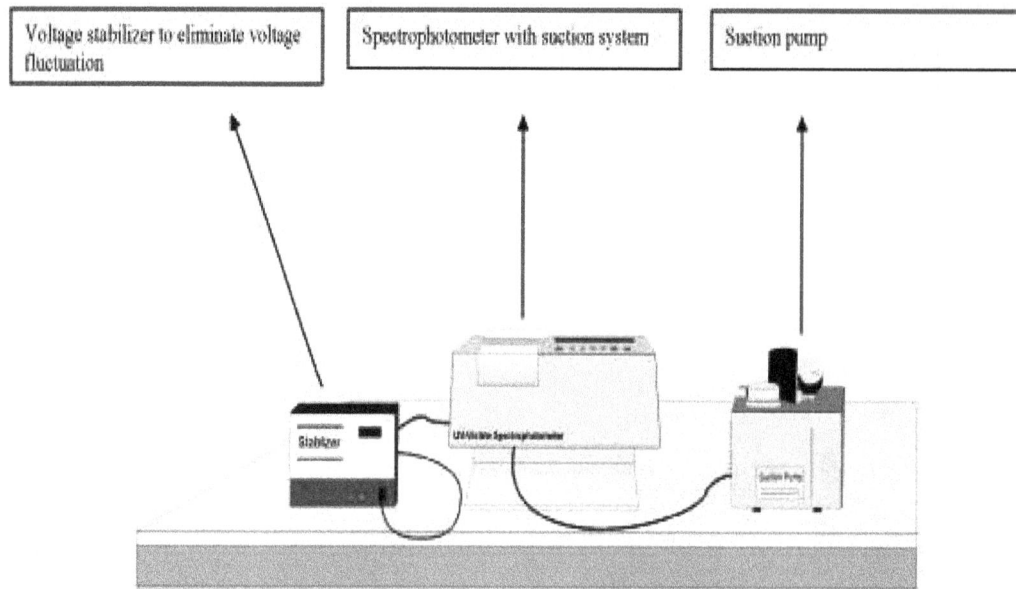

Figure 3. Suction system to rectify fogging problem and stabilizer to control voltage variation.

Table 1. Absorbance of standard solution at different λ range.

Standard Solution (nm)	0.000 ppb Abs.	T %	0.025 ppb Abs.	T %	0.035 ppb Abs.	T %	0.045 ppb Abs.	T %
520	0	0	0.015	96.6	0.026	94.2	0.037	91.8
521	0	0	0.016	96.4	0.026	94.2	0.037	91.8
522	0	0	0.016	96.4	0.027	94	0.038	91.6
523	0	0	0.017	96.2	0.028	93.7	0.039	91.4
524	0	0	0.018	95.9	0.029	93.5	0.041	91
525	0	0	0.019	95.7	0.03	93.4	0.042	90.8
526	0	0	0.02	95.5	0.031	93.2	0.043	90.6
527	0	0	0.021	95.3	0.032	92.9	0.043	90.6
528	0	0	0.022	95.1	0.033	92.7	0.044	90.4
529	0	0	0.023	94.9	0.034	92.5	0.045	90.2
530	0	0	0.023	94.9	0.034	92.5	0.045	90
531	0	0	0.024	94.7	0.035	92.3	0.046	90
532	0	0	0.025	94.4	0.035	92.3	0.046	90
533	0	0	0.025	94.4	0.035	92.3	0.046	90
534	0	0	0.025	94.4	0.035	92.3	0.046	90
535	0	0	0.026	94.2	0.036	92.1	0.047	89.7
536	0	0	0.026	94.2	0.036	92.1	0.047	89.7
537	0	0	0.026	94.2	0.036	92.1	0.047	89.7
538	0	0	0.026	94.2	0.036	92.1	0.047	89.7
539	0	0	0.026	94.2	0.036	92.1	0.047	89.7
540	0	0	0.026	94.2	0.036	92.1	0.047	89.7

Selection of Suitable Wavelength

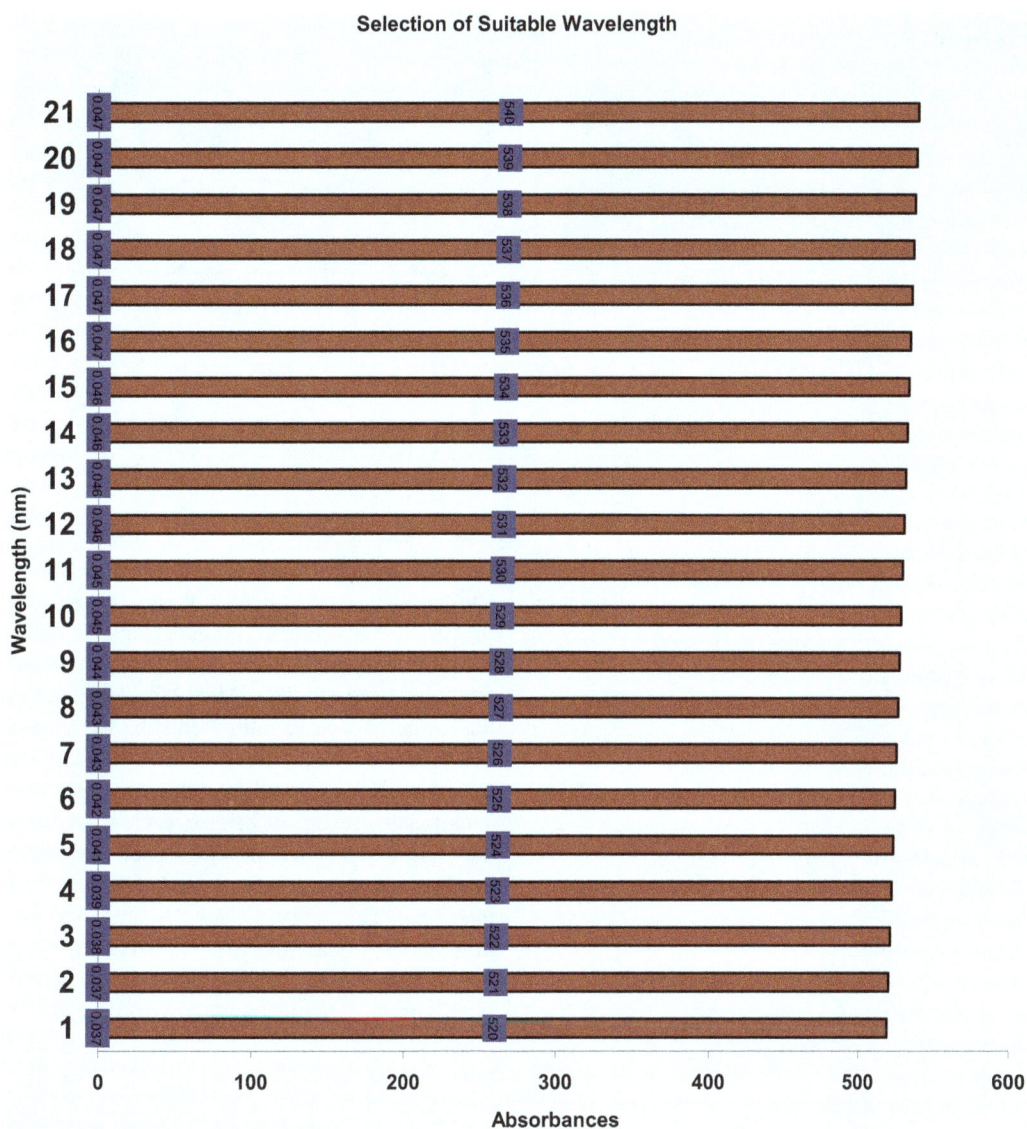

Figure 4. 0.45 ppb arsenic standard at different wavelengths.

Table 2. Arsenic concentration vs. absorbance.

Sr. No.	Concentration (ppb)	Absorbance
1	0	0
2	5	0.008
3	15	0.015
4	25	0.026
5	35	0.036
6	45	0.047

ARSENIC DETERMINATION IN WATER BY SILIVER DIETHYLDITHIOCARBAMATE METHOD

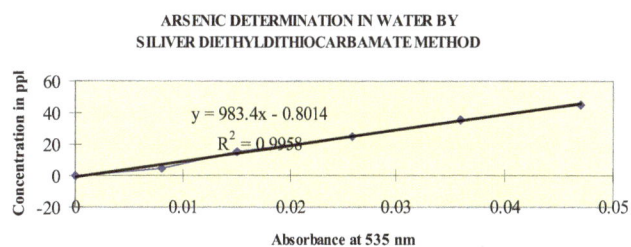

$y = 983.4x - 0.8014$

$R^2 = 0.9958$

Figure 5. Absorbance vs. concentration with straight line.

ARSENIC DETERMINATION IN WATER BY SILIVER DIETHYLDITHIOCARBAMATE METHOD

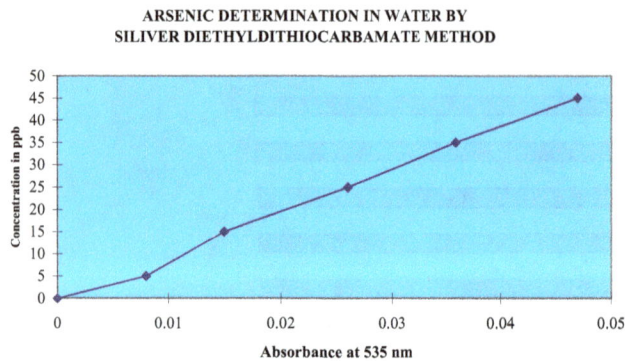

Figure 6. Absorbance vs. concentration without straight line.

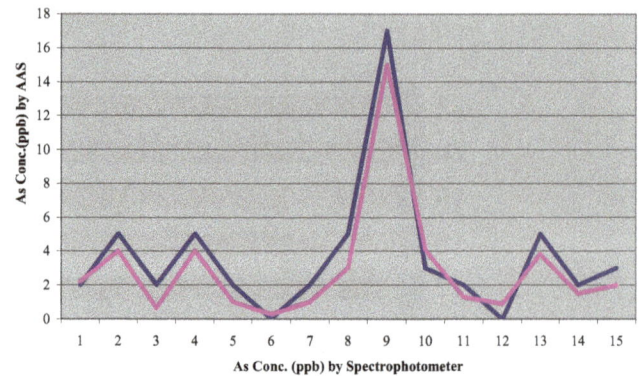

Figure 7. Comparison of arsenic analysis.

Table 3. Spectrometer vs. AAS (As Analysis).

Sr. #	Spectrometer (ppb)	AAS (ppb)
1	2	2
2	5	4
3	2	1
4	5	4
5	2	1
6	0	0
7	2	1
8	5	3
9	16	15
10	3	4
11	2	1
12	0	1
13	5	4
14	2	2
15	3	2

Fifteen water samples were analyzed by pre-calibrated spectrophotometer at 535 nm (λmax.) using the modified method on concentration mode. The same numbers of samples were analyzed by atomic absorption spectrometer. The results can be seen in Table 3 and graphical view is shown in Fig. 7. The difference in concentration is ±1 to 2 ppb, which is not noteworthy at very lower concentrations. The comparison of this modified technique with other commonly used spectrometric techniques, i.e. APHA Method, HACH and WHO, is given in Table 4.

5 Conclusions

The combination of high toxicity and widespread occurrence of arsenic has created a pressing need for effective monitoring and measurement of arsenic in soil and groundwater. Technological advances in a variety of analytical instruments have made improvements in accuracy and detection limit. However, the development of a cost-effective and reliable technique for arsenic determination by using a comparatively inexpensive instrument like spectrometer is what is needed at this time, especially in developing countries facing an arsenic contamination problem. Most laboratories in these countries lack state-of-the-art equipment like the Atomic Absorption Spectrometer (AAS) or the Inductive Coupled Plasma Spectrometer (ICP) to analyze arsenic at low detection levels. This paper presents a brief overview of the scientific literature on existing technologies used for arsenic analysis in the groundwater and also includes research developments in this area. The World Health Organization (WHO) has recommended a guideline value of 10 ppb for arsenic, which is analyzed in most of laboratories on high tech equipment like AAS and ICP due to low this detection limit. These instruments need trained technical manpower to operate and maintain. Many laboratories are unable to analyze arsenic due to lack of such equipment and manpower. Spectrophotometer is a commonly available instrument in most of laboratories. Considering this fact, the modified spectrophotometric method presented here has been developed for the analysis of arsenic at low detection limit. The comparative analytical evaluation on spectrophotometer with modified technique vs. Atomic Absorption Spectrometer reveals almost similar results (±1 to 2 ppb). Hopefully this modified method will be useful to enhance the analytical capabilities with respect to arsenic determination in most laboratories.

Acknowledgement. The authors are grateful to Mr. Ignaz Worm for his help editing of this manuscript.

Table 4. Comparison of modified technique with other spectrometric methods.

Steps	Modified Analytical Technique (2002)	HACH Method (1997)	Standard Methods (APHA, 1992)	Reference Method by WHO (Michael, 1982) and Standard Methods (APHA, 1971)
I	Dampen a cotton ball in 10 % lead acetate solution and put it at appropriate place.	Same	Same	Same
II	25 ml of prepared arsenic absorber solution into absorber tube.	25 ml prepared arsenic absorber solution into absorber tube.	4 ml of SDDC into absorber tube.	Same
III	5 ml of 1-ppm arsenic standard added into 250 ml flask and volume make up with sample.	No standard addition	No standard addition	No standard addition
IV	Volume of sample ≈245 ml	250 ml sample taken.	70 ml sample taken.	35 ml sample taken.
V	Stirring (100–150 rpm) continued up to step-XII.	Stir control set to 5 and heat control to 0.	Stirring continued up to Step-XI.	No stirring.
VI	Added 25 ml conc. HCl into generation flask of 250 ml volume	Added 10 ml acetate buffer and flushed with Nitrogen gas at 60 ml min^{-1}	Added 10 ml acetate buffer and flushed with Nitrogen gas at 60 ml min^{-1}	Added 5 ml conc. HCL into generation flask.
VII	Added 1 ml of stannous chloride.	15 ml of 1 % sodium borohydride injected within 2 min.	15 ml of 1 % sodium borohydride injected within 2 min.	Added 0.40 ml of stannous chloride.
VIII	Added 3 ml of potassium iodide solution.	Additional N$_2$ gas flushing for 15 min.	Additional N$_2$ gas flushing for 15 min.	Added 2 ml. Of potassium iodide solution.
IX	12 min given as "Reaction Time" at 40 °C.	15 min given as reaction time.	–	15 min given as reaction time without heat control
X	6 g zinc added (0.3–1.5 mm or 14–50 mesh size)	6 g zinc added (20 mesh size)	–	Added 3 g of zinc.
XI	12 min given as reaction time at 60 °C.	15 min given as reaction time and heat control set at 3.	–	30 min given as reaction time without heat control.
XII	16 min given as reaction time at 40 °C.	15 min given as reaction time and heat control set at 1.	–	–
XIII	Measurement of blank and sample on spectrophotometer at 535 nm.	Measurement of blank and sample on spectrophotometer at 520 nm.	Measurement of blank and sample on spectrophotometer at 520 nm.	Measurement of blank and sample on spectrophotometer at 535 nm.

References

Andreae, M. O.: Determination of arsenic species in natural waters, Anal. Chem., 49, 820–825, 1977.

APHA, AWWA and WEF: Standard Methods for the Examination of Water and Wastewater, 13th Edn., American Public Health Association, American Water Works Association and Water Environment Federation, Washington, DC, 62–66, 1971.

APHA, AWWA and WEF: Standard Methods for the Examination of Water and Wastewater, 18th Edn., American Public Health Association, American Water Works Association and Water Environment Federation, Washington, DC, 3–50, 1992.

Calvert, C. C.: Arsenicals in animal feeds and wastes, in: Arsenical Pesticides, edited by: Woolson, E. A., Washington, DC, American Chemical Society (ACS Symp. Series No. 7), 1975.

Carapella, S. C.: Arsenic and compounds, in: The encyclopedia of chemistry, third edn., edited by: Hampel, C. A. and Hawley, G. G., New York, Van Nostrand Reinhold Company, 1973.

Christian, G. D. and Feldman, F. J.: Atomic absorption spectroscopy; Applications in agriculture, biology, and medicine, New York, Wiley-Interscience, 188–195, 1970.

Chu, R. C., Barron, G. P., and Baumgarner, P. A.: Arsenic determination at sub-microgram levels by arsine evolution and flameless atomic absorption spectrophotometric technique, Anal. Chem., 44, 1476–1479, 1972.

Clement, W. H. and Faust, S. D.: A new convenient method for determining arsenic (+3) in natural waters, Environ. Lett., 5, 155–164, 1973.

Fishman, M. and Spencer, R.: Automated atomic absorption spectrometric determination of total arsenic in water and streambed materials, Anal. Chem., 49, 1599–1602, 1977.

Guha Mazumder, D. N., Chakraborty, A. K., and Ghose, A.: Chronic arsenic toxicity from drinking tubewell water in rural West Bengal, B. World Health Organ., 66, 499–506, 1988.

Michael, J. S.: Examination of water for pollution control, Vol. 2. First Edn., World Health Organization; 179–186, 1982.

Palmer, P. T.: A Review of Analytical Methods for the Determination of Mercury, Arsenic and Pesticides Residues on Museum Object, Department of Chemistry and Biochemistry, San Francisco State University, San Francisco, California, Collection Forum, 16, 25–41, 2001.

USEPA: Arsenic in Drinking Water: Analytical Methods, US Environmental Protection Agency, United States Environmental Protection Agency, Washington, DC, 1999.

USEPA: List of Drinking Water Contaminants & MCLS.EPA 816-F03-D, United States Environmental Protection Agency, Washington, DC, 2003.

17

Assessment of calculation methods for calcium carbonate saturation in drinking water for DIN 38404-10 compliance

P. J. de Moel[1,2]**, A. W. C. van der Helm**[2,3]**, M. van Rijn**[4]**, J. C. van Dijk**[2,5]**, and W. G. J. van der Meer**[2,6]

[1]Omnisys, Eiberlaan 23, 3871 TG, Hoevelaken, the Netherlands
[2]Delft University of Technology, Faculty of Civil Engineering and Geosciences, Department of Water Management, P.O. Box 5048, 2600 GA, Delft, the Netherlands
[3]Waternet, P.O. Box 94370, 1090 GJ, Amsterdam, the Netherlands
[4]Vitens, P.O. Box 1205, 8001 BE Zwolle, the Netherlands
[5]VanDijkConsulting, Rossenberglaan 9, 3833 BN, Leusden, the Netherlands
[6]Oasen, P.O. Box 122, 2800 AC, Gouda, the Netherlands

Correspondence to: P. J. de Moel (p.j.demoel@tudelft.nl)

Abstract. The new German standard on the calculation of calcite saturation in drinking water, DIN 38404-10, 2012 (DIN), marks a change in drinking water standardization from using simplified equations applicable for nomographs and simple calculators to using extensive chemical modeling requiring computer programs. The standard outlines the chemical modeling and presents a dataset with 10 water samples for validating used computer programs. The DIN standard, as well as the Standard Methods 2330 (SM) and NEN 6533 (NEN) for calculation of calcium carbonate saturation in drinking water were translated into chemical databases for use in PHREEQC (USGS, 2013). This novel approach gave the possibility to compare the calculations as defined in the standards with calculations using widely used chemical databases provided with PHREEQC. From this research it is concluded that the computer program PHREEQC with the developed chemical database din38404-10_2012.dat complies with the DIN standard for calculating Saturation Index (SI) and Calcite Dissolution Capacity (Calcitlösekapazität) or Calcium Carbonate Precipitation Potential (CCPP). This compliance is achieved by assuming equal values for molarity as used in DIN (obsolete) and molality as used in PHREEQC. From comparison with widely used chemical databases it is concluded that the use of molarity limits the use of DIN to a maximum temperature of 45 °C. For current practical applications in water treatment and drinking water applications, the PHREEQC database stimela.dat was developed within the Stimela platform of Delft University of Technology. This database is an extension of the chemical database phreeqc.dat and thus in compliance with SM. The database stimela.dat is also applicable for hot and boiling water, which is important in drinking water supply with regard to scaling of calcium carbonate in in-house drinking water practices. SM and NEN proved to be not accurate enough to comply with DIN, because of their simplifications. The differences in calculation results for DIN, SM and NEN illustrate the need for international unification of the standard for calcium carbonate saturation in drinking water.

1 Introduction

In general, calculation of calcium carbonate saturation in drinking water is performed with a simplification of the processes as shown in Table 1, in which ion pairs are neglected and calcite is assumed to be the determining crystalline phase. The equations for equilibrium constants in Table 1 do not apply to the concentrations of diluted species, but to their (relative) activities which are smaller than the concentrations because of their interaction with each other and with the surrounding water molecules (dipoles).

Table 1. Processes generally used for description of calcium carbonate ($CaCO_3$) saturation in drinking water.

Part	Reaction equations	Equilibrium constants
Dissolution/precipitation of $CaCO_3$	$CaCO_3\,(s) \leftrightarrow Ca^{2+} + CO_3^{2-}$	$K_s = \{Ca^{2+}\} \cdot \{CO_3^{2-}\}$
Dissociation of carbon dioxide $CO_2 + H_2O = H_2CO_3$	$CO_2 + H_2O \leftrightarrow HCO_3^- + H^+$ $HCO_3^- \leftrightarrow CO_3^{2-} + H^+$	$K_1 = \{HCO_3^-\} \cdot \{H^+\}/\{CO_2\}$ $K_2 = \{CO_3^{2-}\} \cdot \{H^+\}/\{HCO_3^-\}$
Ionisation of water	$H_2O\,(l) \leftrightarrow OH^- + H^+$	$K_w = \{H^+\} \cdot \{OH^-\}$ $\lg\{H^+\} = -pH$

$\{X\}$ = relative activity of dissolved species/ion X.

The Saturation Index (SI) is a measure of the thermodynamic driving force to the equilibrium state. The definition of SI for precipitation/dissolution reactions with the equilibrium constants of Table 1 gives:

$$
\begin{aligned}
SI &= \lg\left(\frac{\{Ca^{2+}\}\{CO_3^{2-}\}}{K_S}\right) \\
&= \lg\{Ca^{2+}\} + \lg\{HCO_3^-\} + \lg K_2 - \lg K_S + pH \quad (1)
\end{aligned}
$$

Langelier combined the first four terms in the last part of Eq. (1) into the parameter $-pH_s$ (Langelier, 1936). Therefore SI for calcium carbonate is often called Langelier Index (LI) or Langelier Saturation Index (LSI). The parameter SI is dimensionless, because the activities in Eq. (1) are relative dimensionless values compared to standard conditions (standard molality, $1\,\text{mol kgw}^{-1}$). This is also the case for pH and the K values in Table 1 (Buck et al., 2002; Gamsjäger et al., 2008).

A practical parameter that is associated with the calcium carbonate saturation is the Calcium Carbonate Precipitation Potential (CCPP) (Standard Methods 2330, 2010), which is a generally used measure for the amount of calcium carbonate which theoretically can precipitate.

SI and CCPP are positive for oversaturated water, zero for saturated water, and negative for undersaturated water. Negative CCPP values are also reported as Calcium Carbonate Dissolution Potential, Calcite Dissolution Capacity or Aggressive Carbon dioxide. Calcium carbonate equilibrium or saturation according to Eq. (1) is shown in Fig. 1, assuming $\{HCO_3^-\} = 2\,\{Ca^{2+}\}$. The pH at equilibrium is higher for water with lower calcium content, i.e. for softer water.

In December 2012, a new German standard for the calculation of calcium carbonate saturation in drinking water was released (DIN 38404-10, 2012, hereafter referred to as DIN). According to the DIN, it is no longer allowed to calculate the SI and CCPP with the simplifications described above. The DIN requires these parameters to be calculated with all specified aqueous species, including complexes with sulphate and phosphate. In order to do so, the DIN standard describes the calculation method with the chemical principles and a data set containing 10 water samples for validation of the calculation method or computer program used.

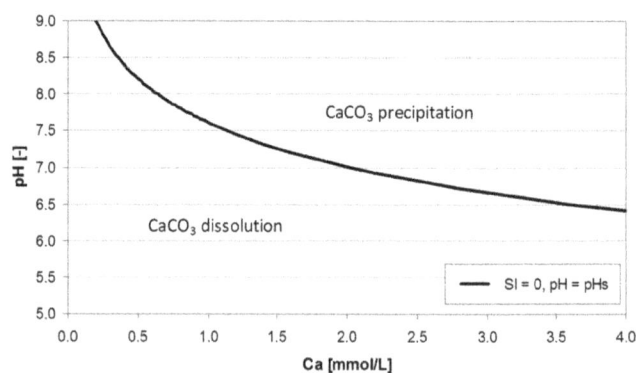

Figure 1. Calcium carbonate equilibrium or saturation ($SI = 0$, at $25\,°C$ and $\{HCO_3^-\} = 2\,\{Ca^{2+}\}$) with over- and under-saturation i.e. calcium carbonate precipitation and dissolution.

The objective of this research is to determine the best calculation method for SI and CCPP in drinking water by testing the calculation methods described in the recent German standard (DIN 38404-10, 2012), the equivalent US standard (Standard Methods 2330, 2010, hereafter referred to as SM), the Dutch standard (NEN 6533, 1990, hereafter referred to as NEN) and a number of widely used databases for calculating chemical equilibria in water, with a validation set given in the DIN standard.

2 Materials and methods

2.1 Calculation method DIN 38404-10

Table 2 gives an overview of all elements, phases and dissolved species in DIN. For both calcium and magnesium this standard distinguishes eight different dissolved species. The standard covers only calcite as the least soluble crystalline form of $CaCO_3$, and applies only to "water for distribution as drinking water". The standard requires that the concentrations of all the elements mentioned are known, as well as pH and temperature. Because of its complexity DIN requires an extensive iterative computer calculation for both SI and CCPP determination, the latter reported in DIN as Calcitlösekapazität (Calcite Dissolution Capacity).

Table 2. Chemical elements, phases and dissolved species in DIN 38404-10 (2012).

Element	Phase	Dissolved species
Ca	$CaCO_3$ (s)	Ca^{2+}, $CaCO_3^0$, $CaHCO_3^+$, $CaOH^+$, $CaSO_4^0$, $CaH_2PO_4^+$, $CaHPO_4^0$, $CaPO_4^-$
Mg		Mg^{2+}, $MgCO_3^0$, $MgHCO_3^+$, $MgOH^+$, $MgSO_4^0$, $MgH_2PO_4^+$, $MgHPO_4^0$, $MgPO_4^-$
Na		Na^+
K		K^+
C*	$CaCO_3$ (s)	H_2CO_3 ($= CO_2.aq$), HCO_3^-, CO_3^{2-}
Cl		Cl^-
N		NO_3^-
S*		HSO_4^-, SO_4^{2-}
P*		H_3PO_4, $H_2PO_4^-$, HPO_4^{2-}, PO_4^{3-}
H	H_2O (l)	H^+, OH^-

* The dissolved species already specified for the cations Ca and Mg are not repeated for the anions.

Chemistry in DIN is based on molarity (mol L^{-1} solution) instead of molality (mol kg^{-1} solvent i.e. water) as used in chemical thermodynamics (Gamsjäger et al., 2008). The values expressed as molarity and molality are equal for a solution with a density of 1 kg L^{-1} while assuming that the mass of the solution equals the mass of water.

2.2 Calculation method Standard Methods 2330

SM gives a set of equations for calculating the value for SI. For CCPP no analytical equation is available and the value can only be obtained by iterative computer calculations. SM distinguishes three crystalline forms of $CaCO_3$ (calcite, vaterite and aragonite) and further only two aqueous species (Ca^{2+} and HCO_3^-), with the assumption that either all other species can be neglected or the reduction to these two aqueous species can be estimated. The influence of other ions is accounted for through the ionic strength and successively in the activity coefficients of Ca^{2+} and HCO_3^-. The standards DIN, SM and NEN have slightly different defined parameters for alkalinity. Therefore, Total Inorganic Carbon (C_T, or TIC) or Dissolved Inorganic Carbon (DIC) is used, which requires calculation of the concentrations of CO_2 and CO_3^{2-} (via K_1 and K_2). Standard Methods gives K_2 as published by Plummer and Busenberg (1982). This study uses K_1 from the same publication. In Table 3 the chemical elements used in SM are shown.

2.3 Calculation method NEN 6533

Just as SM, NEN gives a set of equations for calculating the value for SI. For CCPP no analytical equation is available and the value can only be obtained by iterative computer calculations. NEN considers only Ca^{2+}, HCO_3^- and H^+ and uses only one crystalline form of $CaCO_3$ (calcite), but adjusts the solubility product of $CaCO_3$, in order to take into account the soluble species of $CaCO_3^0$ and $CaHCO_3^+$, by decreasing the pK_S value with 0.037 (at 0 °C) ranging to 0.057 (at 30 °C) based on a combination of K_s values from Plummer and Busenberg

Table 3. Chemical elements, phases and dissolved species in Standard Methods 2330 (2010).

Element	Phase	Dissolved species
Ca	$CaCO_3$ (s)[a]	Ca^{2+}[b]
C	$CaCO_3$ (s)[a]	HCO_3^-, CO_3^{2-}[c]
H	H_2O (l)	H^+, OH^-[c]

[a] three crystalline forms of $CaCO_3$: calcite, aragonite and vaterite

[b] ion pairs $CaHCO_3^+$, $CaSO_4^0$ and $CaOH^+$ by assumption only, for estimation of $[Ca^{2+}]$

[c] CO_3^{2-}, OH^-, and H^+ only in Alkalinity, for estimation of $[HCO_3^-]$

(1982) and Jacobson and Langmuir (1974). Recent literature assesses the approach for K_s used by Jacobson and Langmuir (1974) as incorrect (de Visscher et al., 2012). NEN uses the term aggressiveness for calcium carbonate expressed as aggressive carbon dioxide.

2.4 Calculation software PHREEQC

The computer program PHREEQC, developed by the United States Geological Survey (USGS, 2013) is the de facto international standard for calculating chemical equilibria in groundwater. This program (PHREEQC.exe) solves the mathematical equations that are generated from a chemical database (.dat) and an input file (.pqi), both adjustable by the user (Parkhurst and Appelo, 2013). PHREEQC version 3.0.3 was used to assess the different calculation methods for SI and CCPP according to DIN, SM, NEN and the chemical databases from Table 4. All databases in the upper part of Table 4 are distributed with PHREEQC version 3.0.3 in which "phreeqc.dat" is the default database.

Table 4. Databases in the PHREEQC data format for calculating chemical equilibria in water.

Database	Institution	Program
phreeqc.dat	US-USGS	PHREEQC
wateq4f.dat	US-USGS	WATEQ4F
minteq.v4.dat	US-EPA	MINTEQA2, VISUAL MINTEQ and MINEQL+
llnl.dat	US-LLNL	EQ3/6 and Geochemist's Workbench (GWB)
sit.dat	FR-ANDRA	PHREEQC
pitzer.dat	US-USGS	PHRQPITZ
stimela.dat	NL-Omnisys/Delft UT	PHREEQC/Stimela
din38404-10_2012.dat	NL-Omnisys/Delft UT	PHREEQC/Stimela
sm2330_2010.dat	NL-Omnisys/ Delft UT	PHREEQC/Stimela
nen6533_1990.dat	NL-Omnisys/ Delft UT	PHREEQC/Stimela

2.5 Calculation method with widely used chemical databases

Several widely used computer programs for calculating chemical equilibria in water have been developed by different institutions for different purposes. These programs include their own chemical database, all in their own specific data format. The most prominent databases are also available in the data format for the computer program PHREEQC. The upper part in Table 4 shows a number of databases available in the PHREEQC data format that are able to calculate SI and CCPP, with their institution of origin and the computer program for which they were originally developed.

The computer program PHREEQC and its related database phreeqc.dat is widely used and also is listed in Standard Methods for use of calculation of calcium carbonate saturation indices. The phreeqc.dat database was developed for the calculation of chemical equilibria in groundwater. The calcium carbonate chemistry in the database phreeqc.dat is based on Nordstrom et al. (1990), which is the most recent update of the much-cited publications of Jacobson and Langmuir (1974), Truesdell and Jones (1974) and Plummer and Busenberg (1982). Nordstrom et al. (1990) give equilibrium constants for natural water for a temperature range from 0 to 100 °C, at a pressure of 1 bar. Their dataset is also adopted by Stumm and Morgan (1996). In phreeqc.dat more ion pairs are included than in DIN, such as ion pairs of sodium with sulphate, phosphate, bicarbonate, carbonate and hydroxide.

The database wateq4f.dat is also based on Nordstrom et al. (1990) and therefore almost identical to phreeqc.dat for SI and CCPP calculation. The minteq.v4.dat database has been developed by US EPA, for version 4 of MINTEQA2. The llnl.dat database, compiled by the Lawrence Livermore National Laboratory (Daveler and Wolery, 1992), is by far the most extensive database with respect to thermodynamic equilibrium constants. The databases sit.dat and pitzer.dat are in accordance with the specific ion interaction theory (SIT) of Grenthe et al. (1997) respectively the specific ion interaction model of Pitzer (1973). They were designed to extend the calculation methods for natural water with a high con-

tent of dissolved salts (ionic strength > 500 mmol kgw^{-1}) and they were calibrated on, for instance, seawater and brine. In pitzer.dat, oxygen (dissolved and as gas phase) and all compounds with N and P are not included.

2.6 Calculation method with specific chemical databases

The specifically developed chemical databases are summarized in the lower part of Table 4. The stimela.dat database is developed specifically for water treatment by Omnisys and Delft University of Technology as part of the Stimela modeling environment (van der Helm and Rietveld, 2002). The database is based on phreeqc.dat with extra species and phases to comply to SM and DIN, and with additional redox-uncoupled elements (de Moel et al., 2013). The stimela.dat database will be used in further calculations instead of phreeqc.dat.

The chemical specifications of DIN were converted into a newly developed database for PHREEQC (referred to as "din38404-10_2012.dat"). Starting with phreeqc.dat as a basis, all elements, species and phases were removed which are not mentioned in DIN. An exception was made for the parts of the database that PHREEQC needs in order to run, such as the elements H and O, the gases O_2 and CO_2, and H_2O, alkalinity and the oxidation state (E). Subsequently the numerical values for the equilibrium constants of the reactions (log_k in PHREEQC; lg(K_0) in DIN), the change in enthalpy of the reactions (delta_h in PHREEQC; ΔH in DIN) and the activity coefficients (gamma in PHREEQC; $f(i)$ in DIN with its ion size parameter $g(i)$) were adjusted to the values given in DIN. The last step for building din38404-10_2012.dat was the determination of A_1 to A_6 in the analytical expressions for the equilibrium constants (log_k) in PHREEQC for the reactions with a heat capacity (C_p in DIN):

Table 5. Water quality validation data set from DIN consisting of 10 different drinking water samples.

Parameter		Unit	S1	S2	S3	S4	S5	S6	S7	S8	S9	S10
Temperature		°C	10.0	15.0	10.0	10.0	10.0	15.0	12.0	10.0	15.0	61.0
pH		–	7.34	7.80	7.00	5.60	7.37	7.86	7.59	7.47	7.30	7.30
Calcium	Ca	$mmol\,L^{-1}$	1.40	0.75	3.50	0.15	1.40	0.78	1.30	1.00	2.65	1.00
Magnesium	Mg	$mmol\,L^{-1}$	0.23	0.10	0.70	0.05	0.25	0.10	0.25	0.25	0.20	0.18
Sodium	Na	$mmol\,L^{-1}$	0.30	0.40	2.30	0.30	0.40	0.45	1.60	0.20	0.30	0.20
Potassium	K	$mmol\,L^{-1}$	0.05	0.10	0.30	0.10	0.07	0.05	0.15	0.06	0.06	0.05
Total Inorganic Carbon	C	$mmol\,L^{-1}$	2.737	1.630	6.724	1.375	2.662	1.584	1.159	2.094	4.672	2.057
Chloride	Cl	$mmol\,L^{-1}$	0.25	0.30	2.70	0.34	0.55	0.28	0.85	0.35	0.75	0.10
Nitrate	N	$mmol\,L^{-1}$	0.15	0.03	0.50	0.18	0.20	0.00	0.10	0.05	0.05	0.05
Sulphate	S	$mmol\,L^{-1}$	0.38	0.15	1.20	0.05	0.25	0.15	1.40	0.20	0.55	0.25
Phosphate	P	$mmol\,L^{-1}$	0.00	0.00	0.00	0.00	0.07	0.07	0.00	0.03	0.00	0.03
By DIN calculated validation results:												
SI		–	−0.402	−0.279	−0.121	−4.155	−0.381	−0.221	−0.553	−0.497	0.095	−0.009
CCPP		$mmol\,L^{-1}$	−0.145	−0.032	−0.163	−1.111	−0.140	−0.030	−0.055	−0.115	0.071	−0.003

$$A_1 = \lg(K_0) + \frac{\frac{\Delta H}{T_0} - C_P(1 + \ln(T_0))}{R\ln(10)}$$

$$A_2 = A_5 = A_6 = 0$$

$$A_3 = \frac{C_P T_0 - \Delta H}{R\ln(10)}$$

$$A_4 = \frac{C_P}{R} \tag{2}$$

The temperature related DIN parameters A (constant), B (ion size coefficient), and DK (dielectric constant) in the calculation of the activity coefficients are included in PHREEQC in an equivalent way. DIN assumes no effect of uncharged ions, therefore b in the Debye-Hückel equation as used in WATEQ (Truesdell and Jones, 1974) is set to 0 in din38404-10_2012.dat (the default value used in PHREEQC is 0.1).

Similar to the development of the DIN database for PHREEQC, also databases were newly developed for Standard Methods 2330 (sm2330_2010.dat) and NEN 6533 (nen6533_1990.dat) and therefore a number of issues was resolved. In order for PHREEQC to run, log_k values for ion pairs need to be defined. However, in SM and NEN ion pairs are not included, see Table 3, therefore the log_k values of all reaction equations of these species was set to −100 in "sm2330_2010.dat" and "nen6533_1990.dat". SM uses the Davies equation for the influence of the ionic strength for charged ions according to the default method in PHREEQC, therefore, the gamma option in sm2330_2010.dat has been omitted. NEN uses the WATEQ Debye-Hückel equation for charged ions, thus the gamma option is used in nen6533_1990.dat. For species without charge (including H_2O and CO_2) an activity coefficient of 1.0 has been assumed in both sm2330_2010.dat and nen6533_1990.dat ($b = 0$ in gamma). For all calculations with PHREEQC it is assumed that the oxygen content is al-

ways 100 % saturated at 1.0 atm in dry air with an oxygen content of 20.8 %.

2.7 Drinking water validation data set

For validation of the calculation methods, the data set of 10 water quality validation samples given in DIN was used. The validation data set in DIN contains the measured water quality parameters of the 10 samples. Part of the data is summarized in Table 5, including DIN calculated SI and CCPP values. Actually, in the DIN the Calcitlösekapazität (D_C) in $mg\,L^{-1}$ is given from which the CCPP in Table 5 in $mmol\,L^{-1}$ is calculated with:

$$CCPP = \frac{-D_C}{MW_{CaCO_3}} \tag{3}$$

The water quality dataset in Table 5 ranges from very soft water (sample 4) to hard water with high sulphate content (sample 3). The water temperature of the samples is between 10 and 15 °C (sample 1–9), and 61 °C (sample 10); pH is between 7.00 and 7.86 (sample 1–3 and 5–10), and 5.60 (sample 4). Sample 4 and sample 10 are outside the scope of DIN ("water for distribution as drinking water") because of pH respectively temperature. The validation set lacks samples with pH above 7.86, that is typical for soft and softened drinking water.

According to DIN most water samples in the validation set are slightly calcium carbonate dissolving ($-0.2 < CCPP < 0.0\,mmol\,L^{-1}$), except for sample 4, which is highly calcium carbonate dissolving, and sample 9, which has a small calcium carbonate precipitation potential. The parameter alkalinity is not included in Table 5, because for proper comparison of the DIN, SM and NEN standard the sum of inorganic carbon species in a solution is used. The density of the different samples is not given in DIN, as it is based on molarity. Molarity is converted into molality for PHREEQC by assuming a density of $1.0\,kg\,L^{-1}$ for all

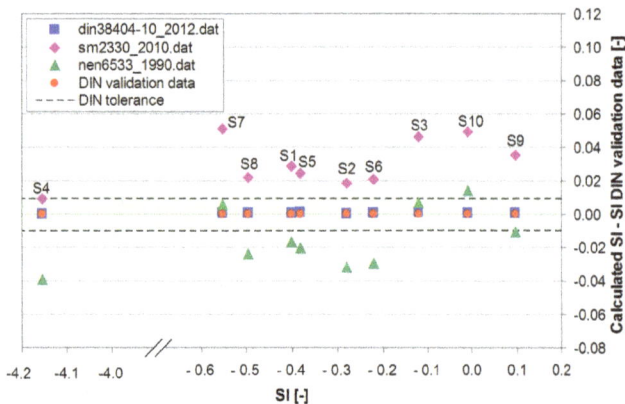

Figure 2. Deviations of the SIs calculated in PHREEQC according to the DIN, SM and NEN standards, from the SIs given in the DIN validation data set for the 10 water quality samples S1 to S10 (see Table 5).

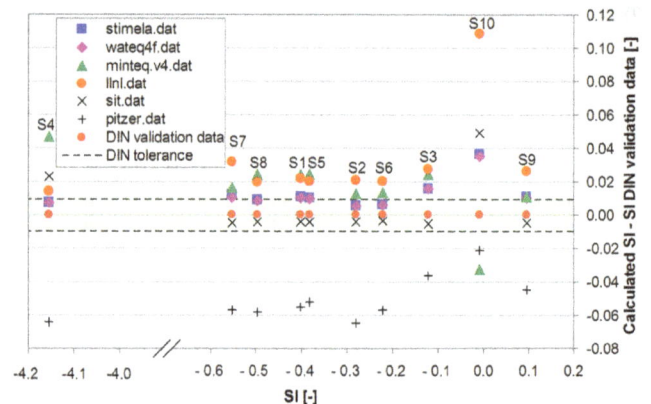

Figure 3. Deviations of the SIs calculated in PHREEQC with widely used chemical databases, from the SIs given in the DIN validation data set for the 10 water quality samples S1 to S10 (see Table 5).

samples, regardless of water temperature. The assumption of a density of $1.0\,kg\,L^{-1}$ is only used for the calculations with DIN, SM and NEN. For the calculations with the other databases from Table 4, the solution density for conversion from concentrations in $mmol\,L^{-1}$ into $mmol\,kgw^{-1}$ is calculated by PHREEQC itself.

3 Results

3.1 SI for calcium carbonate

In Fig. 2 the deviations are shown between the SIs calculated in PHREEQC according to the DIN, SM and NEN standards and the SIs according to the DIN validation data set, see Table 5. Also the tolerance for calculated SI values of 0.01 given in the DIN standard is shown in Fig. 2.

From the data shown in Fig. 2, it is observed that the SI values calculated with din38404-10_2012.dat in PHREEQC have a maximal deviation of 0.0011. The calculation method with din38404-10_2012.dat complies with the DIN standard in which a tolerance of ± 0.01 SI is specified for the calculation results.

SI values calculated with the SM database are on average 0.030 higher than the SIs according to DIN. This is mainly caused by neglecting ion pairs in SM; more specifically, the ion pair $CaSO_4^0$, which leads to an overestimation of the Ca^{2+} concentration. The overestimation is smaller for sample 4 because of the low concentration of ions in the sample and the overestimation is larger for the samples 3, 7 and 9 with high sulphate concentrations. Almost all calculated SI values are out of the DIN tolerance range.

For most samples, the NEN database provides SI values that are more than 0.01 lower than the DIN database because of the higher K_s value used in NEN, except for the samples 3, 7 and 10. The NEN database gives better results for the samples 3 and 7 with higher sulphate concentrations than the SM

database. This is remarkable since the reason for decreasing pK_s value in NEN is to take into account the effect of ion pairs $CaCO_3^0$ and $CaHCO_3^+$, and not $CaSO_4^0$ ion pair. Even though the calculation methods of SM and NEN are similar, the lower pK_s value leads to an almost equal difference of around 0.045 between the two methods. It should be noted that pK_w in NEN is only validated for a temperature range of 0 to 30 °C; therefore, the calculated SI value for sample 10 with a temperature of 61 °C is only illustrative.

In Fig. 3 the deviations are shown between the SIs calculated in PHREEQC according to the widely used chemical equilibrium databases and the SIs according to the DIN validation data set, see Table 5. Also the tolerance for calculated SI values of 0.01 given in the DIN standard is shown in Fig. 3.

Differences between the calculated SI values with stimela.dat and the values according to the DIN validation data are on average 0.012. The differences are caused by small differences in the values of equilibrium constants and activity coefficients, and not by the ion pairs that are present in stimela.dat, but are not present in the DIN standard. For 5 of the 10 samples stimela.dat calculates SI within the tolerance of 0.01 SI as specified in the DIN standard. The larger error observed for sample 10 is caused partly by the fact that DIN neglects the change in density due to the higher temperature.

The SI values calculated with stimela.dat and wateq4f.dat are almost the same for all samples. The SI values with minteq.v4.dat and llnl.dat are, with sample 10 left out, on average 0.022 higher than the SI values according to DIN. This is mainly due to a slight difference in the values of lgK_2-lgK_s for both databases at a temperature between 10 and 15 °C. At 61 °C (sample 10), the difference in SI for these databases is large due to the large differences in K_s. The large difference in sample 4 for minteq.v4.dat is the overall effect of many small differences that reinforce each other, from which half is caused by a difference in $lg\{HCO_3^-\}$.

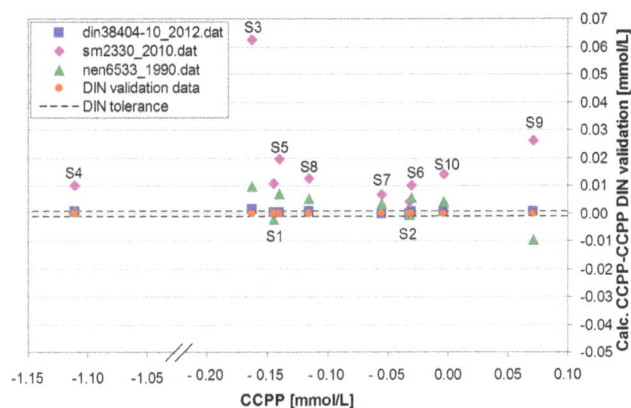

Figure 4. Deviations of the CCPPs calculated in PHREEQC according to the DIN, SM and NEN standards, from the CCPPs given in the DIN validation data set for the 10 water quality samples S1 to S10 (see Table 5).

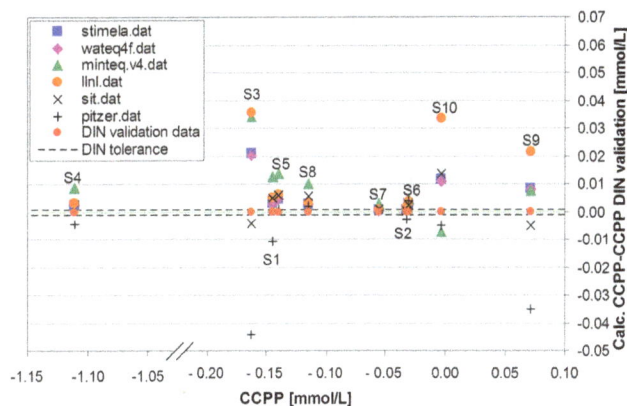

Figure 5. Deviations of the CCPPs calculated in PHREEQC with the widely used chemical databases, from the CCPPs given in the DIN validation data set for the 10 water quality samples S1 to S10 (see Table 5).

The database sit.dat gives for most samples a difference of less than 0.01. The differences are mainly caused by small differences in K_s and K_2. The database pitzer.dat gives large differences up to 0.065. The differences between stimela.dat and pitzer.dat are entirely due to the difference in the activity of Ca^{2+} and HCO_3^-, since K_2 and K_s are the same. It can be observed that the Pitzer model is less suitable for the "low salt" samples given in the DIN standard.

The bandwidth of the calculation results with the comprehensive databases in Fig. 3 is smaller than the bandwidth for the results with the simple calculations according to SM and NEN shown in Fig. 2.

3.2 Calcium carbonate precipitation potential

In Fig. 4 the deviations are shown between the CCPPs calculated in PHREEQC according to the DIN, SM and NEN standards and the CCPPs according to the DIN validation data set, see Table 5. Also the tolerance for calculated CCPP values of 0.001 mmol L^{-1} given in the DIN standard is shown in Fig. 4.

From the data shown in Fig. 4, it is observed that the CCPPs calculated with the DIN database comply with the validation values. The maximum deviation observed is +0.0011 mmol L^{-1} (sample 3), which is not significantly out-of-tolerance.

CCPPs calculated with the SM database give values which are on average 0.018 mmol L^{-1} higher than CCPPs according to DIN, with a peak of 0.062 mmol L^{-1} for sample 3. This is due to neglecting ion pairs in SM. The CCPP calculated with the NEN database always gives lower values than calculated with SM due to difference in pK_s as described for SI. The fixed difference in the pK_s gives variable differences for the CCPP, because of the differences in calcium, carbonate and bicarbonate concentrations in the ten samples. It is noted that pK_w in NEN is only validated for a temperature range of 0

to 30 °C, therefore the calculated CCPP for sample 10 with a temperature of 61 °C is only illustrative.

In Fig. 5 the deviations are shown between the CCPPs calculated in PHREEQC according to the widely used chemical equilibrium databases and the CCPPs according to the DIN validation data set, see Table 5. Also the tolerance for calculated CCPPs of 0.001 mmol L^{-1} given in the DIN standard is shown in Fig. 5.

From the data in Fig. 5 it is observed that for most samples the CCPPs calculated with stimela.dat are 0.000 to 0.005 mmol L^{-1} (0.0 to 0.5 mg $CaCO_3$ L^{-1}) larger than the values in the DIN validation data set. This means that for calcium dissolving water, according to stimela.dat, less $CaCO_3$ might be dissolved than according to the DIN standard. The differences are slightly larger for samples 9 and 10, and for sample 3 the difference is 0.021 mmol L^{-1} (2.1 mg $CaCO_3$ L^{-1}). This difference is caused by accumulation of several small differences, and not by ion pairs which are in stimela.dat but not in din38404-10_2012.dat. For only 2 of the 10 samples stimela.dat calculates CCPP within the tolerance of 0.001 mmol L^{-1} as specified in the DIN standard.

From the calculation with stimela.dat it is observed that the free ion Ca^{2+} forms 90–99 % of the total calcium content in all ten samples, the remainder is present as $CaSO_4^0$, $CaHCO_3^+$, $CaCO_3^0$ and $CaHPO_4^0$ (in order of importance). These ion pairs cause the large deviation as mentioned for sample 3 calculated with sm2330_2010.dat. Other ion pairs for Ca are negligible. From the calculation with stimela.dat it can also be observed that the carbon ion pairs are of less importance since all carbon in the 10 samples is 98–100 % present as CO_2, HCO_3^- or CO_3^{2-}. The CCPP values calculated with wateq4f.dat are almost the same as for stimela.dat for all 10 samples.

It can be observed that the differences between the CCPPs from the DIN validation data and the CCPPs calculated with minteq.v4.dat, llnl.dat, sit.dat and pitzer.dat are generally larger than the differences calculated with stimela.dat and watq4f.dat, and that all values are generally in the range of $\pm 0.01\,\mathrm{mmol\,L^{-1}}$. The causes for the differences of minteq.v4.dat, llnl.dat, sit.dat and pitzer.dat are the same as for the differences observed in SI values. The largest differences occur for samples 3 and 9 due to a large influence of the ion pair $CaSO_4^0$. In addition, a large difference occurs for sample 10, which is mainly caused by the differences in K_s and K_2.

4 Discussion

4.1 SI versus CCPP

SI indicates thermodynamical driving force, while CCPP is total mass $CaCO_3$ reacted to obtain equilibrium. The parameters are not quantitatively related, as shown in Fig. 6, only qualitatively; SI and CCPP are positive for oversaturated water, zero for saturated water, and negative for undersaturated water. In 2003 the German drinking water regulations were changed from SI as the guideline parameter to CCPP ("Calcitlösekapazität"). This approach focuses better on the water quality issue, which is precipitation and dissolution of calcite.

4.2 Tolerance and accuracy

For natural water the DIN tolerance range for SI (0.01) and for CCPP ($0.001\,\mathrm{mmol\,L^{-1}}$) are not equivalent. This can be demonstrated by assuming the allowed deviation in pH for sample 5 (7.38 instead of 7.37) which results in an unaccepted deviation in CCPP of $0.005\,\mathrm{mmol\,L^{-1}}$ (five times to high).

An error of 0.01 in the log-concentration parameter SI corresponds to errors of 2.3 % ($= 10^{0.01}$) in concentration of monovalent ions and 1.2 % ($= 10^{0.01/2}$) for bivalent ions, at equal equilibrium constant. Natural soft water with a calcium concentration of $1.0\,\mathrm{mmol\,L^{-1}}$ would allow for a deviation of $(1.0 \times 1.2\,\% =)\,0.012\,\mathrm{mmol\,L^{-1}}$, which is 12 times larger than the tolerance range for CCPP. The tolerances for SI and CCPP are equivalent at a calcium concentration of $(0.001/1.2\,\% =)\,0.083\,\mathrm{mmol\,L^{-1}}$ $(3.3\,\mathrm{mg\,L^{-1}})$. The DIN dataset has an average calcium concentration of $1.38\,\mathrm{mmol\,L^{-1}}$ and a lowest concentration of $0.15\,\mathrm{mmol\,L^{-1}}$. This shows that in DIN the restrictions for CCPP are by far more stringent than for SI (or pH).

The allowable tolerance for SI and CCPP as defined in DIN is smaller than the deviations caused by the inaccuracy of the measurements of the chemical parameters, which form the input of the calculations. The DIN specifies for pH a tolerance of 0.05 based on determination of p value (acidity), m value (alkalinity), measured pH and calculated pH. Consequently, 0.05 is also the minimal accuracy for SI, which is 5 times larger than the tolerance of the SI calculation of

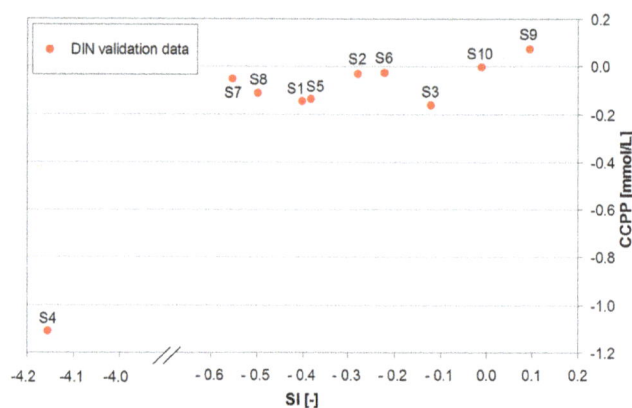

Figure 6. CCPP values plotted against the SI (data from the DIN validation data set, see Table 5).

0.01 defined in DIN. The DIN specifies a minimal accuracy for the ionic balance of 5 % (ionic strength $> 4\,\mathrm{mmol\,L^{-1}}$) to 10 % (ionic strength $< 4\,\mathrm{mmol\,L^{-1}}$) which can be adopted as minimal accurary levels for calcium and alkalinity. Depending on the ionic strength, natural soft water with a calcium concentration of $1.0\,\mathrm{mmol\,L^{-1}}$ would allow for a deviation of $(1.0 \times 5\,\% =)\,0.05\,\mathrm{mmol\,L^{-1}}$ to $(1.0 \times 10\,\% =)\,0.1\,\mathrm{mmol\,L^{-1}}$, which is 50 to 100 times larger than the DIN tolerance range for calculation of CCPP. Figures 2 to 5 show that almost all presented models for the calculation of SI and CCPP comply with a tolerance range of 0.05 for SI and 0.05 to $0.1\,\mathrm{mmol\,L^{-1}}$ for CCPP. It would be preferable to introduce tolerances for SI and CCPP which are consistent with each other and based on practical accuracy of the measurements of at least pH (including the DIN accuracy check) and calcium.

4.3 Molarity versus molality

The equations for equilibrium in DIN are based on the obsolete molarity system ($\mathrm{mol\,L^{-1}}$). Modern chemistry uses molality ($\mathrm{mol\,kgw^{-1}}$). For drinking water with its low salt content and its temperature between 0 and 25 °C the difference is very small. However at higher temperatures, the difference in density is no longer negligible. Above 45 °C the density is below $0.99\,\mathrm{kg\,L^{-1}}$ giving differences in concentration above 1 %. The density of sample 10 in the DIN validation set is $0.983\,\mathrm{kg\,L^{-1}}$, as calculated by PHREEQC with stimela.dat. The differences in SI and CCPP calculated according to DIN due to neglecting changes in density are respectively 0.013 and $0.004\,\mathrm{mmol\,L^{-1}}$. This makes the applicability of DIN for water with a temperature higher than 45 °C unjustified and sample 10 should therefore be omitted from the DIN validation database.

4.4 Warm, hot and boiling water

DIN and NEN are only applicable in lower temperature water. NEN gives as temperature range 0 to 30 °C, while DIN does not give temperature limits but from the calculations performed in this research it is found that it is valid up to a range of 40 to 50 °C. Standard Methods gives a temperature range for K_s and K_2 of 0 to 90 °C, according to Plummer and Busenberg (1982). For drinking water practice the temperature range should be extended up to 100 °C since customers heat up and boil water and scaling of calcium carbonate is a critical factor for in-house drinking water practices. This requires that equilibrium constants and activity coefficients used for calculation of SI and CCPP should be valid in the temperature range of 0 to 100 °C.

In addition, the vapor pressure of water is important, in combination with the concentrations of dissolved gases. At a higher vapor pressure the gas partial pressure of N_2, O_2, CO_2 might result in degassing and therefore in a reduction of CO_2 content. This process occurs during gas bubble formation when heating water, which already occurs below the boiling point of water. The partial pressure of oxygen in contact with wet air is around 0.20 atm for water samples 1 to 9 from the DIN validation set with water temperatures from 10 to 15 °C, and 0.16 atm for sample 10 with a temperature of 61 °C.

The current standards do not fulfil the need from drinking water practice to include calculation of calcium carbonate scaling at high temperatures. In future research, the effect of higher temperatures up to 100 °C on calcium carbonate saturation in drinking water will be assessed. In this research stimela.dat, which is based on phreeqc.dat, will be used since values for equilibrium constants and activity coefficients are valid up to 100 °C and effects of degassing can be taken into account.

4.5 Scientific justification

The chemical databases and standards appear to have significant differences in the values of the equilibrium constants. The influence of the difference in activity coefficients is less significant, with the exception of the unsuitable models (SIT and Pitzer) which have not been calibrated for potable water with a low ionic strength. An international standardization of this basic chemistry is performed within the "IUPAC-NIST Solubility Data Series" of the International Union for Pure and Applied Chemistry (IUPAC) and the National Institute of Standards and Technology (NIST). A volume on alkaline earth carbonates has been published recently (de Visscher et al., 2012; de Visscher and Vanderdeelen, 2012). The values of K_s, K_1, K_2 and K_w in phreeqc.dat and thus in stimela.dat appear to be well in accordance with the presented results.

It is expected that the results will be considered by USGS for future versions of phreeqc.dat and will consequently be adopted in stimela.dat. Pending this scientific justification, the K values in stimela.dat will be used in further research, according to Nordstrom et al. (1990), mainly because of the wide temperature range. This approach is in compliance with Standard Methods 2330 D.

5 Conclusions

The computer program PHREEQC (USGS, 2013) with the developed chemical database din38404-10_2012.dat complies with the new German standard DIN 38404-10 (2012) for calculating SI and CCPP. This standard covers water that is intended for distribution as drinking water with its specific limitations on concentrations and temperature. This compliance is achieved by assuming equal values for molarity as used in DIN (obsolete) and molality as used in PHREEQC. From comparison with widely used chemical databases it is concluded that the use of molarity limits the use of DIN to a maximum temperature of 45 °C. Alternative international standards (Standard Methods and NEN) are not accurate enough to comply with DIN, because of their simplifications. It should be noted that the DIN tolerance range is more stringent than the accuracy of the chemical analyses which serve as input for the computer calculations. On the other hand, the differences in calculation results for DIN, SM and NEN illustrate the need for international unification of these standards. Running extensive chemical models i.e. databases on the DIN validation data set results in deviations outside the DIN tolerance range. None of these widely used models fully complies with the DIN standard. However, it must be noted that all models, including DIN, lack proper scientific justification and/or scientific acceptance. This might be achieved in the near future within the framework of the "IUPAC-NIST Solubility Data Series" of the International Union for Pure and Applied Chemistry (IUPAC) and the National Institute of Standards and Technology (NIST). For application of drinking water supply it is concluded that the standards should also be applicable for hot and boiling water, since scaling of calcium carbonate is a critical factor for in-house drinking water practices. For current practical applications the PHREEQC database stimela.dat was developed within the Stimela platform of Delft University of Technology. This database is an extension of phreeqc.dat focused on water treatment and drinking water applications. This approach is in compliance with Standard Methods 2330 D.

6 Supplementary material

For use of the Supplement the free software phreeqc-3.0.6-7757.msi and IPhreeqcCOM-3.0.6-7757-win32.msi or a higher version should be installed. List of files:

- CalciumCarbonateSaturation_v20131020.xlsm containing din38404-10_2012.dat and the DIN validation data set

- din38404-10_2012.dat

- sm2330_2010.dat

- nen6533_1990.dat

- stimela.dat

– DIN38404-10_compliance.pqi

References

Buck, R. P., Rondinini, S., Covington, A. K., Bauke, F. G. K., Bretts, C. M. A., Camõus, M. F., Milton, M. J. T., Mussini, T., Naumann, R., Pratt, K. W., Spitzer, P., and Wilson, G. S.: Measurement of pH. Definition, standards and procedures (IUPAC Recommendations 2002), Pure Appl. Chem., 74, 2169–2200, 2002.

Daveler, S. A. and Wolery, T. J.: EQPT, A data file preprocessor for the EQ3/6 software package: User's guide and related documentation (Version 7.0), Lawrence Livermore National Laboratory, Livermore, USA, UCRL-MA-110662 PT II, 1992.

de Moel, P. J., van Dijk, J. C., and van der Meer, W. G. J.: Aquatic chemistry for engineers – Volume 1 – Starting with PHREEQC 3, Delft University of Technology, Delft, the Netherlands, 2013.

de Visscher, A. and Vanderdeelen, J.: IUPAC-NIST Solubility Data Series. 95. Alkaline earth carbonates in aqueous systems. Part 2. Ca, J. Phys. Chem. Ref. Data, 41, 023105-1–023105-137, 2012.

de Visscher, A., Vanderdeelen, J., Königsberger, E., Churagulov, B. R., Ichikuni, M., and Tsurumi, M.: IUPAC-NIST Solubility Data Series. 95. Alkaline earth carbonates in aqueous systems. Part 1. Introduction, Be and Mg, J. Phys. Chem. Ref. Data, 41, 013105-1–013105-67, 2012.

DIN 38404-10: German standard methods for the examination of water, waste water and sludge – Physical and physico-chemical parameters (group C) – Part 10: Calculation of the calcit saturation of water (C 10), DIN Deutsches Institut für Normung, Berlin, Germany, 2012.

Gamsjäger, H., Lorimer, J. W., Scharlin, P., and Shaw, D. G.: Glossary of terms related to solubility (IUPAC Recommendations 2008), J. Pure Appl. Chem., 80, 233–276, 2008.

Grenthe, I., Plyasunov, A. V., and Spahiu, K.: Estimations of medium effects on thermodynamic data, Modelling in aquatic chemistry, edited by: Grenthe, I. and Puigdomenech, I., OECD Nuclear Energy Agency, Paris, France, 1997.

Jacobson, R. L. and Langmuir, D.: Dissociation constants of calcite and $CaHCO_3^+$ from 0 to 50 °C, Geochim. Cosmochim. Ac., 38, 301–318, 1974.

Langelier, W. F.: The analytical control of anti-corrosion water treatment, J. Am. Water Works Assoc., 28, 1500–1521, 1936.

NEN 6533: Water – Bepaling van de agressiviteit ten opzichte van calciumcarbonaat en berekening van de verzadigingsindex (Water – Determination of the agressivity to calcium carbonate and the calculation of the saturation index), Nederlands Normalisatie-insituut, Delft, the Netherlands, 1990.

Nordstrom, D. K., Plummer, L. N., Langmuir, D., Busenberg, E., May, H. M., Jones, B. F., and Parkhurst, D. L.: Revised chemical equilibrium data for major water-mineral reactions and their limitations, in: Chemical modeling in aqueous systems II (Symposium Series 416), edited by: Bassett, D. and Melchior, R. L., American Chemical Society, Washington D.C., USA, 1990.

Parkhurst, D. L. and Appelo, C. A. J.: Description of input and examples for PHREEQC version 3 – A computer program for speciation, batch-reaction, one-dimensional transport, and inverse geochemical calculations, US Geological Survey, Denver, USA, 2013.

Pitzer, K. S.: Thermodynamics of electrolytes. I. Theoretical basis and general equations, J. Phys. Chem., 77, 268–277, 1973.

Plummer, L. N. and Busenberg, E.: The solubilities of calcite, aragonite and vaterite in CO_2-H_2O solutions between 0 and 90 °C, and an evaluation of the aqueous model for the system $CaCO_3$-CO_2-H_2O, Geochim. Cosmochim. Ac., 46, 1011–1040, 1982.

Standard Methods 2330: Calcium carbonate saturation (2330), American Public Health Association/American Water Works Association/Water Environment Federation, Baltimore, USA, 2010.

Stumm, W. and Morgan, J. J.: Aquatic chemistry – Chemical equilibria and rates in natural waters (3rd Edn.), John Wiley & sons, New York, USA, 1996.

Truesdell, A. H. and Jones, B. F.: WATEQ, A computer program for calculating chemical equilibria of natural waters, J. Res. US Geol. Survey, 2, 233–248, 1974.

USGS: phreeqc-3.0.6-7757.msi, http://wwwbrr.cr.usgs.gov/projects/GWC_coupled/phreeqc/, last access: 23 October 2013.

van der Helm, A. W. C. and Rietveld, L. C.: Modelling of drinking water treatment processes within the Stimela environment, Wa. Sci. Technol., 2, 87–93, 2002.

18

Conversion of organic micropollutants with limited bromate formation during the Peroxone process in drinking water treatment

A. H. Knol[1]**, K. Lekkerkerker-Teunissen**[1]**, C. J. Houtman**[2]**, J. Scheideler**[3]**, A. Ried**[3]**, and J. C. van Dijk**[4]

[1]Dunea N.V., P.O. Box 756, 2700 AT Zoetermeer, the Netherlands
[2]Het Waterlaboratorium, P.O. Box 734, 2003 RS Haarlem, the Netherlands
[3]Xylem, Boschstrasse 4–14, 32051 Herford, Germany
[4]TU Delft, Stevinweg 1, 2628 CN Delft, the Netherlands

Correspondence to: A. H. Knol (t.knol@dunea.nl)

Abstract. Advanced oxidation with O_3 / H_2O_2 (peroxone) was conducted on pilot plant scale on pre-treated Meuse river water to investigate the conversion of organic micropollutants (OMPs) and the formation of bromate. Fourteen selected model compounds were dosed to the pre-treated river water on a regular basis to assess the efficiency of the peroxone process and to establish the influence of the water matrix.

The ozone dose was the main factor in the conversion of the model compounds, however, the ozone dose was limited because of bromate formation. The hydrogen peroxide dosage had only a minor effect on the conversion, but it limited the bromate formation effectively. In terms of limited chemical consumption, maximal conversion and to comply the strict Dutch drinking water act for bromate of $1 \mu g \, L^{-1}$, a practical peroxone setting was $6 \, mg \, L^{-1}$ hydrogen peroxide and $1.5 \, mg \, L^{-1}$ ozone. During the investigation period, the average conversion of the model compounds was 78.9 %.

The conversion of OMPs was higher at higher water temperatures and lower concentrations of DOC and bicarbonate. The bromate formation also was higher at higher water temperature and lower bicarbonate concentration and proportional with the bromide concentration, above a threshold of about $32 \mu g \, L^{-1}$ bromide. The peroxone process can be controlled on basis of the (derived) parameters water temperature, bicarbonate and DOC.

1 Introduction

All over the world surface water is to some extend contaminated with organic micro pollutants (OMPs). It is expected that concentrations of OMPs will increase, due to growth and aging of populations and global warming. In areas with a shortage of groundwater, drinking water companies use the available surface water as source for drinking water production. They are more and more aware of the fact that conventional treatment technologies, such as coagulation, filtration and activated carbon filtration, are not adequate in removing all OMPs from the surface water. The combination of the increasing concentrations of OMPs in surface waters and the inadequate removal of the polar OMPs with conventional treatment processes, necessitate research on an additional barrier against OMPs.

Drinking water company Dunea, in the western part of the Netherlands, recognizes the threat of OMPs in their source the Afgedamde Maas (Enclosed Meuse), a side branch of the Meuse River, although the drinking water quality still complies the standards of the Dutch Drinking Water Law, including the standards for OMPs. Managed aquifer recharge (MAR) by dune filtration and the dosing of powdered activated carbon (PAC) are the current barriers against these substances.

Only nonpolar OMPs are well removed by MAR and PAC. Polar OMPs are adsorbed less and/or converted (IJpelaar, 2008). Pharmaceuticals and pesticides are main contaminants detected consistently in the Dutch Meuse River (Houtman et al., 2010). However, the effect on human health at low concentrations is judged negligible (Schriks et al., 2010; Houtman et al., 2014).

Membrane filtration and advanced oxidation (AOP) are the two common technologies to reduce concentrations of OMPs in pretreated surface water. After careful consideration, Dunea chose advanced oxidation as the most optimal technique (Abrahamse et al., 2007) and carried out research with this technology. When AOP is installed before MAR, it is expected that these two processes will provide a synergistic hybrid system (Lekkerkerker-Teunissen et al., 2012).

In comparison with other AOP technologies, the O_3 / H_2O_2 (or peroxone) process, is known as energy efficient. The two mechanisms to convert OMPs are direct oxidation by ozone and oxidation by in-situ produced hydroxyl radicals ($\cdot$OH). Although the reaction rate of ozone, depending on the type of compounds, is relatively slow and typically in the range of $k = 1.0$ till $10^6\,M^{-1}\,s^{-1}$ (Gottschalk et al., 2010), and the reaction rate of hydroxyl radicals is much higher and typically in the range of $k = 10^8$ till $10^{10}\,M^{-1}\,s^{-1}$, direct oxidation cannot be neglected when applying peroxone (von Gunten, 2003a). Decomposition of ozone is accelerated by addition of hydrogen peroxide with a reaction rate of $k = 1.1 \times 10^5\,M^{-1}\,s^{-1}$.

The efficiency of peroxone in converting OMPs largely depends on the water quality matrix. Ozone and hydroxyl radicals not only react with OMPs, but also with scavengers as natural organic matter (NOM, mainly humic substances) and bicarbonate (von Gunten, 2003a). Besides that, the water temperature of the Meuse River yearly varies between close to zero to around 25 °C, which influences dissociation coefficients and hydraulic conditions (as mixing energy) in the reactor.

A reaction that is of particular importance is the reaction of ozone with bromide into bromate, since bromate is a suspected carcinogen (Kurokawa et al., 1982). The WHO, EPA and European Drinking Water Directive established a guideline of $10\,\mu g\,L^{-1}$ bromate. Two guidelines are mentioned in the Dutch Drinking Water Act: In case of disinfection with ozone, the appointed guideline is $5\,\mu g\,L^{-1}$ as a 90 % percentile value with a maximum of $10\,\mu g\,L^{-1}$. If ozone is applied for oxidation, the bromate guideline is $1\,\mu g\,L^{-1}$, which means that applying peroxone can be seen as an ultimate challenge. Nevertheless, the negligible risk level (in the Netherlands defined as the concentration at which one additional case of cancer would occur in one million lifelong exposed people; 10^{-6}) is even lower and calculated to be 0.2–$0.6\,\mu g\,L^{-1}$ (WHO, 2005). The target value of this research was an average bromate formation of $0.5\,\mu g\,L^{-1}$ with a maximum of $1\,\mu g\,L^{-1}$. Also taking into account the concentration levelling MAR after applying peroxone, the bromate concentration in drinking water then will not exceed $0.5\,\mu g\,L^{-1}$.

Bromate formation can theoretically be limited by a fast decomposition of ozone and increasing the ratio between the hydrogen peroxide and ozone doses (von Gunten, 2003b).

Bromate formation is thus affected by the varying water matrix parameters pH, water temperature, bromide and bicarbonate. Two other parameters, dissolved organic carbon (DOC, a measurement for natural organic matter) and ammonia, appeared to be not of relevance for the formation of bromate. Bromate is formed proportionally when the ratio of ozone dose and DOC, in $mg\,O_3\,mg\,C^{-1}$, is exceeding 0.4 (Croué et al., 1996; Amy et al., 1993), which is much higher than the applied ratio in this research. The role of ammonia (ammonia can depress the bromate formation) could be neglected, because only in a short period (weeks) the concentration was higher than the detection limit of $0.02\,mg\,NL^{-1}$, which is too low for limiting bromate formation.

The goal of this research was to optimise the use of peroxone, restricted by a bromate target value of $0.5\,\mu g\,L^{-1}$, considering the maximum ozone doses that can be applied, as well as the ozone/hydrogen peroxide ratio and the corresponding conversion of OMPs. The influence of the water matrix, the ozone dose and the hydrogen peroxide concentration on the bromate formation and the conversion of OMPS were investigated in an on-site pilot plant set-up, accompanied by batch experiments and long term duration experiments.

2   Materials and methods

2.1   Peroxone process installation

The pilot plant consisted of an ozone loop reactor (Xylem Wedeco) with sequential injection points (IPs) and sample points (SPs) and degassing chamber, and an ozone generator. A schematic view of the loop reactor is plotted in Fig. 1.

The ozone generator had a capacity between 3.5 and 100 g ozone per hour, produced from oxygen. The minimum dose applicable was $0.7\,mg\,L^{-1}$. Hydrogen peroxide (10 % stock solution) was dosed in the influent water before it entered the loop reactor. Model compounds were dosed before the dosage of hydrogen peroxide. Directly after each dosing point, a static mixer was installed in the pipe. The water velocity in the loop reactor was $1.44\,m\,s^{-1}$ at design capacity of $5.0\,m^3\,h^{-1}$.

The loop reactor was equipped with a multi ozone dosing system. The ozone was dosed in parts by dividing the gas flow over a number of (maximal 6) IPs. A static mixer was installed after each IP. The pressure drop between two IPs was 1.85 mwc at design capacity. The retention time between the IPs was 1.2 s. Between the IPs, SPs were installed, see Fig. 2. The treated water discharged to a degasser, in which the off gas (oxygen and undissolved ozone) and water were separated. The residence time in this contact chamber was about 25 s. The off gas passed a column in which possible residual ozone was catalytically degraded into oxygen and discharged

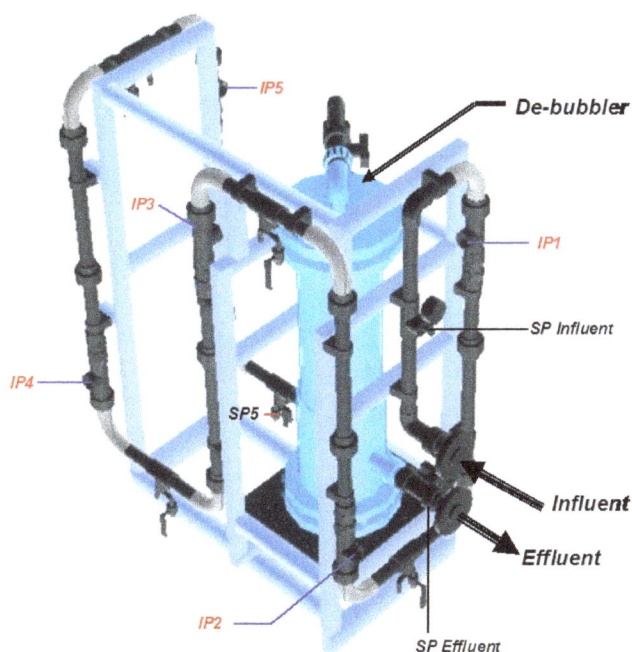

Figure 1. Peroxone loop reactor with injection points (IPs), sample points (SPs) and de-bubbler contact chamber.

Table 1. Minimum, maximum and average values of relevant parameters from RSF.

parameter	unit	minimum	average	maximum
Temperature	$°C$	1.5	12.3	20.8
pH	–	7.81	7.97	8.11
Ammonia	$mg\,NH_4^+\,L^{-1}$	<0.02	<0.02	0.09
DOC	$mg\,L^{-1}$	3.38	4.26	5.26
Bromide	$\mu g\,L^{-1}$	104	119	136
Bicarbonate	$mg\,L^{-1}$	155	175	204

Table 2. Settings of the standard experiments.

parameter	unit	settings
Capacity	$m^3\,h^{-1}$	5.0
Doses peroxide/ozone	$mg\,L^{-1}/mg\,L^{-1}$	6/0.7; 10/0.7; 6/1.0; 10/1.0; 6/1.5; 10/1.5; 6/2.0; 10/2.0
Dosing ozone points	–	IP 3, 4, 5, 6
Dose rate model compounds solution	$m^3\,h^{-1}$	0.068

outside the building by a ventilator. The treated water was collected in a $10\,m^3$ storage tank and treated several times by additional AOP by circulation till the OMP concentrations were lower than the detection limit before discharge.

The ozone content was measured in the feed gas and in the off gas by an BMT964 ozone analyser. In this way the efficiency of the ozone introduction and reaction was determined. Although six IPs were available in the loop reactor, most of the time only four were used (IPs 3, 4, 5 and 6) to limit pressure drop, by bypassing the first two IPs.

2.2 Influent water

The influent water was pre-treated surface water. The source of this water is a dead-end tributary from the Meuse River. In this tributary, with a residence time of several weeks, coagulation is applied. At the intake the water passes micro-strains (mesh width $35\,\mu m$). Afterwards dual media rapid sand filtration is applied. The quality of the intake of river water varies over the seasons as a result of meteorological, biological and hydrological influences. The main water constituents of the rapid sand filtrate (RSF) are provided in Table 1.

2.3 Experiments

Because of the varying water quality, research over a longer period was required to establish the formation of bromate and conversion of the model compounds. The influence of the varying parameters on the peroxone process was investigated on a regular basis with standard experiments, in which

8 hydrogen peroxide/ozone settings were used, see Table 2. In this way, design settings were established to apply the peroxone process.

In addition to the standard experiments, spike experiments were performed to investigate the role of bromide in bromate formation.

2.4 Investigated compounds

A set of 14 compounds were dosed to the RSF to investigate the conversion of model compounds by peroxone, see Table 3.

The model compounds were selected based on their different sensitivity for direct oxidation by ozone and hydroxyl radicals and their representativity for different kind of organic pollutants in river water.

The model compounds were spiked in concentrations from 5 till $30\,\mu g\,L^{-1}$, depending on the detection limit of the compound, dosing at least 95 times the detection limit. The applied OMP concentrations are expected to be sufficiently low to prevent interaction between the OMPs, as well as to assure that the degradation is independent of initial concentrations (Wols et al., 2013). The actual concentration of the model compounds was measured before the loop reactor. The concentration of DOC was only slightly increased by spiking, while the influent varied over the year from 3.5–5.5 mg $C\,L^{-1}$.

2.5 Chemical and physical analyses

The hydrogen peroxide concentration in water was analysed on site with a Hach DRL 2000 spectrophotometer. The measurement is based on the reaction of hydrogen peroxide with titanium(IV)oxysulphate solution, following DIN

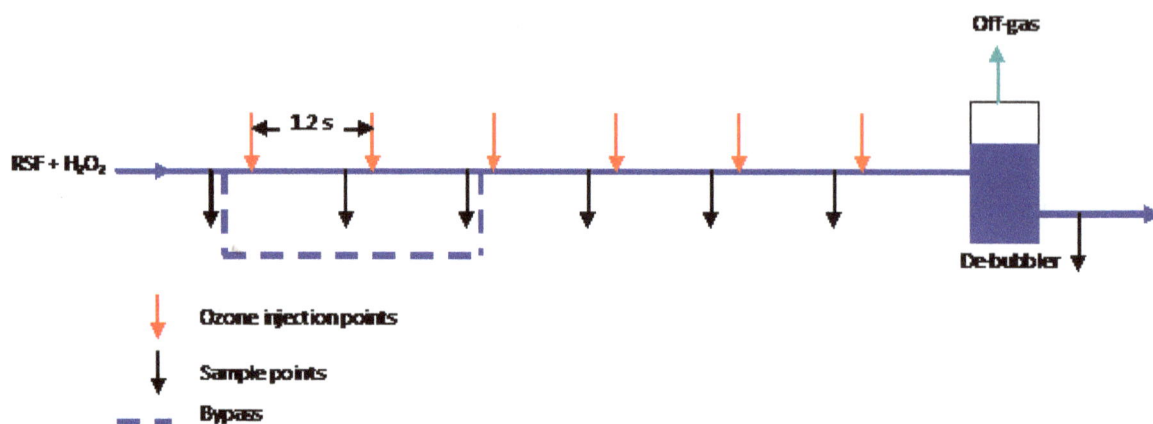

Figure 2. Schematic loop reactor.

Table 3. Average concentration of spiked model compounds in RSF.

Compound	Concentration ($\mu g\,L^{-1}$)
Diglyme	11.0
Bromacil	10.8
Bentazone	11.9
Atrazine	10.1
Isoproturon	10.2
Ibuprofen	18.8
Metformin	4.1
Carbamazepine	5.7
Metoprolol	5.4
Trimethoprim	5.7
Iopromide	1.8
Phenazone	5.8
Diclofenac	31.5
Furosemide	5.4

38409 H15. The water temperature was measured with a PT 100 element in the main influent.

Analysis of the model compounds was performed with fast analysis methods specially developed for this pilot plant research, i.e. they were (1) suitable to efficiently analyse the large number of samples that were generated in the experiments and (2) had large quantification ranges to enable determination of the removal rate of each model compound under the varying process conditions (Lekkerkerker-Teunissen et al., 2012). The methods used an Ultra Performance Liquid Chromatograph (UPLC, Waters Acquity), equipped with a quaternary pump and combined with a Quattro Xevo triple quadrupole Mass Selective Detector (Waters Micromass) with electro spray ionization. $50\,\mu L$ volumes of samples were injected without prior sample preparation. Compounds were measured in three separate runs, slightly differing in eluent composition and gradi-

ents. The first run analysed (quantification ranges given between parentheses) atrazine ($0.05–25\,\mu g\,L^{-1}$), bentazone ($0.10–25\,\mu g\,L^{-1}$), bromacil ($0.05–25\,\mu g L^{-1}$), diglyme ($0.15–25\,\mu g\,L^{-1}$), ibuprofen ($1.5–250\,\mu g\,L^{-1}$) and isoproturone ($0.05–25\,\mu g\,L^{-1}$), all in positive ion mode, except ibuprofen which was measured in negative ion mode. The second run analyzed diclofenac ($0.15–150\,\mu g\,L^{-1}$), furosemide ($0.025–25\,\mu g\,L^{-1}$), metformin ($0.005–5\,\mu g\,L^{-1}$) and phenazone ($0.005–5\,\mu g\,L^{-1}$), of which the first two were measured in negative ion mode and metformin and phenazone in positive ion mode. The third run measured carbamazepine ($0.005–5\,\mu g\,L^{-1}$), iopromide ($0.025–25\,\mu g L^{-1}$), metoprolol ($0.005–5\,\mu g\,L^{-1}$) and trimethoprim ($0.005–5\,\mu g\,L^{-1}$), all in positive ion mode. Quantification was performed using external calibration series of seven concentrations.

Bicarbonate concentrations were determined via titration of hypochloric acid ($0.1\,N$ increments) using methyl orange as indicator.

Nitrate concentrations were determined with continuous flow analysis (Skalar San^{++}). Concentrations of ammonium and nitrite were determined with an automated discrete photometric analyzer (Aquakem). Dissolved organic carbon (DOC) concentrations were determined with Non-Purgeable Organic Carbon Analysis (Shimadzu TOC-V$_{CPH}$). A sample was acidified to a pH of 2–3 with hypochloric acid and the inorganic carbon was subsequently eliminated with purging gas (O_2). The remaining total carbon C was measured and the result is generally referred to as TOC.

UV transmission was measured spectrophotometrically at a wavelength of 254 nm.

Bromate was analysed using ion exchange chromatography followed by conductivity detection (Dionex IonPac AS9SC). The measured bromate concentration was confirmed using a two point calibrated UV absorption measurement at a wavelength of 200 nm. The reporting limit of bromate was $0.5\,\mu g\,L^{-1}$, although values higher than the detection limit ($0.1\,\mu g\,L^{-1}$) were also evaluated to assess the bro-

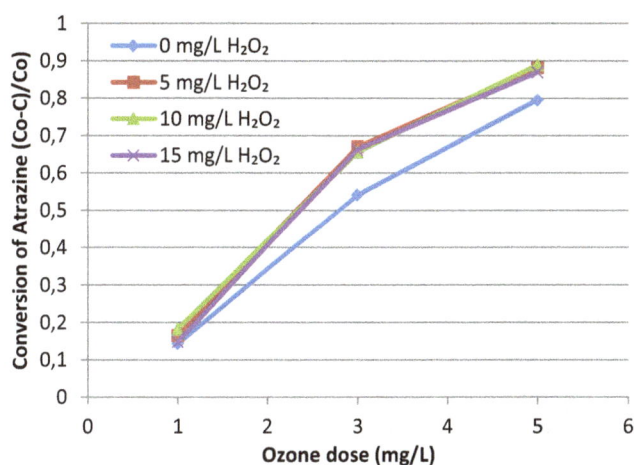

Figure 3. Conversion of atrazine at different settings of the peroxone process (Br$^-$ $124\,\mu g\,L^{-1}$, HCO$_3^-$ $158\,mg\,L^{-1}$, DOC 3,70 mg C L^{-1}, water temperature 11.9 °C).

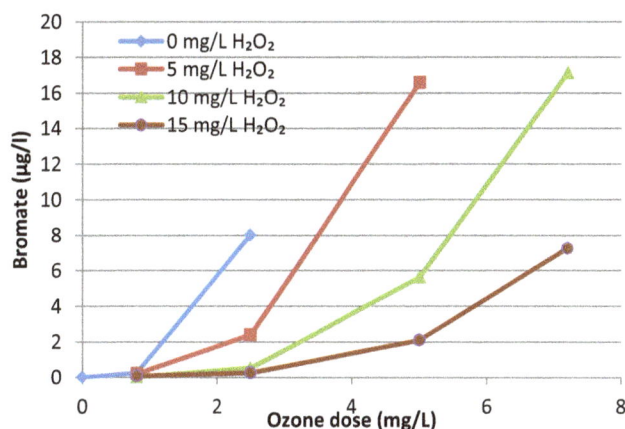

Figure 4. Bromate formation at different settings of the peroxone process (Br$^-$ $126\,\mu g\,L^{-1}$, HCO$_3^-$ $159\,mg\,L^{-1}$, DOC 3,26 mg C L^{-1}, water temperature 9.7 °C).

mate formation, because the reporting limit was equal to the bromate target value.

3 Results and discussion

3.1 Proof of principle

Atrazine was used to investigate the sensitivity for the peroxone process. The conversion increased with the ozone dose, see Fig. 3.

At a dose of $5\,mg\,L^{-1}$ ozone, about 80 % of atrazine was converted. By adding $5\,mg\,L^{-1}$ hydrogen peroxide, the conversion reached almost 90 %. Increase of the hydrogen peroxide dose above $5\,mg\,L^{-1}$ did not influence the conversion.

The energy consumption necessary to produce 5 g ozone to treat $1\,m^3$ water was about 0.045 kWh. To achieve a similar atrazine conversion with a comparable pretreated surface water by applying UV / H$_2$O$_2$, with the same hydrogen peroxide dose, the energy consumption in order to generate UV radiation is at least ten times higher (Lekkerkerker-Teunissen et al., 2013), which demonstrates the energy efficiency of the peroxone process.

Bromate formation also increased with the ozone dose, see Fig. 4.

However, by adding hydrogen peroxide, bromate formation was limited and the limitation was more when the ratio peroxide/ozone was higher. This observation is in line with findings of (von Gunten et al., 2003b). From Figs. 3 and 4 can be concluded, that the ozone dose is the main factor in converting atrazine and that the hydrogen peroxide dose is the main factor in limiting the bromate formation. Almost 90 % of atrazine was converted with $5\,mg\,L^{-1}$ ozone and at least $10\,mg\,L^{-1}$ hydrogen peroxide, without exceeding the WHO guideline for bromate of $10\,\mu g\,L^{-1}$. To comply a bro-

mate concentration below $0.5\,\mu g\,L^{-1}$ at an ozone dose of $5\,mg\,L^{-1}$, even $15\,mg\,L^{-1}$ hydrogen peroxide is not enough.

3.2 Bromate formation

In the period August 2011 up to and including March 2012, every other week peroxone dosing was investigated with standard experiments. The bromate concentrations are shown in Fig. 5. Because the pilot plant settings were similar during the experiments, differences in bromate formation were caused by variations in the water matrix.

As seen before (Knol, 2000), the influent water can contain low background concentrations of bromate. Figure 5 shows the bromate concentrations formed at different ozone/peroxide settings for the standard experiments over the test period. The curves show similar shapes, only the bromate values differed at the same settings. The bromate formation increased with increasing ozone dose and was reduced by increasing the hydrogen peroxide dose at a given ozone dose. Minimum bromate formation occurred in winter season, maximum bromate formation in summer season.

Only the peroxone settings with the high ozone dose of $2.0\,mg\,L^{-1}$, with either 6 or $10\,mg\,L^{-1}$ peroxide, exceeded the value of $0.5\,\mu g\,L^{-1}$ (with a maximum of $1.0\,\mu g\,L^{-1}$). Minimal dose of $6\,mg\,L^{-1}$ hydrogen peroxide combined with ozone doses up to $1.0\,mg\,L^{-1}$ limited the bromate concentration without exception below $0.5\,\mu g\,L^{-1}$. The bromate concentrations at setting 6/1.5 varied over the test period from 0.27 till $0.69\,\mu g\,L^{-1}$, with an average value of $0.41\,\mu g\,L^{-1}$. Thus a safe optimal setting concerning the bromate formation was found as $6\,mg\,L^{-1}$ hydrogen peroxide and $1.5\,mg\,L^{-1}$ ozone.

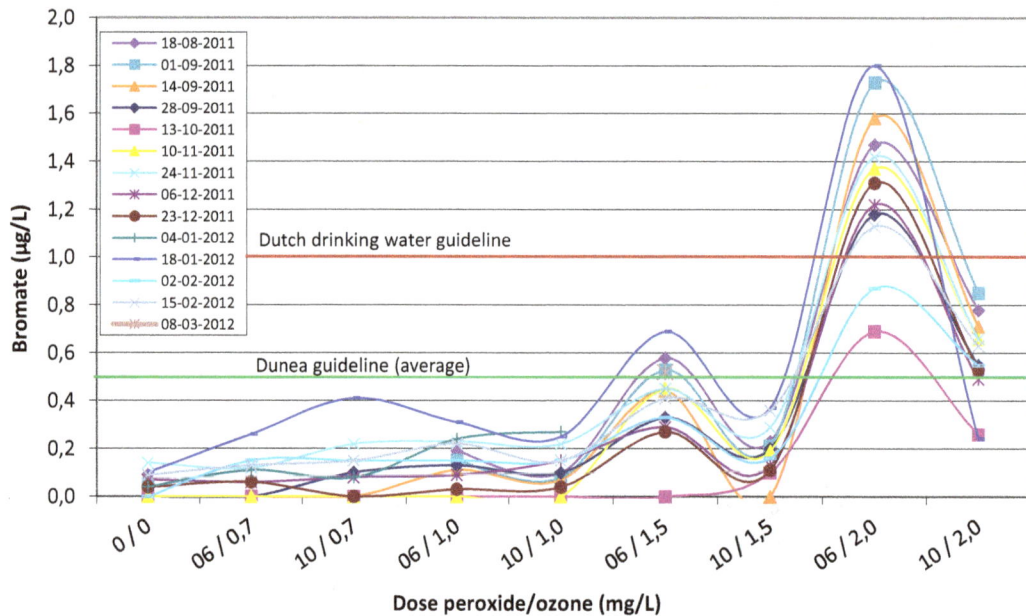

Figure 5. Bromate concentration at standard settings ($n = 14$, August 2011 up to and including March 2012, data points are connected to improve the interpretation).

3.3   Influence water matrix on bromate formation

To indicate the responsible parameter(s) for the variation in bromate concentration, the measured values of water temperature, bromide, bicarbonate and pH of RSF are plotted against bromate of setting 6/2.0 during the test period, Fig. 6.

The correlations in Fig. 6 are not strong, but certainly trends are visible: the bromate concentration increased with increasing water temperature and bromide concentration and decreasing bicarbonate concentration and pH.

Increase of bromate concentrations with increasing water temperature was expected (Croué et al., 1996; von Gunten, 2003b). Croué et al. reported an increase of 10 % per 10 °C, von Gunten even about 20 % per 10 °C. The variation measured in this study was between 17 and 34 % which is in line with previous research, regarding the 17 °C temperature difference between the measured minimum and maximum temperature during the standard experiments. The bromate concentration increased with increasing bromide concentration which is expected at bromide concentrations higher than $20 \mu g L^{-1}$ (Gottschalk et al., 2010).The bromate concentration decreased with increasing bicarbonate concentration. Bicarbonate is a known scavenger of hydroxyl radicals, so probably less radicals are available to react with bromide. Decreasing bromate formation with increasing pH is not in line with theory (von Gunten 2003b). However, the variation in pH during the standard experiments was small with a bandwidth of 0.19 units (7.81–8.00).

Probably the water temperature and the bicarbonate concentration strengthen each other. With increasing water temperature, the bicarbonate concentration decreased, Fig. 7,

both resulting in an increased bromate formation. No other correlations were found between the parameters water temperature, pH and the concentrations of bromide and bicarbonate under tested conditions.

For a better insight in the role of bromide in the bromate formation, concerning the forecasted increase in concentration in the Meuse River in global warming scenarios, this parameter is independently changed by spiking in RSF. The actual bromide concentration in the influent of $130 \mu g L^{-1}$ was increased to 270 and $560 \mu g L^{-1}$, Fig. 8.

Bromate formation strongly increased with increasing bromide concentration. It is therefore of importance to limit industrial spills and discharges of bromide in the main stream of the River Meuse, which are held accountable for about two-thirds of the bromide load in the Meuse (Volz, 2013). Based on the equation in Fig. 8, a provisional bromide threshold of $32 \mu g L^{-1}$ (0.24/0.0075) can be calculated. This value compares to the earlier mentioned $20 \mu g L^{-1}$ as reported by Gottschalk et al. (2010).

3.4   Conversion of organic micropollutants

The average conversion of fourteen model compounds at the four different ozone dosages (0.7 to 2.0 mg L^{-1}) and at a peroxide concentration of $6 \, mg \, L^{-1}$ is shown in Fig. 9. Conversions varied between compounds. Eliminations between less than 10 to over 95 % were observed. Reactivity of organic compounds towards ozone is strongly dependent on the molecular structure of compounds. Electron rich moieties such as aromatic rings and $C = C$ double bonds are main reaction sites at which ozone attacks (Sonntag et al., 2012;

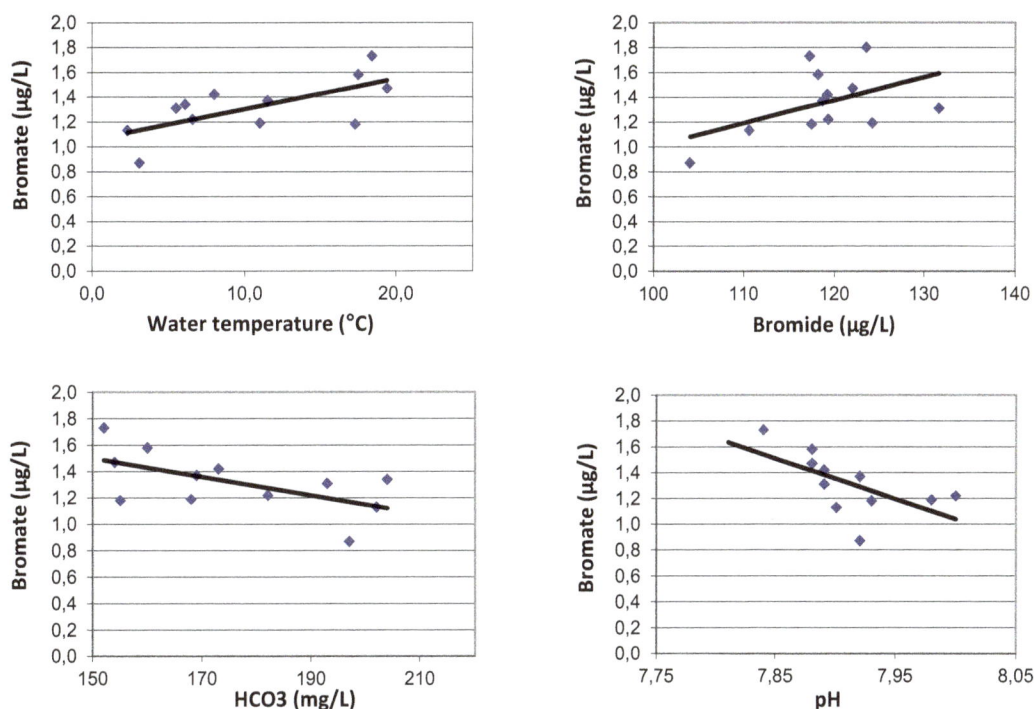

Figure 6. Water temperature, bromide, bicarbonate and pH vs. bromate concentration.

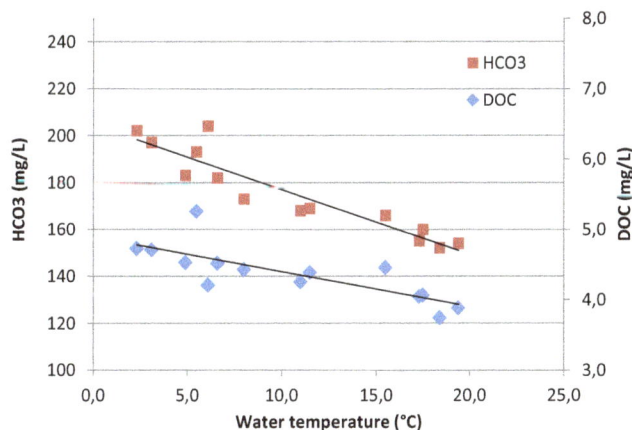

Figure 7. Correlation between water temperature, DOC and bicarbonate concentration.

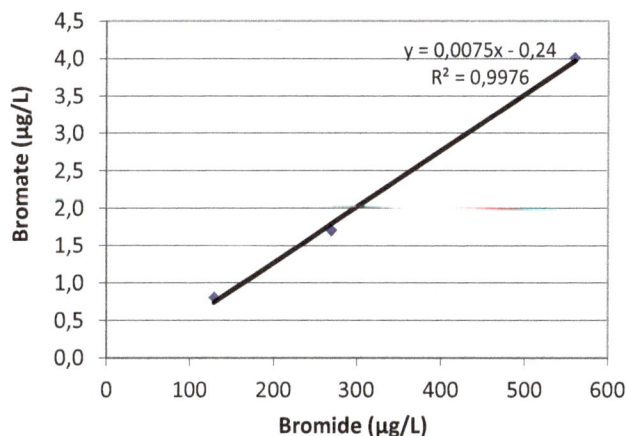

Figure 8. Bromate formation as function of bromide concentration at setting 6/1.5 (water temperature 1.6 °C, bicarbonate 201 mg L^{-1}).

Ternes et al., 2002). This structure dependency was clearly reflected in the results of the model compounds.

Lowest conversions were found for the model compounds metformin, atrazine, iopromide and diglyme, at the left part of Fig. 9. With the exception of iopromide, these compounds all lack aromatic rings and unsaturated C-C bonds. Their limited conversion is in line with results published by Acero et al. (2001) who found that compounds without aromatic and double bond structures, like MTBE, are poorly oxidizable by ozone alone. Iopromide, an X-ray contrast agent also contains one aromatic ring, but this ring is substituted at all six

C positions (a.o. with three jodine atoms) and therefore not easily oxidized.

Higher conversions, between 38 and 68 % at 0.7 mg L^{-1} ozone and between 85 and 99 % at 2.0 mg L^{-1} ozone, were obtained for ibuprofen, metoprolol, bentazone, isoproturon and phenazone.. These five compounds all possess one aromatic ring that is not substituted with halogens at which ozone can attack.

Furosemide, converted for about 95 % at all tested ozone doses, has one aromatic ring in its structure, together with

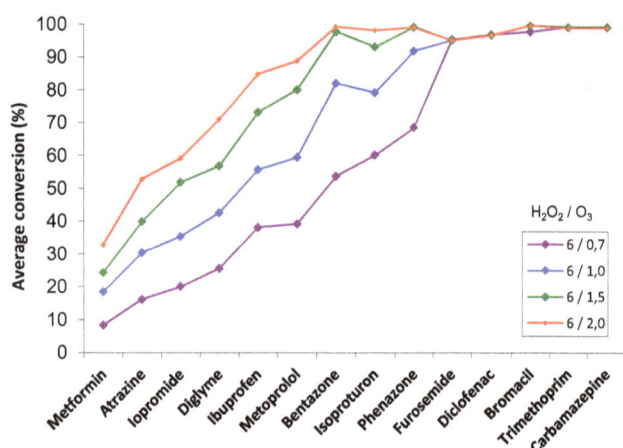

Figure 9. Average ($n = 14$) conversion in percentages of model compounds as function of ozone dosage at a hydrogen peroxide concentration of $6\,\mathrm{mg\,L^{-1}}$ in the period August 2011 till March 2012.

two $C = C$ double bonds. In addition, furosemide has an aromatic amine-N, which also acts as electron rich reaction site for ozone (Lee et al., 2014).

Highest conversions ($\geq 97\%$) were observed for diclofenac, trimethoprim and carbamazepine, due to the presence of two oxidizable aromatic rings in their structures. For diclofenac and carbamazepine, this is in line with Ternes et al. (2002).

The high conversion of bromacil is surprising regarding the fact that this herbicide does not have aromatic rings in its structure. It however does possess a bromine substituted $C = C$ bond. Bromacil was thus directly oxidized by the attack of ozone on this $C = C$ bond and the consecutive loss of bromine (Hapeman et al., 1997). This reaction is energetically favorable due to the fact that bromine acts as a very good leaving group and renders the conversion rate of bromacil comparable to that of the easily oxidizable model compounds with aromatic rings.

The conversion of the model compounds was also assessed at $10\,\mathrm{mg\,L^{-1}}$ peroxide. The influence of an increased hydrogen peroxide concentration at the same ozone dose on the conversion was however found to be small. In fact, in all cases the average conversion was equal or lower, with a maximum of 4% decrease at a dose of $1.5\,\mathrm{mg\,L^{-1}}$ ozone. This is in line with findings published previously for bromacil (Hapeman et al., 1997) and for compounds in hospital waste water effluent (Lee et al., 2014).

Two mechanisms could have been responsible for this phenomenon. Firstly, decomposition of the ozone is accelerated at a higher hydrogen peroxide dosage, which is a disadvantage for compounds that are sensitive for direct oxidation by ozone, and, secondly, excessive peroxide itself can act as a scavenger for hydroxyl radicals.

In general, the conversion of the model compounds was higher with higher ozone dosage. The conversion of the easily oxidizable compounds furosemide, diclophenac, bromocil, trimethoprim and carbamazepine already was maximal at the lowest dosage of $0.7\,\mathrm{mg\,L^{-1}}$. The conversion of the remaining compounds improved considerably by enhancing the ozone dosage to $2.0\,\mathrm{mg\,L^{-1}}$. Although from the perspective of micropollutant conversion a dosage of $2\,\mathrm{mg\,L^{-1}}$ would be optimal, the bromate formation at this setting exceeded the value of $0.5\,\mu\mathrm{g\,L^{-1}}$. The setting of $6/1.5$ combined acceptable bromate formation with an average conversion of the model compounds of 78.9% ($\sigma\ 24.8\%$).

It should therefore be noted that the peroxone process, like all advanced oxidation processes, does not provide a full solution for the problem of OMPs in water sources. Also because oxidation by peroxone leads to the formation of degradation products of OMPs instead of to full mineralisation, e.g. Escher et al. (2009); Sonntag et al. (2012), that may have unwanted toxic properties.

For these reasons, the peroxone process should preferably not be implemented as the only strategy for the removal of OMPs, but in combination with other techniques, in this case followed by biological degradation during MAR and adsorption during PAC filtration.

3.5 Influence of the water matrix on OMP conversion

Using the collected data of the standard experiments, the relevant water quality parameters of the influent are plotted against the conversion of the model compounds for which the observed conversion was less than 90% (Fig. 10).

The conversion of OMPs increased with increasing water temperature and decreasing concentrations of DOC and bicarbonate. The bromate formation increased with increasing water temperature and bicarbonate concentration. Furthermore, the conversion of OMPs and bromate formation increased with ozone dose and the bromate formation was reduced by increasing the dose of hydrogen peroxide. Therefore, the conversion of OMPs and bromate formation can be levelled by adjusting the ozone and hydrogen peroxide doses to water temperature, DOC and bicarbonate: In winter period the conversion of OMPs can be increased by increasing the ozone dose without exceeding the target bromate value and in summer period the bromate formation can be reduced by increasing the hydrogen peroxide dose, without effect on the conversion of OMPs. Controlling peroxone on basis of the online measured or derived parameters as water temperature, bicarbonate and DOC is feasible.

4 Conclusions

Advanced oxidation with O_3 / H_2O_2 was conducted on pilot plant scale on pre-treated Meuse river water to investigate the conversion of 14 selected organic micropollutants and the formation of bromate. The peroxone process effi-

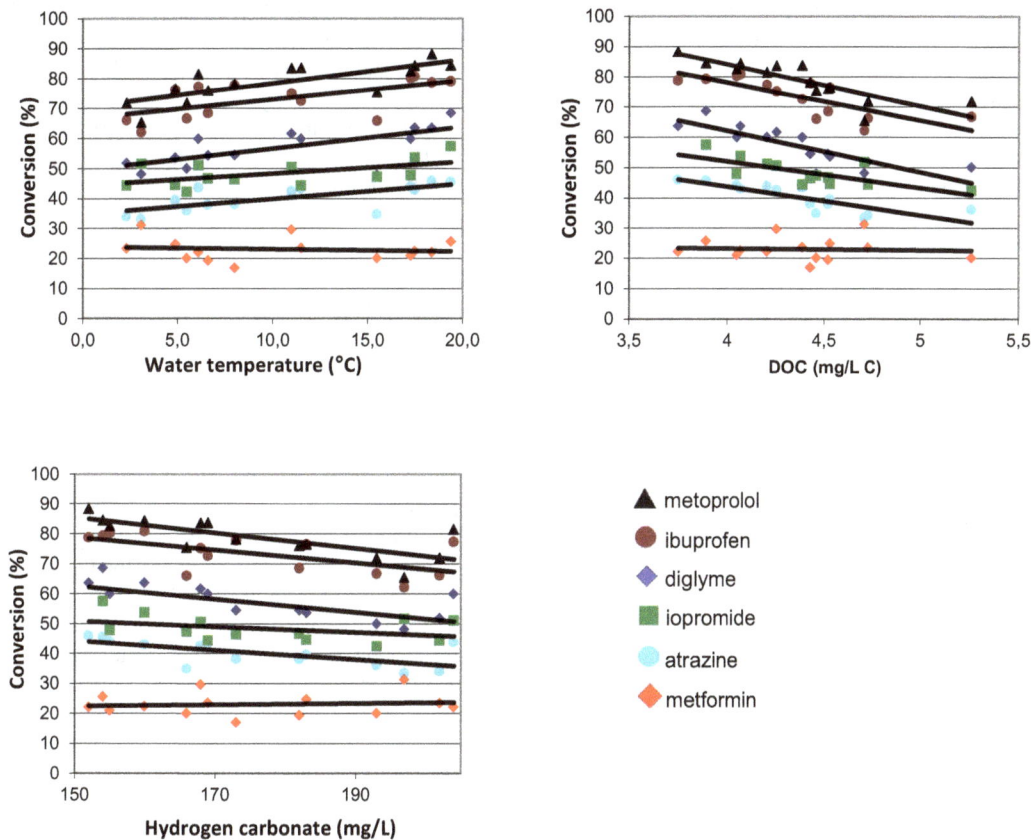

Figure 10. Conversion of compounds versus water temperature and DOC and bicarbonate concentration.

ciently degraded organic micropollutants with aromatic rings and/or unsaturated C-C bonds in their molecular structures. As many OMPs contain these features, the peroxone process might be a useful technique in the purification of contaminated surface water. The dosage of ozone, and with that the conversion of OMPs, is limited due to the bromate formation. Nevertheless an average conversion of 14 model compounds of well over 70 % was achieved with $6 \, \text{mg} \, \text{L}^{-1}$ peroxide and $1.5 \, \text{mg} \, \text{L}^{-1}$ ozone. The applied ozone dose was the main factor in the conversion of the model compounds. The hydrogen peroxide dosage had only a minor effect on the conversion, but limited the bromate formation effectively to levels below $0.5 \, \mu\text{g} \, \text{L}^{-1}$.

The peroxone process can be controlled on basis of the (derived) parameters water temperature, bicarbonate and DOC. Analyses of bromate then can be used to adjust the process.

Acknowledgements. The authors would like to thank the staff of treatment location Bergambacht for their technical support.

References

Abrahamse, A. J., IJpelaar, G. F., and Knol, A. H.: Project Uitbreiding Zuivering DZH, optionele technieken en locatie keuze, KWR, Dunea, 2007.

Acero, J. L., Haderlein, S. B., Schmidt, T. C., Suter, M. J. F., and von Gunten, U.: MTBE oxidation by conventional ozonation and the combination ozone/hydrogen peroxide: efficiency of the processes and bromate formation, Environ. Sci. Technol., 35, 4252–4259, 2001.

Amy, G., SiddiQui, M., Ozekin, K., and Westerho, P.: Treshold levels for bromate formation in drinking water, in: Proc. ISWA/AIDE International Workshop Bromate and water treatment, Paris, 22–24 November 1993, 169–180, 1993.

Croué, J. P., Koudjonou, B. K., and Legube, B.: Parameters affecting the formation of bromate ion during ozonation, Ozone Sci. Engin., 18, 1–18, 1996.

Escher, B. I., Bramaz, N., and Ort, C.: JEM spotlight: monitoring the treatment efficiency of a full scale ozonation on a sewage treatment plant with a mode-of-action based test battery, J. Environ. Monitor., 11, 1836–1846, 2009.

Gottschalk, C., Libra, J. A., and Saupe, A.: Ozonation of Water and Waste Water, a Practical Guide to Understanding Ozone and its Application, 2nd Edn., 2010.

Hapeman, C. J., Anderson, B. G., Torrents, A., and Acher, A. J.: Mechanistic investigations concerning the aqueous ozonolysis of bromacil, J. Agr. Food Chem., 45, 1006–1011, 1997.

Houtman, C. J.: Emerging contaminants in surface waters and their relevance for the production of drinking water in Europe, J. Integr. Environ. Sci., 1–25, 2010.

Houtman, C. J., Kroesbergen, J., Lekkerkerker-Teunissen, K., and van der Hoek, J. P.: Human health risk assessment of the mixture of pharmaceuticals in Dutch drinking water and its sources based on frequent monitoring data, Sci. Total Environ., 496, 54–62, 2014.

IJpelaar, G.: Robuustheid van de Zuivering bij DZH, Management Samenvatting Prioritaire Stoffen, 2005–2007, Kiwa Water Research, 2008.

Knol, A. H.: Gedrag van Bromaat in de Zuivering van Maas Tot Drinkwater, Duinwaterbedrijf Zuid-Holland, 2000.

Kurokawa, Y., Hayashi, Y., Maekawa, A., Takahashi, M., and Kokubo, T.: Induction of renal cell tumors in F-344 rats by oral administration of potassium bromate, A Food Additive, Gann, 73, 335–338, 1982.

Lee, Y., Kovalova, L., McArdell, C. S., and von Gunten, U.: Prediction of micropollutant elimination during ozonation of a hospital wastewater effluent, Water Res., 64, 134–148, 2014.

Lekkerkerker-Teunissen, K., Knol, A. H., Scheideler, J., Ried, A., Verberk, J. Q. J. C., and van Dijk, J. C.: Serial ozone-AOP and UV-AOP for synergistic and effective organic micropollutant treatment and bromate formation control, Sep. Purif. Technol., 100, 22–29, 2012.

Lekkerkerker-Teunissen, K., Knol, A. H., Derks, J. G., Heringa, M. B., Houtman, C. J., Hofman-Caris, C. H. M., Beerendonk, E. F., Reus, A., Verberk, J. Q. J. C., and van Dijk, J. C.: Pilot plant results with three different types of lamps for advanced oxidation, Ozone Sci. Engin., 35, 38–48, 2013.

Schriks, M., Heringa, M. B., van der Kooij, M. M. E., de Voogt, P., and van Wezel, A. P.: Toxicological relevance of emerging contaminants for drinking water quality, Water Res., 44, 461–476, 2010.

Sonntag, C. and von Gunten, U.: Chemistry of Ozone in Water and Wastewater Treatment: From Basic Principles to Applications, IWA Publishing, 2012.

Ternes, T. A., Meisenheimer, M., McDowell, D., Sacher, F., Brauch, H. J., Haist-Gulde, B., Preuss, G., Wilme, U., and Zulei-Seibert, N.: Removal of pharmaceuticals during drinking water treatment, Environ. Sci. Technol., 36, 3855–3863, 2002.

Volz, J.: Bromidebronnen in Het Stroomgebied van de Maas, Een Kwalitatieve en Kwantitatieve Verkenning, 21 Maart 2013.

von Gunten, U.: Ozonation of drinking water: Part I. Oxidation kinetics and product formation, Water Res., 37, 1443–1467, 2003a.

von Gunten, U.: Ozonation of drinking water: Part II. Disinfection and by-product formation in presence of bromide, iodide and chlorine, Water Res., 37, 1469–1487, 2003b.

WHO: Bromate in Drinking Water, Background Document for Development of WHO Guidelines for Drinking-water Quality, (WHO/SDE/WSH/05.08/78), 2005.

Wols, B. A., Hofman-Caris, C. H. M., and Beerendonk, E. F.: Degradation of 40 selected pharmaceuticals by UV/H_2O_2, Water Res., 47, 5876–5888, doi:10.1016/j.watres.2013.07.008, 2013.

19

Online data processing for proactive UK water distribution network operation

J. Machell, S. R. Mounce, B. Farley, and J. B. Boxall

Pennine Water Group, University of Sheffield, Civil and Structural Engineering, Mappin Street, Sheffield S1 3JD, UK

Correspondence to: J. Machell (j.machell@sheffield.ac.uk)

Abstract. Operational benefits and efficiencies generated using prevalent water industry methods and techniques are becoming more difficult to achieve; as demonstrated by English and Welsh water companies' static position with regards the economic level of leakage. Water companies are often unaware of network incidents such as burst pipes or low pressure events until they are reported by customers; and therefore use reactive strategies to manage the effects of these events. It is apparent that new approaches need to be identified and applied to promote proactive network management if potential operational productivity and standards of service improvements are to be realised.

This paper describes how measured flow and pressure data from instrumentation deployed in a UK water distribution network was automatically gathered, checked, analysed and presented using recently developed techniques to generate apposite information about network performance. The work demonstrated that these technologies can provide early warning, and hence additional time to that previously available, thereby creating opportunity to proactively manage a network; for example to minimise the negative impact on standards of customer service caused by unplanned events such as burst pipes.

Each method, applied individually, demonstrated improvement on current industry processes. Combined application resulted in further improvements; including quicker and more localised burst main location. Future possibilities are explored, from which a vision of seamless integration between such technologies emerges to enable proactive management of distribution network events.

1 Introduction

Internationally it is becoming difficult to improve upon the effectiveness and efficiency of network operations such as leakage management because the limits of current techniques are being reached. The English and Welsh regulator Ofwat concluded that all but two water companies have reached an economic level of leakage (Pearson and Trow, 2005) using traditional techniques (House of Lords, 2006). Consequently, there is a need for water companies to be more innovative. The technology required to fully understand, manage, and automate distribution network operation either does not yet exist, is only partially evolved, or has not yet been proven reliable and cost effective for live networks. New technologies and asset management techniques therefore need to be ex-

plored if knowledge and understanding of water distribution, and customer service, is to be improved.

The aim of this paper is to present the practical application of findings from several related research areas. The tools, techniques and methods discussed have previously been presented individually, and verified to different degrees; but they have never been applied in combination to a single case study to demonstrate their combined potential. This paper seeks therefore to highlight the multiplicative benefit of combined application, and the potential for proactive distribution network management that these techniques could enable.

2 Opportunity and case study

2.1 Opportunity

Flow and pressure instrumentation has evolved over many years, and highly robust and accurate off the shelf designs are readily available. The current placement of instrumentation in water distribution networks in the UK is to meet the requirement to monitor and report leakage and low pressure; but measurement locations are not always the most sensitive for this purpose. Automated techniques have therefore been developed to identify these optimal locations (Perez et al., 2009, 2011; Farley et al., 2008, 2010a, b; Rosich et al., 2012; Preis et al., 2011; Goulet et al., 2013) and continue to be improved.

Until recently, the UK water industry norm for data collection from distribution network instrumentation was via manual methods carried out during site visits, often only once a month or even less frequently. One reason for infrequent data collection was cost; data transfer via telephone systems being particularly prohibitive in the past. However, as data gathering and communication technologies steadily improved they became less expensive to own and operate, and water companies began to automate data collection. Data capture rate remained the same, but transfer to central storage was typically increased to once every 24 h. Instrument communication and data transfer is now tending to shift positively towards more advanced General Packet Radio Service (GPRS) or low power radio hopping solutions, with some vendors providing support for the widely used Internet Protocol (IP). For the work reported here, GPRS technologies were employed which enabled the change from data downloads once every 24 h, to once every 30 min; 48 times more frequent than previously. This higher frequency flow and pressure data enables network performance to be analysed in a short time scale previously not possible. Rapid detection of burst pipe and low pressure events allows proactive management of their effects thereby minimising water loss and standards of service failures.

Despite these improvements, some fractions of water use, for example much domestic and small business consumption, remain estimated. The next generations of instruments and communication technologies will make data collection and transfer more efficient and cost effective; even from remote sites. Wireless, m-bus, z-wave, Wi-Fi, EDGE, 4G & 5G, and data analytic techniques are in continuous development, and have the potential to further revolutionise the way water companies gather and hence analyse and use data. Also, in the future, technologies such as Automated Meter Reading (AMR), or Advanced Metering Infrastructure (AMI), will allow water companies to gather high frequency consumption data directly from residential and commercial customers as well as existing sites. This will remove the need to estimate components of leakage calculations, and flow or demand in hydraulic modelling applications; and provide a far better understanding of distribution network flow and pressure dynamics.

Data analytical tools are also evolving rapidly. The need to efficiently manage and analyse increasing numbers of data streams has spawned numerous techniques, many new to the water industry (Srirangarajan et al., 2010; Zan et al., 2011; Romano et al., 2013; Michalak et al., 2012). Artificial intelligence and network hydraulic simulation methods are able to automatically verify, process, and continually analyse large amounts of data. They convert data into information and present it in a time frame completely impossible using manual or semi-automatic systems (Gama, 2010; Preis et al., 2012, Machell et al., 2010).

This study took the opportunity to apply several state of the art methods to a single case study; as individual techniques and then in combination.

2.2 Case study

The water company collaborating in this work had recently implemented a remote communication/data transfer evaluation project in a complex distribution network that was defined by a number of interconnected District Metered areas (DMAs). A DMA is a hydraulically bounded area within a distribution network that typically contains 500–5000 customer properties. The hydraulic operation of the network was not well understood, and relative performance measures such as leakage per property were high; providing significant challenge. The project involved the installation of 490 data loggers to gather data from DMA inlet flow meters, and pressure measurements from incorporated transducers. Flow and pressure data was sampled at 15 min intervals. Measurements were also taken at 1 min intervals from some locations to determine whether a higher data resolution could provide any useful extra information. The loggers were equipped with a GSM modem and were, at the time, unique in the industry in that they communicated via GPRS and transferred data to a mobile telephony provider's data centre. 2 measurements, 15 min apart, were transferred from each instrument every 30 min. From there, the data was relayed to the water company via a high speed broadband connection known as a pipeline, into a central data store.

Access to the water company's project study area, instrumentation and data, provided an ideal opportunity to apply and evaluate the methods used in this work. An overview of the individual state of the art for these methods is now presented.

3 Overview of the state of the art techniques applied

3.1 Instrument location & DMA subdivision

Typically, pressure data is used to identify low pressure events within a DMA, or for their retrospective analysis. The point of highest elevation is often accepted as a pragmatic and reasonably indicative location to monitor pressure. However, a monitoring location may also be selected because of a history of known low pressure events in a specific area of a network. In the UK, this pressure monitoring location is commonly known as the DG2 point, and is primarily used to report against the regulatory minimum pressure levels of service requirement. The current approach does not necessarily identify the most sensitive location for detecting pressure fluctuations. If the most sensitive locations were determined, more fluctuations and events would be detected and awareness of system performance increased; for example, more burst pipes can be identified, and more accurate estimates of the number of customers suffering low pressure, over what duration, can be determined. A method to identify optimal locations for pressure instruments used for detection and location of leak/burst events was developed by Farley et al. (2008, 2010a, b). The approach utilised a methodology that searches a Jacobian sensitivity matrix produced by sequentially modelling leak/burst events at all nodes in a 1-D hydraulic model, and evaluating the change in pressure response at all possible instrument locations. The matrix was then searched (focussing on detection) to maximise overall system sensitivity and to minimise the number of instruments required in a given network; or where appropriate, to complement existing instrumentation to increase detection sensitivity.

To validate the approach, a number of pressure instruments where deployed in selected DMAs and a series of burst events were simulated by opening a number of fire hydrants inside DMAs. Data was collected from the pressure instruments located inside each DMA, including the current DG2 location, and at the most sensitive (or optimal) location previously identified by a Jacobean matrix approach. Analysis of pressure measurements recorded before, during, and after the burst events were used to evaluate the sensitivity of each instrument location to the pressure changes caused by the simulated burst events (Farley et al., 2010a). Figure 1 shows model predicted, and actual sensitivities for 5 simulated bursts at eight instrumentation points in a single DMA highlighting the correlation between predicted (model) and actual (real) sensitivities. Predictions were made using typical UK industry standard hydraulic models with no additional calibration.

From Fig. 1 it is clear that hydraulic models can be used to accurately predict pressure changes caused by burst mains, and can therefore be applied to support the optimal location methodology. Additional analysis of the data demonstrated that data from optimally located pressure instrument(s) can be used to detect more low pressure and leak/burst events within a DMA than data from current DG2 locations, and how to selectively supplement existing instrumentation to improve detection.

The methodology to identify the most sensitive locations for instruments can be extended to provide event location information. Further work (Farley et al., 2013) explored an approach to search the Jacobian sensitivity matrix to provide

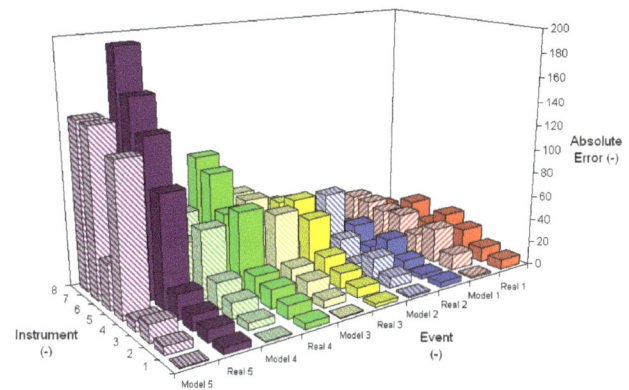

Figure 1. Model predicted vs. actual data sensitivity (for looped DMA).

differential location information, requiring the integration of a Genetic Algorithm (GA) search routine to improve efficiency. Single, and multiple, optimally (providing selected levels of overlapping or different zones of sensitivity) located instruments can be selected to identify sub-areas within a DMA, without the need for network reconfiguration or closed valves. This method has great potential, not only for detecting new leaks/burst events, but also steering leak location efforts.

3.2 Data quality and checking

If maximum benefit is to be obtained from investment in collecting large volumes of data, management, storage and access to both current and historical data sets is required.

The distribution network flow and pressure data quality gathered for this study can generally be classed as "dirty". Dirty data is manifest as large chunks of missing, corrupt or out of range values from faulty loggers and/or the presence of erroneous date and time stamps. A methodology for dealing with these issues was developed as an integral part of the online Artificial Intelligence (AI) system (Sect. 3.3) in order to more easily manage large amounts of data of varying quality. Data checking included statistical tests for stationarity (Mounce et al., 2010).

7-Technology's data warehouse product, Data Manager, was used in this study to pre-process and make data accessible to the Automated Data Analysis system (AI system) and online hydraulic models described later. Data Manager was used to perform the following functions:

- Receive flow, pressure and service reservoir level data

- Perform checks and conversions on the data received (using a pre-processor) and data emulation where measured data was not available for some reason

- Maintain cyclic buffers of weekly profiles for selected instruments

– Make the pre-processed data available for online AI system analysis and hydraulic simulations (historical and current)

– Automatically create and back up a database of historic flow and pressure data

Pre-processing involved automatic data checks for missing or corrupt data and/or data values that fell outside a normal value distribution at each instrument location. This was done using either absolute values or rate of change of values in the data time series. Where a data check failed, error messages were created and, if required, data emulation was performed. Emulation was used to replace missing or corrupt data with a fixed or an average value for the measurement site or, by reference to another site that usually has very similar time series values. In this way, Data Manager was configured to interface the real network measured values with its equivalent in an online AI system (Sect. 3.3) and hydraulic models; the latter used the data as boundary conditions for simulations (Sect. 3.4).

Data quality was a significant issue especially when obtaining long period time series. The quality and continuity of data is essential for online modelling and AI systems. Although a hydraulic model can deal with missing data using substituted values, as the number of substitutions increases the accuracy of the model decreases and, eventually, a point is reached where the resultant error is so large simulation results cannot be relied on to be fit for purpose.

3.3 Automated data analysis for burst detection

The challenge of timely data analysis can be met using automated computation methods. Event detection algorithms work by obtaining data, performing some analysis and then returning outputs such as probabilities or fuzzy values. These are then processed into a binary classification i.e. generate an alarm or not. Despite some software offering profile alarm levels, there has been very little application to date in the water industry. Existing state of the art systems use flat line alarm thresholds for key measurement sites, which are then continuously monitored to enable identification of large bursts. The collaborating water company established and implemented site specific high and low thresholds for DMA inlet flow and pressure flat line alarms, to provide simple data analysis of absolute, individual values. At the time, implementing these flat line alarms was a significant move forward for detecting network events and represented a step change in awareness of network hydraulic performance. In particular, they proved useful for detecting sudden catastrophic bursts, and acting as a failsafe. However, due to continual adjustment of alarm level settings over time the flat-line system started to generate a high number of false alarms (ghosts) that did not correlate with any known events, also many pipe bursts were still not detected and no action was taken. A significant issue with flat-line alarm thresholds is the trade-off

between false alarms (alarms with no identifiable cause) and non-detection of smaller burst events. Although this is an issue for all event detection systems, it is more significant for flat lines where the level has often been set arbitrarily, without any of the positive aspects of data driven methodologies.

More sophisticated data analysis methodologies using Artificial Neural Network (ANN) and Fuzzy Logic (FL) technology were therefore applied. Recorded flow and pressure data from each measurement location was sorted by DMA name and stored in a database. Assembled data sets which passed data quality checks were first pre-processed to deal with any missing data using well proven time series analysis techniques for filling, then normalised by re-scaling to a range required by the ANN, and finally restructured into the format required by the analysis system. A single model was used for each data stream. The pre-processed data sets were used to train a Mixture Density Network (MDN) ANN to make time series predictions based on a lag of past time series values. Importantly, this prediction was not a particular value; the MDN architecture learns the distribution (through a mixture model) for a particular instrument from past data, and assigns a confidence level to the observed flow values. A Fuzzy Inference System (FIS), consisting of a set of fuzzy rules, analysed this mixture model and the observed value over a suitable time window and generated classification fuzzy values (an indication of abnormal events such as burst pipes) such that a % confidence could be assigned to alerts and, in addition, an accurate estimate of likely burst size could be determined. Figure 2 shows how this continually updated model constructs the probability density at time steps into the future horizon.

The algorithms and software employed, as well as further background on system integration, are described in more detail in Mounce et al. (2003) and Mounce et al. (2010). Abnormal classifications by the FIS were entered into an alerts database, and automated email alerts were sent to the water company's control room staff.

A first quarter (three month) evaluation of the system, when leak/burst rates are usually highest in the UK (commonly attributed to freeze/thaw cycles), was conducted using manual data inspection, and correlation to repair information recorded on the water company's Work Management System (WMS) and customer contact database. During this period the online AI system was analysing flow and pressure data from as many as 156 flow and 255 pressure instruments, the actual number being dependent upon changes in data quality over time. A total of 227 alerts generated by the AI system were reviewed, 78 for flow and 149 for pressure. (For comparison it should be noted the flat line system generated 47 853 alarms for the same data streams and the same three month period. It should be stated however, that the flat line alarm thresholds were probably valid at the time of initial set up of. Nonetheless, they were not re-assessed often enough to keep them valid through changes in network operational characteristics and/or changes in demand patterns.)

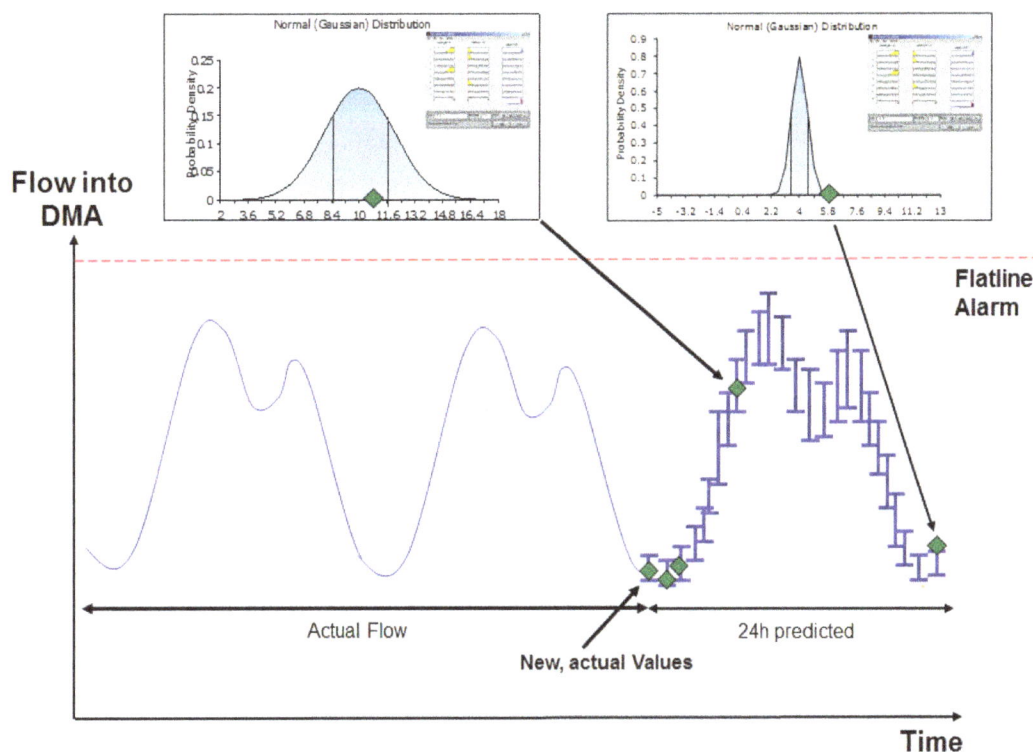

Figure 2. Mixture model for future flow prediction and fuzzy interpretation.

Their classifications are shown in Fig. 3. The class "abnormal" includes all cases where the AI system produced an alarm, and subsequent manual data interpretation confirmed that a noteworthy change in the data stream did occur, but for which there was no correlation with mains repairs or customer contacts i.e. the result of an unknown cause that did not fall into one of the other 5 (normal) categories that correlated with a known causes. The signature of some of these abnormal events such as large industrial demands, or closure and opening of valves, could not necessarily be differentiated from that of bursts using the AI analysis system. However, their detection was still of significant interest to the water company in that it provided important operational information that was previously undetected. Unexpected or unlicensed water use such as for the filling of private fire tanks, increased industrial use for new processes, unauthorised filling of street cleaning equipment and water bowsers, or illegal connections, can all generate abnormal flow events and are all activities that water service providers need to be made aware of if proactive management and control of the network is to be realised. Detections can also be produced by network rezoning activities, changes to valve arrangements or pump schedules and other operational scenarios, thereby providing an additional check of the timing, magnitude and effect of such activities. This information is particularly important when such activities have been outsourced to third party contractors.

From Fig. 3 it can be seen that during the period of analysis a good correlation was found between the AI system generated alerts and company WMS/customer contact information. This was very good for flow where 95 % of flow alerts corresponded to WMS/contact information or known engineered events with only 5 % ghosts. The pressure data analysis was found to produce more ghosts (38 %), which was expected based on previous work and other researcher's findings which highlight the non-stationary nature of the pressure profile over time (Mounce et al., 2011, Ye and Fenner, 2010).

3.4 Combination of optimal location and automated analysis technology

The effectiveness of combining optimally located instruments and the AI event detection system was tested for detection and location of simulated bursts. Simulated bursts were used to negate the delays and uncertainty associated with real events. Optimal instrument locations were identified and instruments installed across 16 DMAs. Hydrants were then flushed by water company Field Teams in several DMAs chosen at random; the research team being unaware of the hydrant flushing locations or times. In this way 6 system blind test events were created in 5 different DMAs; all of which were positively detected by the instrumentation and the AI system; which correctly identify which DMA the event had been created in, and accurately reported the magnitude of the flow of each hydrant flush (simulated burst flow).

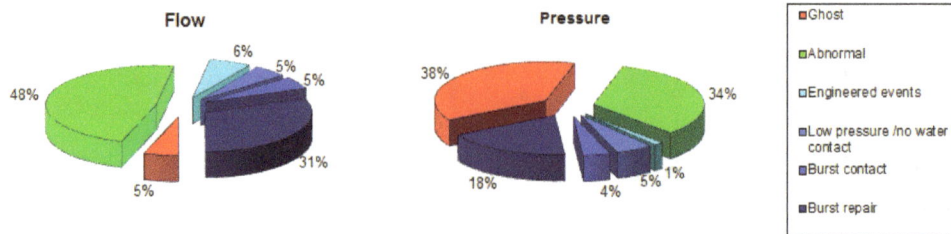

Figure 3. 3-month summary of AI flow and pressure alerts (January to March).

5 of the 6 blind tests were conducted in 4 different DMAs that had additional (optimally located) pressure instrument(s) installed. Results confirmed that the methodology developed to divide a DMA into a number of sub-zones to reduce the search area and improve burst location worked well. The correct sub area within each DMA in which a simulated burst occurred was correctly identified. 3 out of the 5 blind tests produced exact agreement between the model simulated and actual instrument responses; and therefore the correct zone of the burst event could be identified (Farley et al., 2012). In the other 2 DMAs, factors beyond the control of the method, such as instrumentation or logger failure, prevented this.

A successful example result of how pressure instruments can be deployed to provide sub-DMA event location is shown in Fig. 4.

For the case shown in Fig. 4, alerts for the DMA inlet flow meter and DG2 pressure instrument were generated, but not for the additional pressure instrument in the yellow (vertical lined) sub-zone. The pressure instrument at the inlet also did not detect the event, as predicted by the methodology. By reference to the DMA sub-zone sensitivity matrix, it was confirmed that this combination of alerts correctly identified that the hydrant flush had occurred in the blue (horizontal lined) sub-zone, hence demonstrating how this combination of event detection and optimal location can provide sub-DMA location. Field tests also demonstrated the potential for application of the method to multiple DMAs, and that subdivision of DMAs with very differing characteristics was possible. A reduction in search area was achieved for all but one DMA. The number of zones the DMAs were divided into was dependent on their individual characteristics. Results suggested that subdivision of typical DMAs into 4 zones can be achieved through the deployment of only one additional pressure instrument.

An extension of optimal instrument location approaches could be the creation of monitoring areas without the need for closed valves; virtual DMAs. This would enable more open systems with less dead-ends and the inferred water quality issues.

3.5 Online modelling

Aquis, supplied by 7-Technologies A/S, Denmark, was used to build and develop several hydraulic models for 1-D application. These models were used to populate the Jacobean sensitivity matrices for instrument location methods (Sect. 3.1), and to present network flow and pressure information superimposed on a street map background showing network assets and customer properties. Measured flow and pressure inputs and model simulation outputs were used to generate alarms and warnings when flow(s) or pressure(s) anywhere in a network and/or at specific network locations moved outside normal operational values.

A hydraulic network model is most useful when it displays up to date information and, ideally, predictions for the next few hours. The former is governed by the age of available input data. The latter can be achieved by simulation and extrapolation of current state using inbuilt predicted and normalised patterns for non-measured demand components. When measured data is regularly streamed into a model to set simulation boundary conditions the model is termed "online". Online model boundary conditions are continually updated with the most recent data available; in this case the data was 30 min old. An online model is not constrained by fixed (historic) 24 h flow and pressure profiles used as boundary conditions in conventional extended period simulation models. This means that non-diurnal flow and pressure patterns are accounted for and reflected in simulation output.

A 2 DMA model was built and used to demonstrate the effects of changes in network flow and pressure on customers during a real network event when a hydrant was opened by unknown persons in one of the study DMAs (Machell et al., 2010). The effects of changes in flow and pressure could be viewed as they rippled through the network.

A 3 DMA model was used for the evaluation of the optimal pressure instrument number and location methodology and provided flow, pressure, and velocity results for GIS visualisation. Figure 5 shows the extent of this model on a GIS display with results overlaid showing the pressure in each pipe in the network. The time shown in the legend of Fig. 5 is arbitrary. The legend is fully populated to be fully explicit with regard to what is presented in the figure.

A 16 DMA model was configured to simulate every 30 min. The 30 min interval between simulations was appropriate for this work but could have been reduced; the lower limit being dictated by the time taken to capture, transfer, and pre-process raw data. 77 flow and pressure instruments

Figure 4. Sub-DMA location of blind event in the pilot area (after Farley et al., 2012).

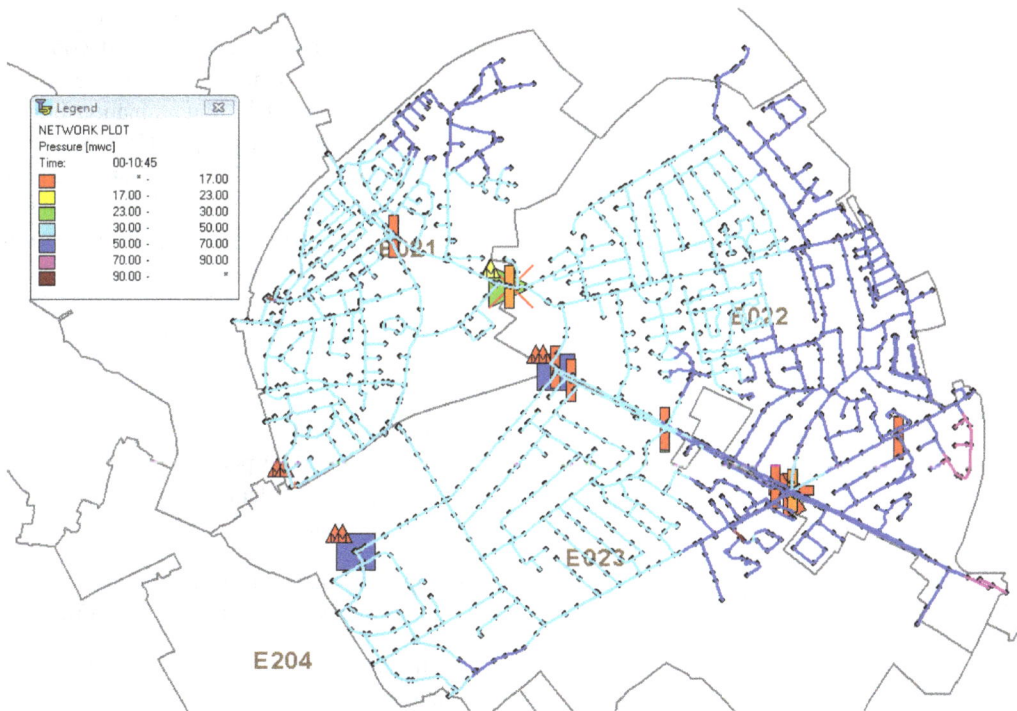

Figure 5. 3 DMA online model showing pressure in pipes at 10:45 GMT.

at the inlets and outlets of 16 DMAs, 4 service reservoirs, 2 pumping stations and a number of DG2 pressure monitors provided boundary condition data for this model. Data Manager pre-processed raw data from these sites and passed it to the model. This model was used to support the DMA subdivision work and instrument/AI system tests (Sect. 3.4) and test alarm and warning functionality of the modelling system.

Pressure data was found to be extremely useful for the online modelling system. It was a powerful resource for model calibration and fault finding, and was used to generate alarms for low or unnecessarily high pressure detections. Availabil-

ity of a stream of continuously updating flow and pressure data enabled calibration to current, rather than historic measurements, and to make it a continuous and iterative process, reflecting ever changing dynamics in the network caused, for example, by changes to valve positions, the timing of pump operations or the turnover rate of service reservoirs.

Simulation results demonstrated that the online models could accurately calculate the magnitude of flow and pressure fluctuations caused by simulated bursts (Sect. 3.1) and hence the effect on standards of service. This functionality can be used to identify faulty instrumentation or corrupt/missing data, unexpected flows or pressures, and pump failure for example. It can also generate warnings when flow or pressure is persistently moving towards an alarm situation. This enables fast response and proactive action to minimise the effect on customers.

On-line simulation of a water distribution network provides a tool that can offer tangible and significant operational benefits for network managers. It allows network operators to progress beyond the reactive and develop a proactive approach to network management. For example, the preliminary effects of a burst main can be detected at an early stage which, in turn, can allow the operators to minimize the later effects and sometimes pre-empt and avoid many standards of service failures by manipulating valves, providing an alternative supply or making other appropriate operational change. Pressure effects on every pipe in a network can be captured to gather knowledge about which customers were affected by an event and for how long – a UK standards of service reporting requirement. With optimally located instrumentation, online models provide a visual overview of all flows and pressures across an entire network. Online models can be used to generate timely warnings of unusual flow and pressure events. They can monitor for persistent changes such as slow but continuous reduction in pressure. Then, in combination with the specialised detection and analysis of the AI system, can be used to help to locate burst pipes and unusual network flow and pressure events.

3.6 Data sampling rate

The unofficial UK industry standard for the temporal resolution of flow and pressure data, at most locations, is 15 min (Mounce et al., 2012). To some extent the use of 15 min hydraulic data is pragmatic (storage space of loggers and receiving systems), but it also reflects a trade-off between the volume of information collected and the detail of flow or pressure fluctuations that can be captured. A good representation of the overall dynamics within a network can be observed with fifteen minute data, although the shape and amplitude of pressure transients cannot be resolved with data points more than a tenth of a second apart. Higher frequency sampling potentially also allows component analysis in order to gain an understanding of the different contributions to the total flow from different types of demand such as domestic and industrial, or flow due to leakage. However, little published work has investigated the benefit of using logging intervals in the sub-fifteen minute range. Data sample rates of 1, 5 and 15 min were evaluated for their suitability when used for flow and pressure event detection by the systems developed in this study; Table 1.

From results shown in Table 1, Mounce et al. (2012) concluded that, at the present time, sampling intervals of 1 or 5 min do not improve event detection sufficiently to justify the extra resources required to gather the data. For example; the increase in power required for battery powered instru-

Table 1. Summary of average improvement in detection times (minutes) over multiple instruments relative to 15 min instantaneous data in each case (Mounce et al., 2012).

	Flow pressure		Flow only		Pressure only	
	Inst.	Avg.	Inst.	Avg.	Inst.	Avg.
15 min	0	44	0	45	0	43
5 min	29	49	38	52	26	48
1 min	53	54	53	53	53	54

ments and loggers (all the loggers used in this study were battery powered), and the data management overheads. Similarly, current online modelling approaches do not yet require data at these higher frequencies. However, this is likely to change in the future as the density of sensing of water distribution system parameters increases due to reducing cost and improving logging capacity and communications options. Pressure transients can cause, and be an indicator of, burst pipes and result in contamination intrusion (Ebacher et al., 2010; Misiunas et al., 2005). Once systems can deal with high frequency, high sample rate data, it will be possible to monitor and analyse transient pressure fluctuations to improve burst detection, identify the cause of transients and prevent contamination events (Jung et al., 2011; Yang et al., 2011).

Data value averaging, inherent in how flow is commonly measured but unusual for pressure, was found to be a useful strategy for both flow and pressure data. A simple, low cost, firmware upgrade to the loggers is recommended so that the data averaging can be performed on the logger in order to reduce the amount of data transmitted via GPRS and provide the following significant benefits:

– Improved detection time compared to using instantaneous data with event detection software

– Eradication of errors associated with short-term variations leading to more accurate hydraulic model calibration and simulations when using values for model boundary conditions and therefore the likely provision of more accurate regulatory compliance data/reporting.

4 Discussion

The goal of this work was to demonstrate how online data processing can benefit proactive water distribution network operation. One of the aims was to free staff from manual interpretation of data, multiple and/or false alarms, and leakage/burst event data. Several individual approaches have been explored, developed and designed to improve some aspect of current industry processes.

Each of the individual methods presented have been shown to provide specific benefits, but their true worth is only realised when all these components are applied in combination. Figure 6 shows an idealised application schematic for all the components.

The system would be initialised by using offline hydraulic models, driven by historic data, to generate pipe sensitivity matrices from which to identify optimal locations for instrumentation within the network. The minimum number of instruments required to meet identification and location sensitivity constraints could then be installed in the field.

The instruments would provide a constant stream of flow and pressure data which would be checked, and preprocessed, before being stored for access by the analytical components of the system. This data would be stored in a "warehouse" and made available to online hydraulic modelling and artificial intelligence based analysis systems, and/or end users.

Once the data was being streamed into the system, the online hydraulic model would be started. The sensitivity matrices could then be recreated using this new, current data, and reviewed to check that instrument locations are still optimal. If they were not, instrument locations could then be moved to the new optimal sites for current network operation. Following network changes such as DMA rezoning, further review would be undertaken to ensure maximum detection sensitivity is maintained at all times; if necessary, by moving, adding or removing instruments. The online system could then be allowed to run continually providing an overview of current flows and pressures in all pipes within the network, and the ability to generate new sensitivity matrices on demand.

The model could then be configured to provide warnings or alarms when unusual flows or pressures were detected by specific instruments. Multiple instruments could be used in combination to generate specific alarms such as pump failure. The impact of a pump failure could be modelled, the effects mapped, and the characteristics used to identify the event should it happen in the future. For example; if flow at A dropped by X%, and pressure at B and C dropped by Y%, display the alarm "The pump at D is not operating within normal parameters". In this way the performance of many dynamic components of a network, for example PRVs and service reservoirs could be continually monitored using a very small number of instruments.

At the same time, data would be streamed into the AI event detection system. This component would automatically analyse large amounts of data, and generate burst pipe (and other hydraulic anomalies) detection alarms and flow estimates, along with a measure of certainty; soon after the event occurred (the actual time to generate an alarm would be dependent upon a number of factors). Combined use of the AI system with dynamic sensitivity matrices generated by the online model would quickly provide location information facilitating quicker repairs.

During an event, the online model would reflect the effect of any event across the entire network highlighting which customers were impacted. If the model was configured to do so, it could identify all pipes with flows and pressures below a definable threshold, record and report them. In this way it would be possible to realise continuous low flow/interruption to supply (DG2/DG3) reporting. The data collected could also be used when investigating an event and its effects on customers.

System performance, and hence the benefits realised, would of course depend upon the performance characteristics of the instrumentation and it being correctly installed. Similarly, online model output would reflect the effort expended in calibration; although tools for this purpose are becoming sophisticated, efficient and cheaper to use. It is not irrational to expect that, with the need for water efficiency coming to the fore, all usage will be metered in the future to minimise loss and waste, and this is reflected in the development of automatic meter reading (AMR), smart metering technologies, and smart water networks. This will benefit hydraulic modelling greatly by reducing uncertainty and making model calibration easier. AI systems will also benefit greatly from this additional measurement data as such systems can be expanded to accommodate almost any number of time series inputs enabling very accurate network flow mapping which will in turn provide better detection and location information. The loop is completed when this information is assimilated into hydraulic models enabling more accurate calibration and the creation of improved sensitivity matrices. The whole process becomes iterative resulting in a stable, very accurate, and sensitive detection and location system.

The vision, that is the natural progression of the work presented here, is seamless integration between a variety of instrumentation types, automated data gathering and online processing, analysis and interpretation/presentation technologies, allied with decision support systems to provide a truly proactive management and control capability. Water quality performance monitoring and reporting would be integrated and automated hydraulics would be programmed to minimise residence time to maintain residual disinfectant and maximise water quality. Automatic remote control of dynamic assets and valves would ensure supply and route it optimally to support unexpected demand from burst events, and to further ensure high water quality.

The vision would be particularly enabled by emerging technologies such as Smart Sensors (Frank, 2013), and Smart Pipes (Metje et al., 2011; Cheng et al., 2006) which will include built in, smart flow, pressure, transient and stress, leakage detection, and water quality parameter measurement and monitoring. A substantial improvement in the availability and ownership cost of instrumentation, data collection, and communications, and the way they are applied will also impact on other distributed infrastructure; for the water sector, this includes sewerage systems. Such work is already underway, for example, on CSO analytics (Guo and Saul, 2011; Mounce et al., 2014) and GA optimised fuzzy logic pump control (Ostojin et al., 2011).

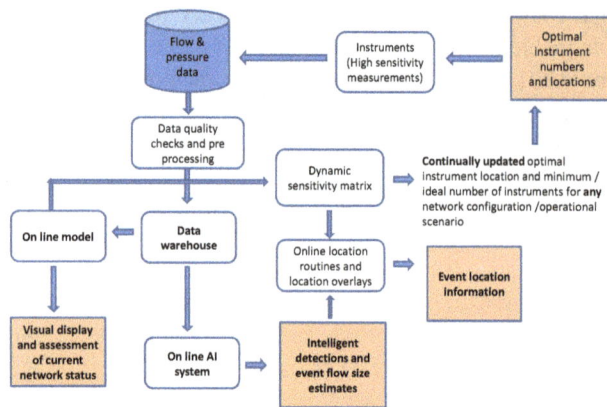

Figure 6. Schematic of idealised combined application.

5 Conclusions

This paper summarises the development, application and verification of a number of different methods/approaches designed to obtain value/benefit from measurements of flow and pressure within water distribution networks.

The individual techniques have shown how network data can be gathered from optimally located instrumentation and automatically checked, analysed and presented, to provide timely information for network operation decision making, and for flow and pressure event detection and location. Each of the methods presented can improve current distribution network knowledge, and are valuable steps towards improving network management.

When the individual methods explored are applied in combination, as shown in Fig. 6, the composite system enables a step change in proactive information network management, including the potential to generate improvements in network performance, customer standards of service, and the economic level of leakage.

In future, entire water supply (and sewerage) networks should be proactively managed from source to tap, using state of the art measurement and control technologies backed by data analysis and decision support systems; much of which will be programmed and analysed by artificial intelligence methods, making life much simpler for system operators and industry decision makers.

Acknowledgements. This work was supported by the EPSRC (grant EP/E003192/1) and industrial collaborators. The authors would like to thank Yorkshire Water Services for their assistance with models and data. Particular thanks are due to R. Patel and L. Soady. In addition, the authors wish to acknowledge 7-Technologies for provision and support of the hydraulic modelling software.

References

Aquis 7-Technologies A/S, Bistruphave 3, 3460 Denmark, http://www.7t.dk (last access: 26 August 2013).

Cheng, J., Wu, X., Li, G., Taheri, F., and Pang, S.-S.: Development of a smart composite pipe joint integrated with piezoelectric layers under tensile loading, Int. J. Solids Struct., 43, 5370–5385, 2006.

Ebacher, G., Besner, M.-C., Lavoie, J., Jung, B. S., Karney, B. W., and Prévost, M.: Transient modeling of a full-scale distribution system: comparison with field data, J. Water Res. Pl.-ASCE, 137, 173–182, 2010.

Farley, B., Boxall, J. B., and Mounce, S. R.: Optimal locations of pressure meter for burst detection, in: Proceedings of the 10th Water Distribution System Analysis Symposium, South Africa, 17–20 August (ASCE online), 2008.

Farley, B., Mounce, S. R., and Boxall, J. B.: Field Testing of an Optimal Sensor Placement Methodology for Event Detection in an Urban Water Distribution Network, Urban Water J., 7, 345–356, 2010a.

Farley, B., Mounce, S. R., and Boxall, J. B.: Field validation of optimal instrumentation method- ology for burst/leak detection and location, in: Proceedings of the 12th Water Distribution System Analysis Symposium, Tucson, Arizona (ASCE online), 2010b.

Farley, B., Mounce, S. R., and Boxall, J. B. Development and Field Validation of a Burst Localisation Methodology, ASCE Water Resources Planning and Management, 139, 604–613, 2013.

Frank, R.: Understanding smart sensors. Artech House – Technology and Engineering, ISBN-13: 978-1-60807-507-2, http://books.google.co.uk/books?id=v4G9jKBCghMC&source=gbs_navlinks_s (last access: 30 March 2014), 2013.

Gama, J.: Knowledge Discovery from data streams. Chapman & Hall/CRC, Business & Economics – http://books.google.co.uk/books?isbn=1439826110 (last access: 30 March 2014), 237 pp., 2010.

Goulet, J.-A., Coutu, S., and Smith, I. F. C.: Model falsification diagnosis and sensor placement for leak detection in pressurized pipe networks, Adv. Eng. Inform., 27.2, 261–269, 2013.

Guo, N. and Saul, A. J.: Improving the operation and maintenance of CSO structures, in: Proceedings of 12th International Conference on Urban Drainage, Porto Alegre, Brazil, 2011.

House of Lords: Water Management, Volume I: Report, House of Lords Science and Technology Committee, HL Paper 191-I, House of Lords, London, UK, 2006.

Jung, B. S., Boulos, P. F., and Altman, T.: Optimal transient network design: A multi-objective approach, J. AWWA, 103, 118–127, 2011.

Machell, J., Mounce, S. R., and Boxall, J. B.: Online modelling of water distribution systems: a UK case study, Drink. Water Eng. Sci., 3, 21–27, doi:10.5194/dwes-3-21-2010, 2010.

Metje, N., Chapman, D. N., Cheneler, D., Ward, M., and Thomas, A. M.: Smart Pipes – Instrumented Water Pipes, Can This Be Made a Reality?, Sensors, 11, 7455–7475, 2011.

Michalak, S., DuBois, A., DuBois, D., Vander Wiel, S., and Hogden, J.: Developing systems for real-time streaming analysis, J. Comput. Graph. Stat., 21, 561–580, 2012.

Misiunas, D., Vítkovský, J., Olsson, G., Simpson, A., and Lambert, M.: Pipeline break detection using pressure transient monitoring, J. Water Res. Pl.-ASCE, 131, 316–325, 2005.

Mounce, S. R., Khan, A., Wood, A. S., Day, A. J., Widdop, P. D., and Machell, J.: Sensor-fusion of hydraulic data for burst detection and location in a treated water distribution system, International Journal of Information Fusion, 4, 217–229, 2003.

Mounce, S. R., Boxall, J. B., and Machell, J.: Development and Verification of an Online Artificial Intelligence System for Burst Detection in Water Distribution Systems, ASCE Water Resources Planning and Management, 136, 309–318, 2010.

Mounce, S. R., Mounce, R. B., and Boxall, J. B.: Novelty detection for time series data analysis in water distribution systems using Support Vector Machines, J. Hydroinform., 13, 672–686, 2011.

Mounce, S. R., Mounce, R. B., and Boxall, J. B.: Identifying sampling interval for event detection in water distribution networks, ASCE Water Resources Planning and Management, 138, 187–191, 2012.

Mounce, S. R., Shepherd, W., Sailor, G., Shucksmith, J., and Saul, A. J.: Predicting CSO chamber depth using Artificial Neural Networks with rainfall radar data, Water Sci. Technol., 69, 1326–1333, 2014.

Ostojin, S., Mounce, S. R., and Boxall, J. B.: An artificial intelligence approach for optimising pumping in sewer systems, J. Hydroinform., 13, 295–306, 2011.

Pearson, D. and Trow, S. W.: Calculating the Economic Levels of Leakage, Leakage 2005 Conference Proceedings, 2005.

Perez, R., Puig, V., Pascual, J., Peralta, A., Landeros, E. and Jordanas, L.:. Pressure sensor distribution for leak detection in Barcelona water distribution network, Wa. Sci. Technol., 9, 715–721, 2009.

Pérez, R., Puig, V., Pascual, J., Quevedo, J., Landeros, E., and Peralta, A.: Methodology for leakage isolation using pressure sensitivity analysis in water distribution networks, Control Eng. Pract., 19, 1157–1167, 2011.

Preis, A., Whittle, A., and Ostfeld, A.: Multi-objective optimization for conjunctive placement of hydraulic and water quality sensors in water distribution systems, Wa. Sci. Technol., 11, 166–171, 2011.

Preis, A., Obaid, T., Allen, M., Iqbal, M., and Whittle, A. J.: Real-time hydraulic modelling of a water distribution system in Singapore, WDSA 2012: 14th Water Distribution Systems Analysis Conference, 24–27 September 2012 in Adelaide, South Australia, Engineers Australia, 2012.

Romano, M., Kapelan, Z., and Savic, D. A.: Evolutionary Algorithm and Expectation Maximisation Strategies for Improved Detection of Pipe Bursts and Other Events in Water Distribution Systems, J. Water Res. Pl.-ASCE, doi:10.1061/(ASCE)WR.1943-5452.0000347, in press, 2013.

Rosich, A., Sarrate, R., and Nejjari, F.: Optimal Sensor Placement for Leakage Detection and Isolation in Water Distribution Networks, IFAC Symp. Series, 8, 776–781, 2012.

Srirangarajan, S., Allen, M., Preis, A., Iqbal, M., Lim, H. B., and Whittle, A. J.: Water main burst event detection and localization, Proceedings of 12th Water Distribution Systems Analysis Conference (WDSA'10), 2010.

Yang, J., LeChevallier, M. W., Teunis, P. F. M., and Xu, M.: Managing risks from virus intrusion into water distribution systems due to pressure transients, J. Water Health, 9, 291–305, 2011.

Ye, G. and Fenner, R.: Kalman Filtering of Hydraulic Measurements for Burst Detection in Water Distribution Systems, ASCE Journal of Pipeline Systems Engineering and Practice, 2, 14–22, 2010.

Zan, T. T. T., Wong, K. J., Lim, H. B., and Whittle, A. J.: A frequency domain burst detection technique for water distribution systems, Sensors IEEE, 1870–1873, 2011.

20

Present challenges for future water sustainable cities: a case study from Italy

L. Bonzanigo and G. Sinnona

Sustainable Development Programme, Fondazione Eni Enrico Mattei (FEEM), Isola di San Giorgio Maggiore, 30124 Venezia, Italia

Correspondence to: L. Bonzanigo (laura.bonzanigo@feem.it)

Abstract. The global population is increasingly concentrated in cities. Cities and urban areas face many challenges – economic, social, health and environmental – which are often exacerbated by an increase in the frequency of natural disasters. Together, these challenges call for a shift towards sustainable cities which reduce their impact on the surrounding environment, whilst at the same time succeeding to make resources available to their increasing number of inhabitants. This study explores the state of the art of water management practices of the highly urbanised Northern Italian region and plans and scope for the future development of water management. Although the region is at present not under severe water stress, recently some cities faced water scarcity problems and were forced to implement water rationing. We assessed the vulnerability of Parma and Ferrara to a water crisis, together with the regular and emergency adaptation measures already in place, and the forecast for the near future. In two workshops, the authors adapted the Australian concept of water sensitive urban design for the Italian context. Although the population remains generally unaware of the impact of the two latest severe drought events (2003 and 2006/7), many adaptation measures towards a more sustainable use of the water resource are already in place – technically, institutionally, and individually. Water managers consider however that the drastic and definite changes needed to integrate the urban water management cycle, and which minimise the ecological footprint of urban spaces, lay far in the future.

1 Introduction

The effects of global warming because of climate change is already increasing vulnerability of several urban areas around the world, through raising sea levels, inland floods, more frequent droughts, periods of increased heat, and the spread of diseases. In 2003, more than 70 000 people died in Europe from a severe heat wave (The World Bank, 2010). These kinds of extreme events will increase in coming years. Due to climate change, the access to basic urban services, especially those linked to water supply, is expected to become more difficult, worsening citizens' quality of life. Urban planned need cities' infrastructure to be able to contrast these threats and ensure that the main services continue to be provided to its inhabitants (i.e. water supply and transport, amongst others).

Cities are uniquely equipped to deal with the challenges they face (C40 Cities, 2011). In some places, like Australia, the widespread realisation of the significance of climate change stimulates citizens' aspirations to ensure that cities reduce their ecological footprint in order to become more sustainable, and to improve their structure and function to make them more liveable in. Yet, change seems slow: many cities still face ongoing investments in the conventional way.

Climate change affects all continents to different extents. Although the focus is primarily on vulnerable cities in developing countries, cities in Europe are also affected by climate change related disasters. The frequencies of both droughts and severe flooding are increasing. In the last decade, 2003 was the hottest summer since 1850, with temperatures 20–30 % higher than in normal years. 40 000 people died because of the heat; alpine glaciers lost 10 % of their total mass, power interruptions occurred frequently, many crops failed, and so forth. Similarly, in 2006/7, in Southern Europe precipitation fell by 89 % in January and 50 % in February. Italy is not immune either and suffers from slight, but important climate variations that lead to the need for a modification of the standard approach towards the management of natural resources, stimulating the implementation of new adaptation measures.

This article explores the state of the art of water management practices of a highly urbanised Northern Italian region. The research was conducted as part of the IWRMnet Water2Adapt project[1]. Although Emilia-Romagna does not generally suffer from severe water stress, recently some cities faced water scarcity problems and at times they were forced to implement water rationing. Moreover, water quality is increasingly affecting water availability. In the Po river basin, pollution is a critical issue for many groundwater aquifers, which constitute the primary water source for several urban areas (Carrera et al., 2013). In the two case studies, we assessed their vulnerability to water crises (past and future), together with ordinary and emergency adaptation measures already in place and the forecast for the near future. The next section describes the case studies and justifies their choice; and then the methods section discusses how the authors explored the vulnerability of the two cities and possible measures to ensure water sustainability. Thirdly, the article examines the outcomes of the workshops.

2 Materials and methods

This article focuses on droughts in the Emilia-Romagna region, both because of the region's differentiated climate between east and west, and for its fast growing urban areas. Emilia-Romagna has the largest number of cities in Italy ranked in terms of population: it has nine cities with more than 100 000 inhabitants, seven of the most populous cities in Italy and ten amongst the first fifty. The case studies in the Water2Adapt project were based on two "medium-size" cities in the region, namely, Parma (West) and Ferrara (East), which present similar historical characteristics whilst having very dissimilar water management practices. We studied their reactions to the two most serious droughts of the last ten years, namely 2003 and 2006/7, and their preparations for longer periods of water scarcity expected in the future due to climate change patterns.

These cities were chosen because of (i) the dependency of each city on a single water source, namely groundwater and the Po river respectively; (ii) their different categories and levels of vulnerability, with Ferrara at risk primarily due to the shortage of water, even during short term droughts, whereas Parma is more sensitive to water quality degeneration and is not so vulnerable to brief water scarcity events.

Ferrara is located near the Po river delta. It has 135 369 residents and an average population density of 334.9 per km^2 (ATO6 Ferrara, 2006). Dependence of Ferrara on the Po River ensures on the one hand nearly unlimited water availability, except during an exceptional drought event (such as in 2003), whilst on the other hand however, it requires advanced water treatment plants, especially due to Ferrara's position along the river, downstream of other significant urban areas. In addition, due to the local topography, pumps are needed to distribute water in the network. The reliance on pumping increases the vulnerability of the network to a varying abstraction of water: when the river levels drop, the pressure in the pipes cannot be maintained and they can be damaged.

The Municipality of Parma has 179 000 residents and a population density of 699 inhabitants km^{-2}. The city acquires nearly all its water from groundwater sources, some of which at present have an annual deficit. Of the two well fields that supply Parma, the southern field is already polluted with nitrates of agricultural origin (up to 50–60 mg L^{-1}) (ATO2 Parma, 2010). The best water originates from city wells, whose quality is protected both by law and by the southern well field, which absorbs any pollution before the water reaches the city wells (Ferrari and Spina, 2011). Parma water managers are assessing the feasibility of diverting water from the Taro and Cerno rivers, which are of acceptable quality according to the regional Water Laws and could potentially supply 280 000 users with 800 L s^{-1}.

Given that water comes primarily from aquifers, it is difficult to estimate the sustainability of present pumping patterns, particularly when there is independent abstraction for industrial and/or agricultural uses, which often compete with requirements for domestic consumption. Recent studies by the Regional Environmental Agency however, underline a yearly water deficit of 1 221 800 m^3 at a provincial level (Provincia di Parma, 2003). Although withdrawals of about 46.3 Mm^3 in 2008 by the main water suppliers' withdrawals accounted for 33% of total water available, the study estimates that the system is not able to respond in a sustainable manner to a potential increase in water needs.

The study aimed to both assess the vulnerability of the two cities to drought, and measures towards a more sustainable and resilient water consumption, which in turn may reduce the exposure of a city to water scarcity. For this paper, the authors adopt the IPCC definition for vulnerability. "The degree to which a system is susceptible to, or unable to cope with, adverse effects of climate change, including climate variability and extremes. Vulnerability is a function of the character, magnitude, and rate of climate variation to which a system is exposed, its sensitivity, and its adaptive capacity" (IPCC, 2001).

In order to explore vulnerability, the authors built on the work of Milman and Short (2008). In their assessment of the sustainability of urban water systems, they identify a new indicator, Water Provision Resilience (WPR), which defines

---

[1]http://www.feem-project.net/water2adapt/

the vulnerability of a water system to providing water, and thus gives useful information for solving the basic shortcomings of the system. It is beyond the scope of this paper to calculate the WPR for the two case studies. However, the authors adopted Milman and Short's identification of six critical aspects of urban water management for the measurement of WPR: supply, infrastructure, service provision, finances, water quality and governance. In the authors' opinion, consideration of these factors introduces resilience to the analysis of sustainability (Milman and Short, 2008). After a preliminary assessment of the two case studies, the authors clustered the six indicators in the following categories: water source, structural measures, institutional and political dimensions, and a sound communication plan. As a consequence, the most important factors for assessing past, present, and future vulnerabilities future vulnerability are: diversification of water sources, infrastructural measures such as desalinization, water reuse, long distance pipe-lines from remote sources, institutional and political framework to regulate the price of water and the related investments to reduce losses, and communication strategies to increase awareness of a proper use of domestic water.

As a framework for the analysis of policies and actions taken by the two cities towards sustainable water management, we applied the Australian concept of water sensitive urban design (WSUD), so far, little explored in European contexts. WSUD is a "holistic approach to the planning and design of urban development that aims to minimise impacts on the natural water cycle and protect the health of aquatic ecosystems" (McAuley and McManus, 2009). In other words, WSUD seeks to address the deficiencies in traditional urban water management practices by minimising the hydrological impacts of urban development in the surrounding environments. WSUD in Australia has evolved from its early association with stormwater management to provide a broader framework for sustainable urban water management. It provides a common and unified method for integrating the interactions between the built form (including urban landscapes) and the urban water cycle (Wong, 2006). In Australia, WSUD is increasingly practiced in new urban greenfield development areas and urban renewal developments linked to a broader Ecologically Sustainable Development agenda. Ecologically Sustainable Development in Australia can be described as going beyond the protection of the environment from the impacts of pollution, to protecting and conserving natural resources (Wong, 2006).

Key transition factors for WS cities are: (i) inter-organisational collaboration and coordination community participation; (ii) uniform regulatory framework and processes; (iii) organisational capacity; and (iv) organisational commitment (Brown and Farrelly, 2007). This article focuses primarily on public sector development and major subdivisions. At the subdivision scale, there is indeed the opportunity, for instance, to undertake large-scale potable wa-

ter substitution, for example using wastewater recycling or stormwater harvesting and reuse.

In order to achieve these objectives, the project organised one workshop in each city with political and technical representatives, where weaknesses of their organisation were assessed through brainstorming sessions based on the methods described above. During the meetings, discussion revolved around how to make the four transition factors operational, by focusing on priorities of water use, social risks of interruptions to the water supply, categories of intervention already in place in the two cities, and future actions planned and/or envisaged (Fig. 1). The main outcome of the workshops was the development of a new water management policy to address present shortcomings through the best practices identified.

3 Results and discussion

Neither Parma nor Ferrara experienced significant water shortages during the water crises in 2003 and 2006–2007. However, the short droughts of 2003 and 2006/7 revealed vulnerability of both cities to major and lasting drought events. Moreover, there emerged the interconnection of two concepts we initially thought as being separate. It appeared that water managers are increasingly implementing some of the WSUD principles regarding future crises, in order to reduce the cities' vulnerability. For instance, because of the antiquity of the centres of the majority of Italian cities, rainwater harvesting is only possible in new areas, and even there storage is a problem due to lack of space. However, the rehabilitation of the wells in Parma demonstrates a clear move towards the minimisation of water consumption patterns and the maximisation of water re-use, whenever possible, which would improve the city's ecological footprint. Nevertheless, a specific application of WSUD principles needs a very critical level of water scarcity, as has occurred in Australia, in order to prompt people to support this approach. In the Po river, droughts are still marginal to citizens' lives and hence the concept is harder to justify in political terms.

Below, there follows a description of the implications of the four selected factors for the management of previous water crisis events, the strengths and weaknesses of the factors, and the future direction of their development.

Firstly, the cities' water source seems to have been the main determining factor. The public water supply in Parma comes from groundwater; thus the effects of drought are delayed, and short-lasting drought events do not lead to immediate water shortages. In Ferrara too, the amount of water diverted from the Po river (av. $1 \, \mathrm{m}^3 \, \mathrm{s}^{-1}$) is negligible compared to the average river flow ($1540 \, \mathrm{m}^3 \, \mathrm{s}^{-1}$) even under drought conditions (environmental flow: $240 \, \mathrm{m}^3 \, \mathrm{s}^{-1}$), and hence it is unlikely that the city will run out of water. However, experts in both cities admit that their reliance on a single water source would significantly increase the risk should a severe water crisis materialise (whether drought or pollution).

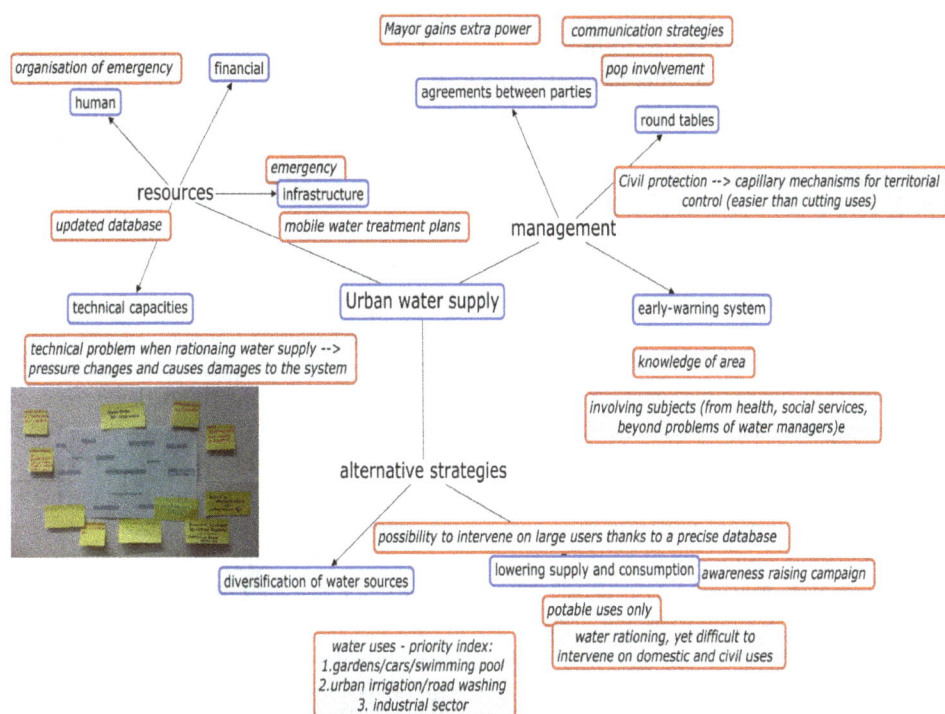

Figure 1. Conceptual map of Parma water supply's vulnerability as emerged during the workshop.

Parma's aquifers are increasingly polluted, which would hinder the city's supply in the future unless action is taken. Ferrara would also suffer, primarily from a water quality crisis. Therefore, the differentiation of water sources remains a priority for both municipalities. In Ferrara, participants suggested that they should invest more in a connection between networks, which would drastically reduce the city's vulnerability to drought as it would add another source to that of the Po river. Parma's water managers have several plans to manage emergencies, which include the diversification of water sources using an aqueduct, which would take water to Parma from a source 35 km away.

Secondly, structural measures in both the long- and short-term would increase the flexibility of the two water systems. Recent droughts, although they did not affect the water supply severely, initiated a whole set of adaptation measures to reduce the cities' long-term vulnerability through good management. In Parma, after the 2006/7 droughts, the Municipality approved a plan for the rehabilitation of old, disused wells for non-potable water uses (including urban irrigation, road and sewage washing, fire fighting, construction sites, and so forth). On the one hand, this led to water and economic savings (such as energy for pumping and treating water for non-potable uses), whilst on the other hand, it marginalised the risk of pollution spillage to the deep aquifer through these wells, which are often the drivers of pollution diffusion. The diversification of the water sources mentioned above remains top of the list of long-term measures. Effort would focus on

the construction of dual systems for rainwater and black water, in order to maximise the potential for reuse, as well as reducing losses in the network up to a maximum of 20 %.

If water supply to the city of Ferrara suddenly failed, it would be necessary to use 5–6000 tank lorries per day to limit the impact on the population! In Ferrara, during the drought event of 2003, the river level fell below the abstraction point. In both the 2003 and 2006/7 droughts, water suppliers installed floating rafts, provided by Civil Protection, in the centre of the river to ensure sufficient water supply despite the critical water levels. Since then, such structures have been utilised twice in summer periods, although they do not constitute a permanent solution for a falling river level. In 2007 the pumps of one of the two diversion structures were extended so that water can be withdrawn at lower levels too.

Thirdly, our stakeholders all agreed that the institutional and political dimensions played a crucial role in avoiding the impacts of severe drought. During water crises, the Mayor acquires emergency powers and can impose restrictions on certain water uses. Both cities agree on the order in which certain uses should be eliminated during critical events: (i) private gardening, car washing, swimming pools, as they did in Parma in August 2007 in southern suburbs of the town; (ii) irrigation of urban areas, road washing; (iii) industrial uses (although large industries that consume more than $100\,000\,\mathrm{m}^3\,\mathrm{year}^{-1}$ usually have their own wells. One example is Barilla, worldwide one of the largest pasta producers). In case of a water emergency, the population would be involved through a concerted awareness raising campaign, while social services, medical personnel and so forth would be mobilised. The problem would gain an interdisciplinary dimension, beyond the water management company's sole authority. It would be feasible to intervene in the water consumption of large users (industries) thanks to the existence of a precise database of all industrial water consumption, which belongs to the Region. In these instances, industries with a private well would have to stop production and make their water available for civil uses. According to experts present at the workshop, these measures would be feasible and sufficient to minimise the impacts of a short and intense drought.

Moreover, together with Civil Protection, when the main water source fails, mayors try to acquire water elsewhere, if possible. This is a preferred solution than to cut usage altogether, which often damages treatment plants and the whole supply network due to large variations in pressure. Yet, it is only possible if the various users and political actors work together in a coordinated fashion, regulated by a legitimate authority – in this instance, the "Cabina di Regia" (emergency decision-making board) called for by the River Basin Authority. A detailed analysis of the decision-making process during an emergency lies beyond the scope of this paper, yet the "Cabina di Regia" is unanimously recognised as a determining factor for the management of water crises. Water authorities and supply companies are drafting new emergency plans, which include who and what organisations to contact (provinces, government, River Basin Authorities, ...) and short-term interventions at the local level: such as tank-lorries, plastic bags containing water, and so forth.

Finally, a sound communication plan is crucial, including both prevention and reaction mechanisms. For example, awareness-raising campaigns are crucial for cutting water consumption and stimulating a more sensitive behaviour on the part of not only citizens, but other urban users too (commercial and industrial organisations, urban agriculture, ...). The local administrators are aware of the fact that all technical adaptation and mitigation measures will not have a significant impact if they do not come with awareness and information campaigns directed at citizens and local users, in order to change their consumption patterns. In Parma, several campaigns are aimed at reducing individual consumption from 200 to less than $150 \, \text{L} \, \text{d}^{-1}$ through events involving children, advertising posts, and events focused on water issues. As a reaction mechanism, much attention focuses on early-warning systems, crucial for a prompt response to a crisis – particularly if linked to a sound institutional structure that can take prompt action. Although these systems are more prominent for floods and pollution crises, drought sensors are currently being implemented by the regional environmental agency, which has a dedicated website on its monitoring system. Stakeholders, despite their overall satisfaction with present communication strategies, have highlighted the need to exploit TV and radio more, although these – and especially the former – require the intervention of the national govern-

ment,. Better cooperation amongst "colleagues" in the water sector is also important, as this has not been satisfactory in the past, according to the institutions themselves. Nevertheless, all participants claim that Emilia-Romagna still remains the leader and coordinator in Italy due to its tradition of research and investments at district level.

4 Conclusions

This study explores the state of practice of water management of the highly urbanised Northern Italian region and plans and scope for the future development of water management. Although the region is at present not under severe water stress, recently some cities faced water scarcity problems and were forced to implement water rationing. We assessed the vulnerability of Parma and Ferrara to a water crisis, together with the regular and emergency adaptation measures already in place, and the forecast for the near future. In two workshops, we adapted the Australian concept of water sensitive urban design for the Italian context. The workshops highlighted that, in order to avoid the extreme consequences of events like droughts and floods cause, it is important to mitigate their impacts not only in emergencies but also on ordinary water management.

Interestingly, although the population is generally unaware of the impacts of the two latest severe drought events, these did stimulate a whole set of adaptation measures working towards a more sustainable use of water resources at municipal and regional level; that is, from the technical dimension to governance and from the individual to the community level. The role of citizens in planning has been increasing, and their responsible behaviour in terms of water consumption patterns is now valued as technical and technological measures. Local administrators are aware of the fact that all technical adaptation and mitigation measures would not have a significant impact if they did not come with awareness and information campaigns directed to citizens and local users in order to change their consumption patterns.

However, despite a move towards greener cities, drastic and definite changes in water management practices according to the concept of the integrated urban water management cycle and in line with WSUD principles have not been fully adopted yet, even though they would drastically minimise the ecological footprint of urban spaces.

Nevertheless, this article demonstrates how ordinary water management, which includes amongst other aspects, standard maintenance works to increase the network efficiency coupled with awareness raising campaigns for more sustainable consumption patterns involving citizens, is an effective preventive measure to significantly reduce the vulnerability of cities to extreme water crises.

The same drought, whose impact was marginal in Parma and Ferrara, caused havoc elsewhere, such as in tourism affecting the coast of Romagna which has water supplied by the Ridracoli dam. In 2007 the reservoir experienced a critical level of exploitation. Hence, the fact that the two cities in the study did not experience severe crises does not itself exclude the fact that other neighbouring areas had to adopt drastic emergency measures. It is important to consider the peculiar characteristics of each context before transferring lessons learnt from one area to another. In addition, this exercise highlights the importance of linking the water sector to other aspects of risk and vulnerability at city level, leading to a holistic WSUD approach.

Nevertheless, the best practice identified in both Parma and Ferrara presents high potential for replication in other cities. Future research should examine how to make WSUD operational in other urban contexts. Given that Ferrara and Parma are not positioned at the extreme end of the vulnerability spectrum for which the WSUD principles were developed, a next step will be to posit a methodology that would be appropriate for different parts of the urban water vulnerability spectrum.

Acknowledgements. The Authors wish to acknowledge the collaboration of the Emilia-Romagna Region and its Environmental Agency and the extensive support of the reviewer Prof. L. Rietveld. This research has been funded by the Water2Adapt project (http://www.feem-project.net/water2adapt/) with the financial support of the Italian Institute for Environmental Protection and Research (ISPRA).

References

ATO2 Parma: Piano di Conservazione della Risorsa Idrica (proposal), ATO Parma, 2010.

ATO6 Ferrara: Il Piano di Conservazione della Risorsa, ATO Ferrara, 2006.

Brown, R. R. and Farrelly, M. A.: Advancing the Adoption of Diverse Water Supplies in Australia: A Survey of Stakeholder Perceptions of Institutional Drivers and Barriers, Report No. 07/04, National Urban Water Governance Program, Monash University, 2007.

Carrera, L., Mysiak, J., and Crimi, J.: Siccità nel bacino del Po: cause, impatti e possibili rimedi, Review of Environment, Energy and Economics, Fondazione Eni Enrico Mattei, 2013.

C40 Cities: www.c40cities.org (last access: 22 August 2011), 2011.

Ferrari, G. and Spina, A.: Personal communication of the authors with Gruppo Iren and ATO2 Parma, 31/05/2011, Parma, Italy, 2011.

IPCC: Glossary of Terms used in the IPCC Fourth Assessment Report, WG II, http://www.ipcc.ch/pdf/glossary/tar-ipcc-terms-en.pdf, 2001.

McAuley, A. and McManus, R.: Water sensitive urban design planning guide. Darwin Harbour WSUD Strategy. Northern Territory Department of Planning and Infrastructure, Australia, 2009.

Milman, A. and Short, A.: Incorporating resilience into sustainability indicators: An example for the urban water sector, Global Environ. Chang., 18, 758–767, 2008.

Provincia di Parma: Piano Territoriale di Coordinamento Provinciale, available at: http://ptcp.provincia.parma.it/ (last access: 30 September 2011), 2003.

The World Bank: Cities and climate change: an urgent agenda, World Bank, available at: http://siteresources.worldbank.org/INTUWM/Resources/340232-1205330656272/CitiesandClimateChange.pdf (last access: 7 April 2014), 2010.

Wong, T. H. F.: An Overview of Water Sensitive Urban Design Practices in Australia, Water Practice and Technology, 10, 2006.

21

Pump schedules optimisation with pressure aspects in complex large-scale water distribution systems

P. Skworcow, D. Paluszczyszyn, and B. Ulanicki

Water Software Systems, De Montfort University, The Gateway, Leicester LE1 9BH, UK

Correspondence to: P. Skworcow (pskworcow@dmu.ac.uk)

Abstract. This paper considers optimisation of pump and valve schedules in complex large-scale water distribution networks (WDN), taking into account pressure aspects such as minimum service pressure and pressure-dependent leakage. An optimisation model is automatically generated in the GAMS language from a hydraulic model in the EPANET format and from additional files describing operational constraints, electricity tariffs and pump station configurations. The paper describes in details how each hydraulic component is modelled. To reduce the size of the optimisation problem the full hydraulic model is simplified using module reduction algorithm, while retaining the nonlinear characteristics of the model. Subsequently, a nonlinear programming solver CONOPT is used to solve the optimisation model, which is in the form of Nonlinear Programming with Discontinuous Derivatives (DNLP). The results produced by CONOPT are processed further by heuristic algorithms to generate integer solution. The proposed approached was tested on a large-scale WDN model provided in the EPANET format. The considered WDN included complex structures and interactions between pump stations. Solving of several scenarios considering different horizons, time steps, operational constraints, demand levels and topological changes demonstrated ability of the approach to automatically generate and solve optimisation problems for a variety of requirements.

1 Introduction

Water distribution networks (WDN), despite operational improvements introduced over the last 10–20 yr, still lose a considerable amount of potable water from their networks due to leakage, whilst using a significant amount of energy for water treatment and pumping. Reduction of leakage, hence savings of clean water, can be achieved by introducing pressure control algorithms, see e.g. Ulanicki et al. (2000). Amount of energy used for pumping can be decreased through optimisation of pumps operation. Optimisation of pumping and pressure control are traditionally studied separately; in water companies pump operation and leakage management are often considered by separate teams.

Modern pumps are often equipped with variable speed drives; hence, the pump outlet pressure could be controlled by manipulating pump speed. If there are pumps upstream from a pressure reducing valve (PRV) without any intermediate tank, the PRV inlet pressure could be reduced by adjusting pumping in the upstream part of the network. Further-

more, taking into account the presence of pressure-dependent leakage whilst optimising pumps operation may influence the obtained schedules. Therefore, for some WDNs it is beneficial to consider pump operation optimisation in conjunction with pressure control. However, even pump operation optimisation on its own is not an easy task due to significant complexity and inherent non-linearity of WDNs, as well as due to number of operational constraints and interactions between different network elements. For example, in our past studies (Skworcow et al., 2009a) the obtained optimal pumping schedules were not intuitive; whilst the tank levels were far from their limits, some pumps did not operate at their maximum capacity during the cheapest tariff, instead they also operated (albeit at significantly lower speed) during the most expensive tariff. Closer examination revealed that further increase of pumping in the cheapest tariff period and reduction of pumping during the more expensive tariff would in fact increase the overall cost, due to pumps operating further from their peak efficiency. Furthermore, as highlighted in Bunn and Reynolds (2009) pumps usually do not operate in isolation; it is typical that any change in the operating duty of one pump may affect the suction or discharge pressure of other pumps in the same system.

Some authors consider optimisation of pump operation as a part of the network design, but the considered case studies are rather small; see e.g. Farmani et al. (2006) and Geem (2009). This paper focuses on optimisation of pump operation in an existing water network. Optimised pump control strategies can be based either on time schedules, see e.g. Ulanicki et al. (2007), or on feedback rules calculated offline, see e.g. Abdelmeguid and Ulanicki (2010). In this paper time schedules approach is considered. The majority of WDN optimisation approaches reported in the literature use a hydraulic simulator or simplified mass-balance models as a key element of their optimisation process and usually consider small scale water distribution systems as case studies, see e.g. Fiorelli et al. (2012) and Lopez-Ibanez et al. (2008). Commercial optimisation packages such as BalanceNet from Innovyze (2013) are able to suggest improvements in operation of complex large-scale WDN, but they typically use mass balance models.

The operational scheduling problem when considering in its full complexity is non-linear and mixed integer and for large scale systems requires huge computational resources. Known approaches try to obtain a suboptimal solution by using simplifying assumptions. Evolutionary algorithms are the most generic search methods and they work efficiently if a simulator of the considered system is available. The simulator can be called tens of thousands of times during the search and in order to reduce the calculation time simplified simulation models are employed, such approach was used for instance by Salomons et al. (2007) and is used by Darwin Scheduler from Bentley Systems (2014). The approach presented by Derceto Aquadapt from Derceto (2014) relies on preparing a highly specialised model of the considered system which is solved using linear and non-linear programming combined with advanced heuristics, the technical details about the algorithm are not available in the literature. To overcome problems with the non-linearity of the hydraulic model Price and Ostfeld (2013) proposed the iterative linearisation procedure, the approach is quite efficient but it solves only continuous version of the optimisation problem in the current formulation. Additional complexity is added to the scheduling problem when the maximum demand charge is considered, this requires the optimisation problem to be formulated over a long time horizon typically 1 month and application of stochastic methods as illustrated in McCormick and Powell (2003).

The approach presented in this paper uses a hydraulic model in the EPANET format as an input, but does not require the EPANET simulator to produce a hydraulically feasible solution. Instead, hydraulic characteristics of the WDN are formulated within the optimisation model itself. Such inclusion of hydraulic characteristics allows taking into account pressure dependent leakage and subsequently including the leakage term in the cost function, thus minimising energy usage and water losses simultaneously. The optimisation model can be automatically adapted to structural changes in the network, such as isolation of part of the network due to pipe burst or installation of additional pumping station, as well as to operational constraints changes, such as allowing lower minimum tank level or higher maximum pump speed. Furthermore, the optimisation model can be generated and solved automatically for different time horizons and different time steps.

The remainder of this paper is structured as follows. Section 2 describes the overall methodology and the developed software. In Sects. 3 and 4 details about obtaining and solving the optimisation model are given. Section 5 describes application of the methodology to a complex large-scale WDN. Finally, conclusions are provided in Sect. 6.

2 Methodology and implementation overview

2.1 Methodology

The proposed method is based on formulating and solving an optimisation problem, similarly to Skworcow et al. (2009b, 2010). However, in this paper the considered network is of significantly higher complexity compared to our previous work, which required some changes to the modelling approach when the optimisation model is formulated, and resulted in a more general method applicable to a wider range of WDNs.

The method involves utilisation of a hydraulic model of the network with pressure dependent leakage and inclusion of a simplified PRV model with the PRV set-points included in a set of decision variables. The cost function represents the total cost of water treatment and pumping. Figure 1 illustrates that with such approach an excessive pumping contributes to a high total cost in two ways. Firstly, it leads to high energy usage. Secondly, it induces high pressure, hence increased leakage, which means that more water needs to be pumped and taken from sources. Therefore, the optimizer attempts to reduce both energy usage and leakage by minimising the total cost.

An optimisation model is automatically obtained from a hydraulic model in the EPANET format and from additional files describing operational constraints, electricity tariffs and pump station configurations. In order to reduce the size of the optimisation problem the full hydraulic model is simplified using module reduction algorithm. In the simplified model all reservoirs and all control elements, such as pumps and valves, remain unchanged, but the number of pipes and nodes is significantly reduced. It should be noted that the connections (pipes) generated by the module reduction algorithm may not represent actual physical pipes. However, parameters of these connections are computed such that the simplified and full models are equivalent mathematically.

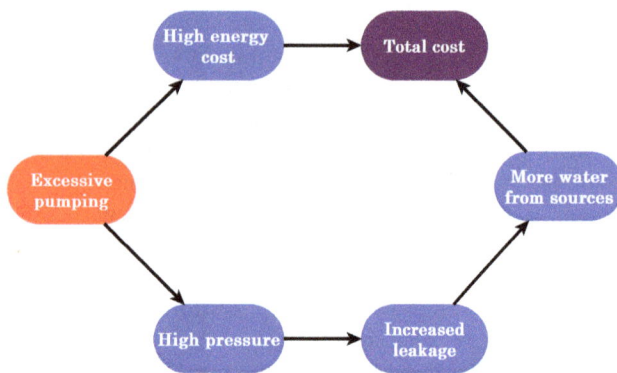

Figure 1. Illustrating how excessive pumping contributes to high total cost when network model with pressure dependent leakage is used.

Details about the model reduction algorithm are given in Paluszczyszyn et al. (2013) and Alzamora et al. (2014).

Some decision variables of the considered optimisation problem are continuous (e.g. water production, pump speed, valve opening) and some are integer (e.g. number of pumps switched on). Problems containing both continuous and integer variables are called mixed-integer problems and are hard to solve numerically, particularly when the problem is also non-linear. Continuous relaxation of integer variables (e.g. allowing 2.5 pumps switched on) enables network scheduling to be treated initially as a continuous optimisation problem solved by a non-linear programming algorithm. Subsequently, the continuous solution can be transformed into an integer solution by manual post-processing, or by further optimisation. For example, the result "2.5 pumps switched on" can be realised by a combination of 2 and 3 pumps switched over the time step. Note that an experienced network operator is able to manually transform continuous pump schedules into equivalent discrete schedules. In this work the main focus is on obtaining the continuous schedules; however, two simple schedules discretisation approaches are also presented in Sect. 4, one fully-automatic and one interactive.

2.2 Implementation

The main software module has been implemented in C# and .NET 4.0. Using a simplified hydraulic model of network in the EPANET format and additional files the optimisation problem is automatically generated by the main software module in a mathematical modelling language called GAMS (Brooke et al., 1998). Subsequently, a non-linear programming solver called CONOPT is called to calculate a continuous optimisation solution. An optimal solution is then fed back from CONOPT into the main software module for analysis and/or further processing and/or export of the results. Specific details of the software functions are as follows:

1. Loads input files required to formulate the optimisation problem (details are given below).

2. Validates the model, i.e. ensures that e.g.: no control rules are associated with pumps or pipes, pressure at leakage nodes is positive, tanks are not emptying or overflowing.

3. Generates GAMS code, runs GAMS (which calls CONOPT), retrieves GAMS results.

4. Handles manipulation of the EPANET model which is required to: (i) use initial schedules (if required) from external time-series files, (ii) manipulate schedules for the purpose of interactive discretisation described in Sect. 4, (iii) produce EPANET file with optimal pump and valve schedules. Note that due to EPANET limitations (Rossman, 2000) valve schedules are implemented as time-based control rules.

5. Handles manipulation of xls files for the purpose of interactive discretisation.

6. Produces time-series files with optimal pump and valve schedules and the resulting tank level trajectories.

Complete information of the WDN and other data required to formulate the optimisation problem is obtained from the following sources:

1. EPANET input file (inp format),

2. EPANET binary simulation results file (bin format) produced by calling the simulator,

3. time-series files (csv format) describing initial schedules; when the scheduler is employed in an on-line receding-horizon environment (Skworcow et al., 2010), the schedules from the previous time step can be used as an initial condition for the current time step,

4. electricity tariffs (csv format),

5. configuration files (txt format) describing the following:

   – lengths of time step and optimisation horizon,

   – configuration of pump stations: (i) fixed or variable speed, (ii) which pump in EPANET belongs to which pump station, (iii) hydraulic curve and power curve coefficients, (iv) constraints: min. and max. number of pumps switched on, min. and max. speed, max. flow,

   – min. and max. flow in pipes and valves,

   – min. and max. pressure at connection nodes,

   – tank level constraints (which are not necessarily equal to the physical limits described in the EPANET model) and inflow/outflow constraints.

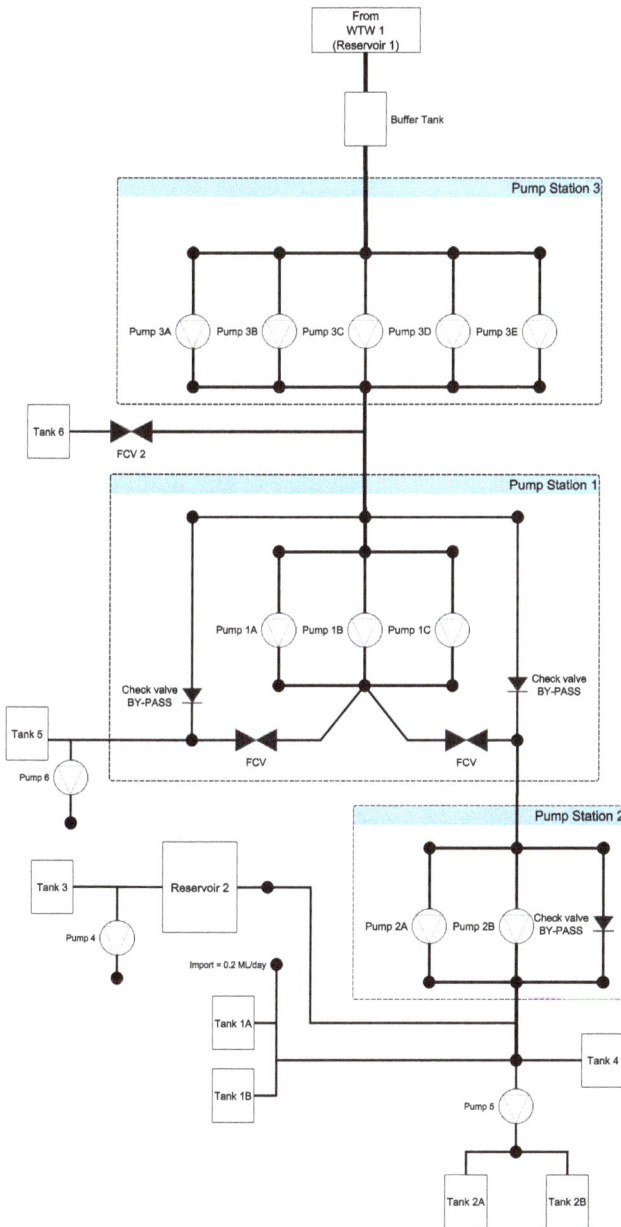

Figure 2. A schematic of the network.

3 Water distribution network scheduling: continuous optimisation

In this section details on formulating and solving a continuous optimisation problem are given. Initial conditions for all variables (flows, pressures etc.) are obtained directly from the EPANET output file from which the network structure was loaded. The optimisation problem has the following three elements, which are described in details in the following subsections: (i) hydraulic model of the network, (ii) objective function, (iii) constraints. The problem is expressed in discrete time with k denoting time step.

3.1 Modelling of WDN for optimisation in GAMS

Each network component has a hydraulic equation. Tanks and pump stations are represented by standard models see e.g. Brdys and Ulanicki (1994) and pipes are represented by the Hazen–Williams formula. A pump station model requires also an additional hydraulic equation and an electrical power characteristic equation. For valves simplified equations are used; details concerning pumps and valves modelling are given below.

3.1.1 Connection nodes

For connection nodes, mass-balance equation is employed; however, since leakage is assumed to be at connection nodes, the standard mass balance equation is modified to include the leakage term:

$$\Lambda_c q(k) + d_c(k) + l_c(k) = 0 \tag{1}$$

where Λ_c is a node branch incidence matrix, q is a vector of branch flows, d_c denotes a vector of demands and l_c denotes a vector of leakages calculated as:

$$l_c(k) = p^\alpha(k)\kappa \tag{2}$$

with p denoting a vector of node pressures, α denoting a leakage exponent and κ denoting a vector of leakage coefficients, see Ulanicki et al. (2000) for details. Note that in the GAMS implementation the variables describing pressure at nodes with non-zero leakage coefficient κ are constrained to be positive, whilst the leakage term in Eq. (1) is zero for nodes with zero leakage coefficient κ.

3.1.2 Pump stations

It is assumed that all pumps in any pump station have the same characteristics as in Brdys and Ulanicki (1994). In addition to the standard hydraulic equation which forces the pump station to operate along its head-flow curve the following equation for each pump station is added:

$$\Delta h(k)u(k) \geq 0 \tag{3}$$

where Δh denotes head increase between inlet and outlet and u denotes number of pumps switched on.

When some pump stations are connected in series without intermediate tanks and/or have by-passes with check-valves (see example in Fig. 2), Eq. (3) prevents a pump station from operating at negative head increase when it is switched on. However, at the same time Eq. (3) allows negative head increase between the pump station inlet and outlet nodes when it is off and the water flows through the by-pass. Note that for networks with pump stations connected in series, if Eq. (3) was not present in the optimisation model, a negative head increase could potentially occur even for a pump station being turned on. This could happen due to the solver choosing to produce a large head increase on the upstream pump station and a negative head increase on the upstream pump station, such that the total head increase (from both pump stations) would still satisfy other constraints and equations. Consequently, Eq. (3) is required for networks with pump stations connected in series to ensure physical feasibility of the solution.

To model electricity usage, instead of using a pump efficiency equation a direct modelling of pump station power is employed, as discussed in Ulanicki et al. (2008). However, the equation is rearranged to allow zero pumps switched on, without introducing *if-else* formulas:

$$P(k)u(k)^2 = Eq(k)^3 + Fq(k)^2 u(k)s(k) +$$
$$Gq(k)u(k)^2 s(k)^2 + Hu(k)^3 s(k)^3 \qquad (4)$$

where E, F, G, H are power coefficients constant for a given pump station, q is flow, P is consumed power, s is speed normalised to a nominal speed for which the pump hydraulic curve was obtained. Additionally it is imposed for all pump stations that $P(k) \geq 0$, so when all pumps in a given pump station are switched off (i.e. $u(k) = 0$) the solver (due to minimising the cost) assigns $P(k) = 0$ for this pump station. Finally, since the coefficients E and F are small compared to G and H, to make a large-scale model easier to solve it is assumed that $E = 0$ and $F = 0$, i.e. the consumed power depends linearly on the pump station flow.

3.1.3 Valves

There are different types of valves in WDN that can be controlled remotely and/or according to a time-schedule; for some, valve opening is controlled directly, while for others pressure drop or flow across the valve is controlled. In the approach proposed in this paper all controllable valves are assumed to be PRVs (control variable is PRV outlet pressure) or FCV (control variable is valve flow). Actual implementation of the control variables in the physical WDN depends on valve construction and is not considered here.

Since head-loss across the valve can be regulated for both FCV and PRV and their direction of flow is known, to reduce the nonlinearity of the model it is proposed to express both FCV and PRV as two simple inequalities:

$$h_{in}(k) > h_{out}(k) \quad q(k) \geq 0 \qquad (5)$$

with the difference between both valve types being their control variables: flow for FCV and outlet pressure for PRV. Consequently, valve flow is defined by other network elements and the mass-balance equation.

Check-valves (non-return valves) are described by the following equation:

$$q(k) = \max\left(0, \frac{|\Delta h(k)|}{R^{0.54}} \text{sign}(\Delta h(k))\right) \qquad (6)$$

where R is a constant valve resistance. Such formulation ensures that valve head-loss is positive if and only if valve flow

is greater than zero; when the flow is zero (i.e. check valve is closed) the head-loss can take any negative value, i.e. inlet and outlet pressures are defined by other network elements. Note that in the Hazen–Williams formula $|\Delta h|^{0.54}$ is used, while here to reduce the nonlinearity of the model it is proposed to use $|\Delta h|$. The justification for such simplification is that head-loss across an open check-valve is relatively small compared to head-loss in other elements, hence such simplification has negligible effects on obtained results. To avoid unnecessary discontinuities, the term $\text{sign}(\Delta h)$ in Eq. (6) is actually implemented as:

$$\text{sign}(\Delta h) \approx \frac{\Delta h}{|\Delta h| + 10^{-14}} \qquad (7)$$

3.2 Objective function

The objective function to be minimised is the total energy cost for water treatment and pumping. Pumping cost depends on the consumed power and the electricity tariff over the pumping duration. The tariff is usually a function of time with cheaper and more expensive periods. For given time step τ_c, the objective function considered over a given time horizon $[k_0, k_f]$ is described by the following equation:

$$\phi = \left(\sum_{j \in J_p} \sum_{k=k_0}^{k_f} \gamma_p^j(k) P_j(k) + \sum_{j \in J_s} \sum_{k=k_0}^{k_f} \gamma_s^j(k) q_s^j(k) \right) \tau_c \qquad (8)$$

here J_p is the set of indices for pump stations and J_s is the set of indices for treatment works. The function $\gamma_p^j(k)$ represents the electricity tariff. The treatment cost for each treatment works is proportional to the flow output with the time-dependent unit price of $\gamma_s^j(k)$. The term P_j represents the electrical power consumed by pump station j and is calculated according to Eq. (4).

3.3 Operational constraints

In addition to constraints described by the hydraulic model equations defined above, operational constraints are applied to keep the system-state within its feasible range. Practical requirements are translated from the linguistic statements into mathematical inequalities. The typical requirements of network scheduling are concerned with tank levels in order to prevent emptying or overflowing, and to maintain adequate storage for emergency purposes:

$$h_{min}(k) \leq h(k) \leq h_{max}(k) \quad \text{for} \quad k \in [k_0, k_f] \qquad (9)$$

Similar constraints must be applied to the heads at critical connection nodes in order to maintain required pressures throughout the water network. Another important constraint is on the final water level of tanks, such that the final level is not smaller than the initial level; without such constraint least-cost optimisation would result in emptying of tanks.

The control variables such as the number of pumps switched on in each pump station, pump speeds or valve flow, are also constrained by lower and upper constraints determined by the features of the control components.

It is evident from the above equations that the overall optimisation model is nonlinear. Furthermore, GAMS recognizes that the model is non-smooth due to the term $|\Delta h|$ in Eq. (7). Hence, the overall optimisation model is of the form Nonlinear Programming with Discontinuous Derivatives (DNLP).

4 Discretisation of continuous schedules

The main focus of this paper is on the continuous optimisation, hence only two simple discretisation approaches are discussed: (i) a fully-automatic discretisation algorithm which does not rely on the EPANET simulation engine but uses GAMS and simple heuristics, and (ii) an interactive discretisation which uses EPANET simulation engine. Both approaches assume that the discretisation time step length is shorter than the continuous optimisation time step length, so for example continuous "2.5 pump switched on for 2 h", can be discretised as "3 pumps on for 1 h and then 2 pumps on for another hour".

4.1 Automatic discretisation

The algorithms progresses through the following steps:

1. Load continuous optimisation results produced by GAMS/CONOPT.

2. For each pump station round the continuous pump control (i.e. the number of pumps switched on) to an integer number, while calculating an accumulated rounding error at each time step. The accumulated rounding error is used at subsequent time steps to decide whether the number of pumps switched on should be rounded up or down, using user-defined thresholds.

3. Generate a new GAMS code where the number of pumps switched on for each pump station and at each time step are fixed, i.e. as calculated in step 2. Initial conditions for all flows and pressures in the network are as calculated by GAMS/CONOPT during the continuous optimisation. Note that in this GAMS code the number of pumps switched on for each pump station and at each time step are no longer decision variables but forced parameters. However, the solver (CONOPT) can still change pump speed and can adjust valve flow to match the integer number of pumps switched on. The cost function to be minimised and the constraints are the same as in the continuous optimisation.

4. Call GAMS/CONOPT and subsequently load the results of integer optimised solution.

5. During the continuous optimisation, pump station flow can be zero only when all pumps in this station are off. However, in the integer optimisation over a long time horizon it may happen that pump station control is forced to have e.g. 1 pump switched on during a particular time step, but this pump is unable to deliver the required head at that time step, hence the pump flow is zero. If such event occurs, the above steps 3 and 4 are repeated, but at the time steps when the resulting pump station flow was zero, the number of pumps switched on is forced to be zero.

4.2 Interactive discretisation

The interactive discretisation approach involves the use of the EPANET simulation engine and a spreadsheet software. The role of the user is to manipulate the discrete schedules initially proposed by the scheduler, by modifying at which time steps the number of pumps switched on is rounded up or down. For networks with flow control valves (FCV) diverting the flow from one pump station into multiple branches, the user may also need to modify the FCV control to match the modified discrete pump schedule. For example, if the continuous pump control at a particular time step is 2.5 pump switched on for 2 h, and it is discretised as 3 pumps on for 1 h and then 2 pumps on for another hour, then the required FCV flow which was calculated during the continuous optimisation needs to be modified to account for increased flow during the first hour and decreased flow during the second hour. The interactive discretisation process progresses iteratively through the following steps, note that the points 1 and 2 in both automatic and interactive discretisation are the same:

1. Load continuous optimisation results produced by GAMS/CONOPT.

2. For each pump station round the continuous pump control (i.e. the number of pumps switched on) to an integer number, thus generating initial discrete schedules.

3. Automatically update the EPANET model with new discrete schedules, simulate the model and retrieve the hydraulic results.

4. Automatically generate an xls file with continuous and discrete schedules, hydraulic results and costs. The file also contains tariffs, plots and other features to simplify the analysis and schedules manipulation.

5. The user modifies at which time steps the number of pumps switched on is rounded up or down for each pump station and may also change FCV schedules. The goal is to: (i) match discrete pump and valve flows (calculated by EPANET and averaged over continuous time step) with continuous pump and valve flows, (ii) match discrete and continuous tank level trajectories and (iii) if possible avoid frequent pump switching.

6. Automatically load the updated discrete schedules from the xls file into the scheduler and go to point 3. The process described in points 3–6 is repeated until the user decides that the results obtained from the discrete schedules and from the continuous optimisation are sufficiently close. For small networks or a short time horizon (24 h) only few iterations are required. For large, complex networks and a long time horizon (7 days) more than ten iterations may be required.

5 Case study: large-scale WDN

This section describes application of the proposed method to optimise operation of a large-scale WDN. The study was based on real data concerning an actual WDN being part of a major water company in the UK.

5.1 Network overview

The considered WDN consists of 12 363 nodes, 12 923 pipes, 4 (forced-head) reservoirs, 10 (variable-head) tanks, 13 pumps in 6 pump stations and 315 valves. The average demand is $451 \, \mathrm{L \, s^{-1}}$ ($39 \, \mathrm{ML \, day^{-1}}$). The system is supplied from 1 major source (water-treatment works) and 2 small imports (under $0.2 \, \mathrm{ML \, day^{-1}}$). The model was provided in the EPANET format. The considered WDN includes complex structures and interactions between pump stations, e.g. pump stations in series without an intermediate tank, pump stations with by-passes, mixture of fixed-speed and variable-speed pump stations, valves diverting the flow from one pump station into many tanks, PRVs fed from booster pumps or a booster pump fed from a PRV.

Due to the network complexity only its schematic with configuration of pump stations is illustrated in Fig. 2. Due to pump station by-passes, when the demand between two pump stations connected in series is low (i.e. at night), one of the pump stations can be turned off and the water will still reach the downstream part of the network with a sufficient pressure.

5.2 Hydraulic model preparation and simplification

Before the automatic model reduction algorithm was applied some manual model preparation was carried out; this included:

1. The model was converted from the Darcy–Weisbach formula to the Hazen–Williams formula, using an operating point when most of the pumps were switched on, i.e. when the flow in pipes was high.

2. Two reservoirs were connected to the system via permanently closed pipelines; these reservoirs were removed.

3. Two connected tanks that follow a similar pressure trajectory were merged into one tank with a suitably chosen diameter.

4. Around 200 permanently closed isolation valves were removed.

5. Several valves that had fixed opening (i.e. throttle control valves (TCV) without any control rules assigned) were replaced with pipes of an equivalent resistance.

6. A TCV to which an open-close control rule was assigned was replaced with an equivalent FCV.

7. A pipe to which an open-close control rule was assigned was replaced with an equivalent valve (FCV) to ensure that only control elements are actually controlled in the model.

The above modifications enable further reduction in the number of network elements; for example, if the isolation valves were not removed, the automatic model reduction algorithm would treat them as control elements, thus retaining them in the reduced model. Table 1 presents functions of valves in the original model and actions performed during the model reduction. Subsequently, the automatic model reduction algorithm was applied; the scale of reduction is shown in Table 2. The model reduction algorithm requires an operating point around which the model will be linearised; in such complex WDN selection of the operating point might present a challenge. However, keeping in mind that the operating point should be representative for normal operation of the network and should be chosen for average demand conditions while keeping at least one pumping unit working at each pumping station (Alzamora et al., 2014), the operating point was chosen at 12:30 h.

To validate how the reduced model replicates the hydraulic behaviour of the original model a goodness of fit in terms of R^2 was calculated for flow trajectories of pumps/valves and for head trajectories of reservoirs/tanks. It was found that the reduced model adequately replicates the hydraulic behaviour of the original model. The R^2 for pump and valve flows was 0.94 in the worst case, 0.99 for most cases and 1.0 for some elements. The R^2 for reservoirs and tanks was 0.5 in the worst case, 0.91 in the second-to-worst case, and between 0.98 and 1.0 for all other reservoirs and tanks. The largest discrepancy was at a small tank which was the furthest from the main source and was empty (according to the original model) at around 18:00 h. Typical performance (i.e. with accuracy obtained for most elements) of the reduced model is illustrated in Figs. 3 and 4. Detailed analysis revealed that the most significant errors were introduced due to the conversion from the Darcy–Weisbach formula to the Hazen–Williams formula.

5.3 Example scheduling results and discussion

The optimisation algorithm was run for over ten scenarios with different constraints on the allowed tank level and on the allowed number of pumps switched on, with two different

Table 1. Function of each valve in the original model and actions performed during the model reduction. (*) not classified as a valve in the total valves count in the original model.

Type	Status	# in original model	Action	# in reduced model
PRV	permanently closed	3	removed	0
	active	39	retained	39
FCV	active	1	retained	1
TCV	close-open control rule	1	converted to FCV	0 (1)
	isolation valve	51	removed	0
	constant opening	220	converted to pipe	0
Pipe*	close-open control rule	1	converted to FCV	0 (1)
Total		315		39 (42)

Table 2. Number of elements in the original and the reduced model.

Elements	Original model	Reduced model	Percentage of reduction
Junctions	12 363	164	99%
Reservoirs	4	2	50%
Tanks	10	9	10.0%
Pipes	12 923	336	97.4%
Pumps	13	13	0.0%
Valves	315	42	86.7%

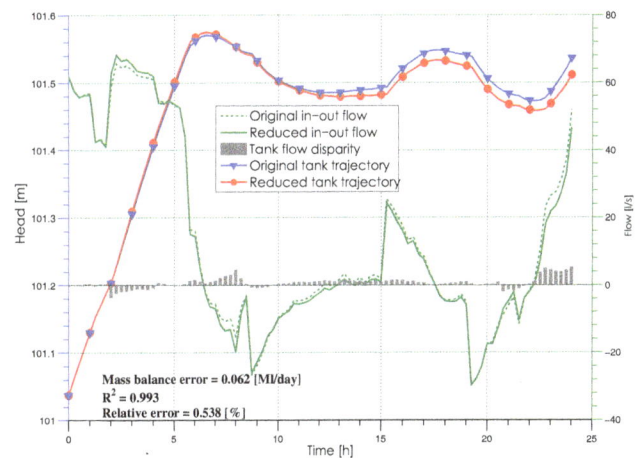

Figure 3. Typical discrepancy in tank level in the original and simplified models.

horizons (24 h and 7 days), with/without pressure dependent leakage and with different demand levels (scaled for different seasons). In all considered scenarios the initial tank level for each tank was assumed to be as in the provided EPANET model. Pressure and flow constraints in different elements were either provided by the water company or assumed and were kept constant for all scenarios. In each case a GAMS code was automatically generated and CONOPT managed to find an optimal continuous solution. However, the automatic discretisation required several trials with different thresholds mentioned in Sect. 4.1. The automatic discretisation algorithm particularly struggled for scenarios with pressure dependent leakage; for these scenarios the interactive discretisation approach was employed.

Subsequently, it was decided to extend the boundaries of the model and include an additional pump station and a tank. After the changes were made in the simplified EPANET model and in an additional file describing pump station constraints, the scheduler successfully generated and solved an updated optimisation model without the need of any changes to the algorithm. Optimisation for 24 h horizon with 1 h time-step and for 7 days horizon with 2 h time-step took around 5 min and 1 h, respectively, on a standard office PC.

It was observed that for all 24 h horizon scenarios it was not possible to fully utilise the allowed capacity of the large tanks and their levels were far from the allowed limits. This was due to the restriction that the final tank level must be

at least as the initial tank level. However, for scenarios with 7 days horizon most tanks hit their upper or lower allowed limits. An example schedule for the largest pump station and an example tank level trajectory for one 7 days scenario are illustrated in Fig. 5 and in Fig. 6, respectively. The tank level increases due to an increased pumping during the cheapest tariff and decreases during the peak tariff. In all considered scenarios it has also been observed that this particular tank is slowly emptying up to the middle of the week and then starts to fill up, since the final level has to be at least as the initial level. These observations suggest that, if allowed by other policies, to reduce the operation cost this tank should operate at lower level than its initial level in the provided EPANET model. However, for another tank which was at the network boundary an opposite tendency was observed: slowly filling up to the middle of the week (with oscillations due to varying daily demand and tariff) and then emptying to finish close to the initial level. This behaviour was due to the fact that the import to the tank was modelled as forced inflow (without any pump), so the inflow head was "free" from the

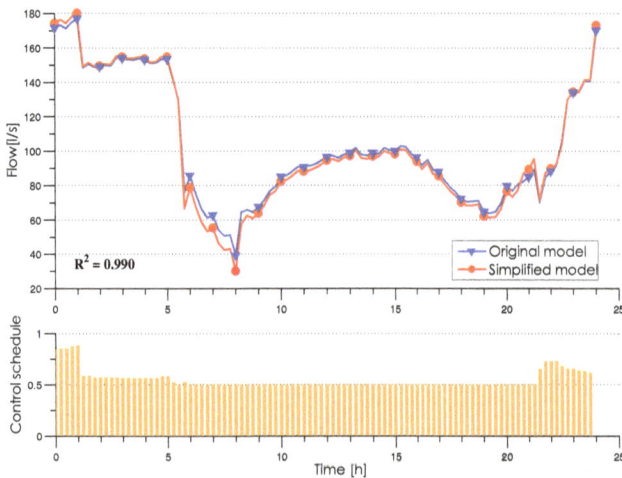

Figure 4. Typical discrepancy in performance of a pump station in the original and simplified models.

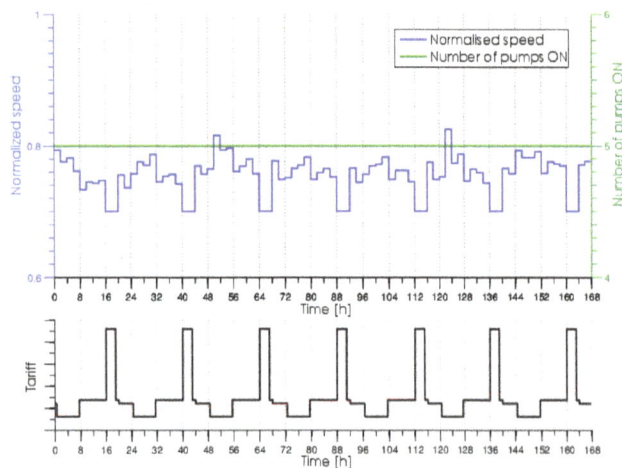

Figure 5. An example schedule for the largest pump station.

optimisation point of view. Therefore, maximising the level in that particular tank enabled small reduction in pumping effort on the downstream pumping station.

Note that the current and optimised operations are not compared, since the provided data considered only one day of operation and on that particular day the final tank levels were far from the initial ones for most tanks. However, the costs for different scenarios were compared against each other. This allowed to formulate several conclusions useful for the water company. For example, two scenarios named A and B considered identical constraints, demands, leakage and topology, but in scenario A a pump station in the middle of the network was fixed speed (as is at present in the physical system) and in scenario B this pump station was equipped with variable speed drive, with the hydraulic and power curves for this pump station being identical in both scenarios A and B. It was found that in scenario B the pump

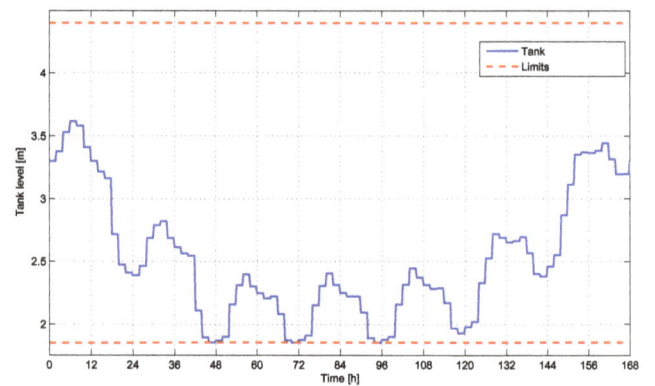

Figure 6. An example tank level trajectory.

station still operated at 100 % speed for majority of time, even when the initial condition was given as 70 %, and the reduction in cost was minimal compared to scenario A. Thus actual installation of a variable speed drive in this pump station in the physical system would not reduce the pumping cost; this demonstrates how the proposed approach can be used to evaluate cost-effectiveness of potential investment in assets related to pumping.

6   Conclusions

Pump operation optimisation is a difficult task due to significant complexity and inherent non-linearity of WDNs. In this paper a time-schedules optimisation is considered and simultaneous optimisation of pumps and valves schedules is employed. An optimisation model is automatically generated in the GAMS language from a hydraulic model in the EPANET format and from additional files describing operational constraints, electricity tariffs and pump station configurations. In order to reduce the size of the optimisation problem the full hydraulic model is simplified using a model reduction algorithm. A nonlinear programming solver CONOPT is used to solve the continuous optimisation problem. Subsequently, the schedules are converted to a mixed-integer form using a simple heuristic.

The proposed approached was tested on a large-scale WDN being part of a major UK water company and provided in the EPANET format. The considered WDN included complex structures and interactions between pump stations. Solving of several scenarios considering different horizons, time steps and operational constraints, and also with topological changes to the hydraulic model demonstrated ability of the approach to automatically generate and solve optimisation problems for a variety of requirements. However, further work is required to improve the current discretisation approaches.

References

Abdelmeguid, H. and Ulanicki, B.: Feedback rules for operation of pumps in a water supply system considering electricity tariffs, in: Water Distribution Systems Analysis, 1188–1205, 2010.

Alzamora, F., Ulanicki, B., and Salomons, E.: A Fast and Practical Method for Model Reduction of Large Scale Water Distribution Networks, J. Water Res. Pl.-ASCE, 140, 444–456, 2014.

Bentley Systems: Darwin Scheduler, http://www.bentley.com/en-GB/Products/WaterCAD/Darwin-Scheduler.htm/, last access: 17 April 2014.

Brdys, M. and Ulanicki, B.: Water Systems: Structures, Algorithms and Applications, Prentice Hall, UK, 1994.

Brooke, A., Kendrick, D., Meeraus, A., and Raman, R.: GAMS: A user's guide, GAMS Development Corporation, Washington, DC, USA, 1998.

Bunn, S. and Reynolds, L.: The energy-efficiency benefits of pump-scheduling optimization for potable water supplies, IBM J. Res. Dev., 53, 5:1–5:13, 2009.

Derceto Inc.: Derceto Aquadapt, http://www.derceto.com/Products-Services/Derceto-Aquadapt/, last access: 17 April 2014.

Farmani, R., Walters, G., and Savic, D.: Evolutionary multi-objective optimization of the design and operation of water distribution network – total cost vs reliability vs water quality, J. Hydroinform., 8, 165–179, 2006.

Fiorelli, D., Schutz, G., Metla, N., and Meyers, J.: Application of an optimal predictive controller for a small water distribution network in Luxembourg, J. Hydroinform., 15, 625–633, 2012.

Geem, Z.: Harmony search optimisation to the pump-included water distribution network design, Civil Eng. Environ. Syst., 26, 211–221, 2009.

Innovyze: BalanceNet, http://www.innovyze.com/products/balancenet/, last access: 30 December 2013.

Lopez-Ibanez, M., Prasad, T., and Paechter, B.: Ant colony optimization for optimal control of pumps in water distribution networks, J. Water Res. Pl-ASCE, 134, 337–346, 2008.

McCormick, G. and Powell, R.: Optimal Pump Scheduling in Water Supply Systems with Maximum Demand Charges, J. Water Res. Pl.-ASCE, 129, 372–379, 2003.

Paluszczyszyn, D., Skworcow, P., and Ulanicki, B.: Online simplification of water distribution network models for optimal scheduling, J. Hydroinform., 15, 652–665, 2013.

Price, E. and Ostfeld, A.: Iterative Linearization Scheme for Convex Nonlinear Equations: Application to Optimal Operation of Water Distribution Systems, J. Water Res. Pl.-ASCE, 139, 299–312, 2013.

Rossman, L.: EPANET Users manual, US: Risk Reduction Engineering Laboratory, Office of Research and Development, US Enviromental Protection Agency, Cincinnati, Ohio, USA, 2000.

Salomons, E., Goryashko, A., Shamir, U., Rao, Z., and Alvisi, S.: Optimizing the operation of the Haifa-A water-distribution network, J. Hydroinform., 9, 51–64, 2007.

Skworcow, P., AbdelMeguid, H., Ulanicki, B., and Bounds, P.: Optimal pump scheduling with pressure control aspects: Case studies, in: Integrating Water Systems: Proceedings of the 10th International Conference on Computing and Control in the Water Industry, 2009a.

Skworcow, P., AbdelMeguid, H., Ulanicki, B., Bounds, P., and Patel, R.: Combined energy and pressure management in water distribution systems, in: 11th Water Distribution Systems Analysis Symposium, Kansas City, USA, 2009b.

Skworcow, P., Ulanicki, B., AbdelMeguid, H., and Paluszczyszyn, D.: Model predictive control for energy and leakage management in water distribution systems, in: UKACC International Conference on Control, Coventry, UK, 2010.

Ulanicki, B., Bounds, P., Rance, J., and Reynolds, L.: Open and closed loop pressure control for leakage reduction, Urban Water J., 2, 105–114, 2000.

Ulanicki, B., Kahler, J., and See, H.: Dynamic optimization approach for solving an optimal scheduling problem in water distribution systems, J. Water Res. Pl.-ASCE, 133, 23–32, 2007.

Ulanicki, B., Kahler, J., and Coulbeck, B.: Modeling the efficiency and power characteristics of a pump group, J. Water Res. Pl.-ASCE, 134, 88–93, 2008.

22

Removal of paraquat pesticide with Fenton reaction in a pilot scale water system

C. Oliveira[1], K. Gruskevica[2], T. Juhna[2], K. Tihomirova[2], A. Alves[1], and L. M. Madeira[1]

[1]LEPABE – Laboratory for Process Engineering, Environment, Biotechnology and Energy, Department of Chemical Engineering, Faculty of Engineering, University of Porto, R. Dr. Roberto Frias, s/n, 4200-465 Porto, Portugal

[2]Department of Water Engineering and Technology, Riga Technical University, 16, Azenes street, Riga, Latvia

Correspondence to: T. Juhna (talis.juhna@rtu.lv) and L. M. Madeira (mmadeira@fe.up.pt)

Abstract. Advanced oxidation processes, such as the Fenton's reagent, are powerful methods for decontamination of different environments from recalcitrant organics. In this work, the degradation of paraquat (PQ) pesticide was assessed (employing the commercial product gramoxone) directly inside the pipes of a pilot scale loop system; the effect of corroded cast iron pipe and loose deposits for catalysing the process was also evaluated. Results showed that complete degradation of paraquat ($[PQ]_0 = 3.9 \times 10^{-4}$ M, $T = 20$–$30\,°C$, $pH_0 = 3$, $[H_2O_2]_0 = 1.5 \times 10^{-2}$ M and [Fe (II)] = 5.0×10^{-4} M,) was achieved within 8 h, either in lab scale or in the pilot loop. Complete PQ degradation was obtained at pH 3 whereas only 30 % of PQ was degraded at pH 5 during 24 h. The installation of old cast iron segments with length from 0.5 to 14 m into PVC pipe loop system had a significant positive effect on degradation rate of PQ, even without addition of iron salt; the longer the iron pipes section, the faster was the pesticide degradation. Addition of loose deposits (mostly corrosion products composed of goethite, magnetite and a hydrated phase of FeO) also catalysed the Fenton reaction due to presence of iron in the deposits. Moreover, gradual addition of hydrogen peroxide improved gramoxone degradation and mineralization. This study showed for the first time that is possible to achieve complete degradation of pesticides in situ pipe water system and that deposits and corroded pipes catalyse oxidation of pesticides.

1 Introduction

Contamination of raw waters with pesticides is recognised as a problem in many countries. Even in trace amounts, pesticides may pass water treatment plants and over long periods accumulate in water distribution pipes (Klamerth et al., 2010; Kralj et al., 2007; Sanches et al., 2010). Moreover, during an accident or deliberate contamination large concentrations may enter the system. Due to sorption in biofilm or on the surfaces of the pipes their removal by network flushing is not efficient. Advanced oxidation processes (AOP) are well-known for generating highly reactive and non-selective hydroxyl radical species, which are used to degrade (and mineralize into water, carbon dioxide and mineral salts) most of organics present in water and wastewater – cf. Reactions (R1) and (R2) (Bautitz and Nogueira, 2007; Hassan et al., 2012; Homem and Santos, 2011; Klamerth et al., 2009; Walling, 1975).

Fenton's reaction is one type of AOP which generates the hydroxyl radicals ($^•OH$) but in this case through the reaction between ferrous ions (Fe^{2+}) and hydrogen peroxide (H_2O_2) according to Reaction (R3) (Andreozzi et al., 1999; Neyens and Baeyens, 2003; Pérez et al,. 2002; Ramirez et al., 2007; Venny et al., 2012; Walling, 1975).

$$\text{Organic matter} + HO^• \rightarrow \text{Oxidation intermediates} \quad (R1)$$

$$\text{Oxidation intermediates} + HO^• \rightarrow CO_2 + H_2O \quad (R2)$$

$$Fe^{2+} + H_2O_2 \rightarrow Fe^{3+} + HO^- + HO^• \quad (R3)$$

Numerous studies have been reported about removal of organic contaminants through Fenton's reaction. Among them, it is worth mentioning the decontamination of wastewaters from dyes (Kuo, 1992; Duarte et al., 2009), chlorophenols (Pera-Titus et al., 2004), or pharmaceuticals (De la Cruz et al., 2012). Decontamination of pesticides like chlorfenvinphos, paraquat, carbofuran, among others by the Fenton's process has also been addressed, this AOP showing to be a very efficient technology (e.g., complete elimination of paraquat was reached after 200 min (Santos et al., 2011), and 90 % of carbofuran elimination after only 5 min of reaction, Ma et al., 2010).

Apart from the classic Fenton to eliminate paraquat from waters, there are other AOPs that present also good efficiencies. Among them are the photocatalytic degradation, using TiO_2 as catalyst (complete degradation after 120 min has been reached) or the direct photolysis (60 % of paraquat was destroyed in less than 3 h) (Moctezuma et al., 1999); the catalytic wet peroxidation (based on heterogeneous oxidation) has also been tested recently – with 92 % of chemical oxygen demand removal after 12 h (Dhaouadi and Adhoum, 2010). Removal of paraquat from water was also the focus of recent work by adsorption over deposits from water networks (Santos et al., 2013); however, in this case there is no degradation of the pesticide, rather simple transfer to another phase (the adsorbent).

Because the Fenton's process requires dissolved iron (homogeneous oxidation), corrosion process in cast iron networks should not be detrimental for the Fenton reaction. Moreover, as an alternative to the homogeneous oxidation, heterogeneous catalysis can be applied by using a solid matrix to support the iron species (Duarte et al., 2009; Herney-Ramirez et al., 2010; Navalon et al., 2010). So, pipe material or corrosion product present in old networks can potentially be used as catalysts of this reaction. Recently it was found that pipe deposits can also act as catalysts in this AOP (Oliveira et al., 2012). This way, such deposits can be reused (upon cleaning/maintenance operations in water networks), or used as catalysts for in situ water treatment, in case of a contamination event.

The objective of this study was (i) to test whether with Fenton reaction it is possible to degrade pesticides (gramoxone) directly in a water distribution system and (ii) to assess the influence of length of cast iron pipe sections and amount of loose deposits on the catalytic process. The study was carried out in a pilot loop to simulate a real water supply system.

Up to the author's knowledge, this is the first report dealing with the use of the Fenton reaction at pilot scale, in a loop reactor, for in situ decontamination of water.

Table 1. Composition of the tap water used along this study.

Tap water properties	Value/ concentration
Al, mg L^{-1}	0.07
Ammonium, mg L^{-1}	0.045
Fe, mg L^{-1}	0.022
Electrical conductivity, $\mu S\,cm^{-1}$	325 ± 45
Chlorides, mg L^{-1}	9
Mn, mg L^{-1}	0.02
Na, mg L^{-1}	5.5
pH	6.8 ± 0.1
Sulphates, mg L^{-1}	71 ± 7

2 Materials and methods

2.1 Reagents

Commercial Paraquat – Gramoxone (GMX) – 32.5 % (w/w) in paraquat dichloride was supplied by Syngenta. Gramoxone was used instead of analytical standard paraquat due to economical reasons. In addition, in case of a deliberate contamination event of a drinking water distribution system, it is more likely that such available commercial products are used instead of expensive analytical ones, even if other organics are also introduced in the network. Two loose deposits samples, one obtained from a local water distribution system (deposit A), and another taken from the tower of a water distribution system (deposit B), both from the city of Riga, were tested as catalysts of the Fenton-like reaction. H_2O_2 solution (30 % v/v), $FeSO_4$ heptahydrate (99.5 %) and anhydrous Na_2SO_3 were purchased from Merck (Darmstadt, Germany). H_2SO_4 (96 %) was from José M. Vaz Pereira, Lda (Lisbon, Portugal), while NaOH (98.7 %) was from José Manuel Gomes dos Santos, Lda (Odivelas, Portugal). Heptafluorobutyric acid (HFBA) was from Sigma Aldrich, and acetonitrile (HPLC grade) from Prolabo.

2.2 Standards preparation

Experiments were conducted with paraquat (PQ) solutions of 100 mg L^{-1}, which were prepared diluting the appropriate amount of gramoxone (GMX) in current tap water (composition data presented in Table 1). This high concentration was chosen to simulate a deliberate contamination of a drinking water system. All gramoxone solutions were stored at 4 °C in polypropylene containers, in which adsorption does not occur. These solutions are stable when exposed to the used conditions (room temperature and while stored at 4 °C or frozen).

2.3 Experimental setup

2.3.1 Stirred batch reactor (Lab scale)

Oxidation reactions done in lab scale were carried out in a stirred batch reactor with 250 mL of capacity. The temperature and the pH of the reaction mixture were respectively measured by a thermocouple and a pH electrode (WTW, SenTix 41 model), connected to a pH-meter from WTW (model Inolab pH Level 2). The temperature was kept constant in the desired value by a Huber thermostatic bath (Polystat CC1 unit) that ensured water recirculation through the jacket of the reactor. No temperature variations higher than $\pm 1\,^\circ\text{C}$ were observed. After temperature stabilisation, the pH of the gramoxone solution was adjusted to the desired value by adding small amounts of $2\,\text{M}$ H_2SO_4 or NaOH aqueous solutions. The start of the oxidation process was remarked by the addition of the catalyst (solid deposit or iron salt, the later for homogenous experiments) and the oxidant agent (hydrogen peroxide). During the reaction, samples were withdrawn, filtered with a $0.2\,\mu\text{m}$ pore size PTFE syringe filter, and analysed as described in Sect. 2.4. To stop the homogenous reaction in the vial samples an excess of Na_2SO_3 was used to instantaneously consume the remaining hydrogen peroxide.

2.3.2 Recirculation tubular reactor (Pilot loop)

The pilot loop used in these experiments (Fig. 1) is made of polyvinyl chloride (PVC) pipes; however, some sections were replaced (see Table 2), if necessary, by iron pipes. The total length of the pilot loop is 28 m (l) and internal diameter is 75 mm (d_i). Iron pipes were obtained from inner heat supply system. However these pipes were made accordingly to the same standard as drinking water system pipes. The reagents (GMX, H_2SO_4, H_2O_2 and ferrous iron) were introduced into the pilot loop using the manual pump coupled to the system.

The pilot loop was filled with tap water that passed through the filter before the experiments (composed of a water softening with manganese dioxide being the colloidal and dissolved iron removed by ion exchange resins). Water was recycled in the loop for 1 h to reach equilibrium conditions. Afterwards gramoxone was introduced into the system to reach the final concentration in paraquat of $100\,\text{mg L}^{-1}$ ($3.9 \times 10^{-4}\,\text{M}$), followed by the addition of the necessary amount of concentrated H_2SO_4 to achieve the desired initial pH.

After the addition of these reagents, the solution was recycled in the system for around one hour to assure the homogeneous mixing. Afterwards, the ferrous iron (or loose deposits – see Table 2) was added to the loop followed by the addition of the H_2O_2 to reach the final concentration of $1.5 \times 10^{-2}\,\text{M}$; this corresponds to time zero of the reaction. Samples were taken along the reaction time and filtrated with a $0.2\,\mu\text{m}$ pore size PTFE syringe filters, for further analysis, as described in the following section. All the experiments were carried out at room temperature ($20 \pm 2\,^\circ\text{C}$).

Experiments performed are described in Table 2.

2.4 Analytical methodology

The paraquat degradation was followed by HPLC-DAD (High Performance Liquid Chromatography with Diode Array Detection), as described previously (Santos et al., 2011). The HPLC-DAD is a Hitachi Elite LaChrom that consists in an L-2130 pump, an L-2200 auto sampler and an L-2455 diode array detector (DAD). Chromatographic analysis of paraquat was performed by direct injection of $99\,\mu\text{L}$ of sample. The chromatographic separation was achieved by a RP C18 Purospher® STAR column ($240\,\text{mm} \times 4\,\text{mm}$, $5\,\mu\text{m}$) reversed phase, supplied by VWR, using a mobile phase of 95 % (v/v) of 10 mM HFBA in water and 5 % (v/v) of acetonitrile, at isocratic conditions, with a flow rate of $1\,\text{mL min}^{-1}$. The spectra acquisition was recorded from 220 to 400 nm and paraquat was quantified at 259 nm, characterized by a retention time of 3 min. The calibration curve for paraquat in water was performed by direct injection of 9 standards, from 0.1 to $100\,\text{mg L}^{-1}$ of paraquat. The coefficient of determination obtained was 0.9999 and the tests revealed an excellent linearity. A detection limit of $0.05\,\text{mg L}^{-1}$ was reached.

Dissolved organic carbon (DOC) measurements were performed in a TOC-5000A Analyzer with an auto sampler ASI-5000 (Shimadzu Corporation, Kyoto, Japan). The methodology is based on a standard method (LVS EN 1484, 2000). Each sample was tested in duplicate and the mean values were calculated (CV ≤ 2 %). The blank and control solutions were analysed with each series of samples in order to verify the accuracy of the results obtained by the method. The minimal detection limit (MDL) was $380\,\mu\text{g L}^{-1}$.

The metals (namely iron) in the solution were determined using a UNICAM 939/959 flame atomic absorption spectrophotometer.

2.5 Solids characterization

Chemical composition of the loose deposit samples was determined by wavelength dispersive X-ray fluorescence (WDXRF) in a Bruker S8 TIGER spectrometer. Samples were analysed in helium atmosphere without previous treatment. The analyses were performed in Full Analysis mode. Results of measurements are expressed in oxide formula units. Mathematical data processing was carried out with integrated Spectra plus software.

To obtain the XRD (X-ray diffraction) diffractograms, a PANalytical model X'Pert PRO with a X'Celerator detector was used. The energy used to produce de X-rays was of 40 kV and 30 mA. Data acquisition was based in the geometry Bragg-Brentano, between $15^\circ \leq 2\Theta \leq 70^\circ$.

d_i = 75 mm
l = 28 m

2,8 m

11,7 m

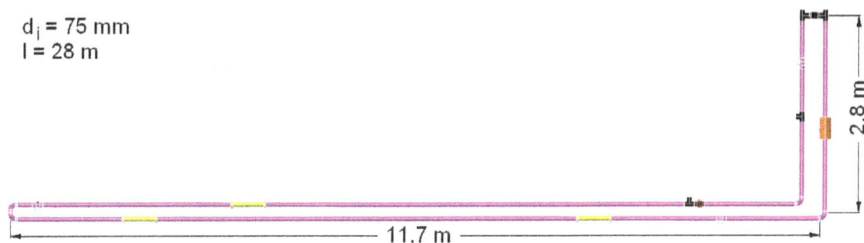

Figure 1. Illustration of the pilot loop used.

3 Results and discussion

3.1 Homogeneous process of PQ degradation

3.1.1 Batch scale vs. pilot loop

A previous parametric study in a stirred batch reactor was done to establish the best conditions for mineralization of PQ in water. The work is described elsewhere (Santos et al., 2011) but briefly the best conditions were: $T = 30\,°C$, $pH_0 = 3$, $[H_2O_2]_0 = 1.5 \times 10^{-2}\,M$ and $[Fe(II)] = 5.0 \times 10^{-4}\,M$, for $[PQ]_0 = 3.9 \times 10^{-4}\,M$ ($100\,mg\,L^{-1}$). This study was performed using the same conditions, but not the temperature because it is not controllable neither in the pilot loop nor in a real situation. The operating conditions used for each experiment in the pilot loop are presented in Table 2, where parameters changed in each run are highlighted in bold.

The first step of this study was to find out if the degradation process in the pilot loop would be similar to the one in the lab scale, so that it could be easily scaled up to the size of a real water network. This possibility would represent a main novelty and step forward as a decontamination in-situ. The same operational conditions were then applied in both lab and pilot scale, and compared in terms of paraquat degradation and gramoxone mineralization.

Figure 2a presents the changes of paraquat concentration in both reactors, i.e., the evolution of the pesticide concentration along reaction time, while Fig. 2b refers to the dissolved organic carbon data. It can be seen that the performance of the process, for both degradation and mineralization, was very similar in both reactors, although their dimensions and mode of operation are considerably different, as well as initial DOC (because in the pilot loop reactor commercial GMX was used, as described above). Besides, it is shown that the decrease of temperature (20 °C in the loop instead of 30 °C) does not influence significantly the process, representing an important advancement in water decontamination in-situ, avoiding the concerns about high temperatures. It should be noted that after 8 h, in the pilot loop, paraquat was completely degraded, even using lower temperatures than in the lab scale; it can thus be concluded that the process can be easily scaled up, keeping the good performance.

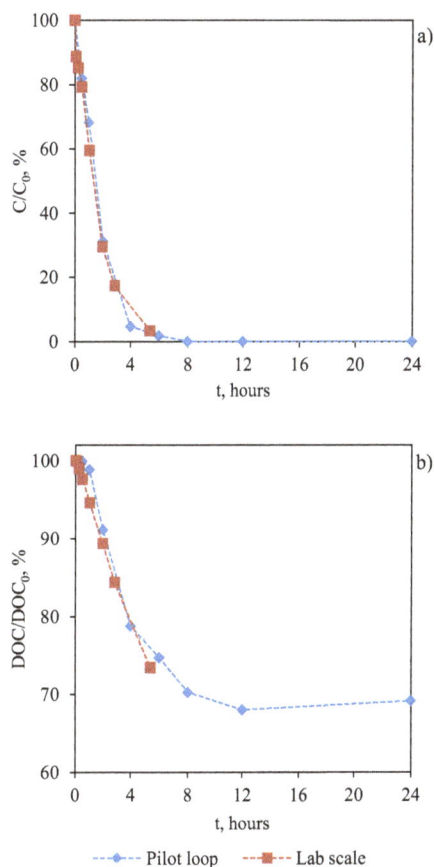

Figure 2. Comparison of PQ degradation (**a**) and GMX mineralization (**b**) between experiments in lab scale ($T = 30\,°C$) and in the pilot loop ($T = 20\,°C$, for the other conditions refer to Table 2, Run #1).

3.1.2 Effect of the initial pH

As it is relatively hard to control the pH in the real water distribution system, the effect of the initial pH in the performance of paraquat degradation was evaluated. Experiments were performed with the initial pH of 3 (Run #1) and the initial pH of 5 (Run #2) – see Table 2 – which is closer to the pH of tap water, and would allow decreasing acid consumption.

From Fig. 3a and b it can be seen that the initial pH has a significant impact in the catalytic process. Indeed, at the

Table 2. Operating conditions for each experiment.

	Run #1	Run #2	Run #3	Run #4	Run #5	Run #6	Run #7	Run #8
$[H_2O_2]_0$, M	1.5×10^{-2}	1.5×10^{-2}	1.5×10^{-2}	1.5×10^{-2}	1.5×10^{-2}	1.5×10^{-2}	1.5×10^{-2} [a]	1.5×10^{-2} [b]
$[FeSO_4 \cdot 7H_2O]_0$, M	5.0×10^{-4}	5.0×10^{-4}	5.0×10^{-4}	–	–	–	5.0×10^{-4}	5.0×10^{-4}
$[Fe]_{salt}$, mg L^{-1}	139	139	139	–	–	–	139	139
[deposit], mg L^{-1}	–	–	–	–	128	744	–	–
$[Fe]_{deposit}$, mg L^{-1}	–	–	–	–	94	606	–	–
$[PQ]_0$, M	3.9×10^{-4}	3.9×10^{-4}	3.9×10^{-4}	3.9×10^{-4}	3.9×10^{-4}	3.9×10^{-4}	3.9×10^{-4}	3.9×10^{-4}
pH_0	**3.0**	**5.0**	3.0	3.0	3.0	3.0	3.0	3.0
$T_{average}$, °C	≈ 20	≈ 20	≈ 20	≈ 20	≈ 20	≈ 20	≈ 20	≈ 20
Pipes types	PVC	PVC	**PVC + iron, 0.5 m**	**PVC + iron, 13.6 m**	PVC	PVC	PVC	PVC
Iron source	Iron (II) salt	Iron (II) salt	Iron (II) salt + iron pipe	Iron pipe	**Loose deposit A**	**Loose deposit B**	Iron (II) salt	Iron (II) salt

[a] Gradual addition of H_2O_2 (32 mL × 5).
[b] Gradual addition of H_2O_2 [(15 mL × 2) + (30 mL × 2) + 70 mL].

initial pH of 5 no mineralization was achieved and only 30 % of the pesticide degradation was reached in 24 h, while for the initial pH of 3 the degradation was complete after 8 h of reaction, and after 12 h the mineralization reached a plateau of 30 %.

The evolution of pH in these runs is presented in Fig. 3c. In both cases there is a pH decrease for short reaction times, which is typical when oxidation occurs with formation of organic acids. The increase of pH for long reaction times can be due to presence of small and unknown particles present along the pipes that are released to the liquid phase and can also be related with the release of the dissolved CO_2 present in the water. It should be noted that in case of initial pH of 5 even after pH dropped to 3.5 within first 1 h of the reaction it did not promote GMX mineralization.

Figure 3d and e shows, respectively, the evolution of soluble and total iron concentration along reaction time, for both experiments. These experiments have the same source of iron – iron (II) salt – and also the same initial load (Table 2); thus, the evolution of total iron concentration is similar for both experiments; however, there are important differences in the soluble iron concentration (Fig. 3d), once the initial pH is quite different and it affects the solubility of the iron species. In fact, it can be seen that much less soluble iron is present in the reaction using an initial pH of 5 (RUN #2) because at high pH values part of the iron present is converted into Fe^{3+}, which precipitates, becoming not available to react with the oxidant and catalyze the process. Therefore, degradation of gramoxone is much worse in such conditions (Fig. 3a and b).

3.2 Heterogeneous Fenton reaction

3.2.1 Iron pipe as catalyst – influence of the length

To evaluate the ability of distribution system water pipes in catalysing the Fenton's reaction, some sections of the PVC from the pilot loop were replaced by used iron pipes (see Table 2). The installed iron pipe sections represented approximately 2 and 50 % of the total length of the loop (RUN #3 and RUN #4, respectively).

It should be highlighted that previous experiments without iron and H_2O_2 (either in PVC or PVC + iron pipes) were performed to assess the possibility of occurring sorption of the pesticide in the pipes. Any significant variation of the pesticide concentration was found. Results presented in Fig. 4a show that the paraquat degradation rate, in the experiment where no iron pipes were installed, was slower than the experiment where 2 % iron pipes were installed (0.5 m long iron pipe), in the presence of $FeSO_4$. This shows that the presence of the iron pipe can promote a faster production of hydroxyl radicals and thus increase the rate of paraquat degradation. As it could be predicted, the greater part of the iron pipes, the faster is the pesticide degradation; this can be confirmed by the analysis of the data from Fig. 4a, from the experiment where 50 % of PVC pipes were replaced by iron pipes (and no iron salt was used), which shows that the degradation rate is the fastest among tested. This can be considered as a great advantage in case of a decontamination demand, once it allows operating in-situ and avoiding the use of some chemicals once the pipes can promote the catalysis.

It should be noted that the fastest and highest mineralization was also achieved in the experiment where 50 % of the loop consisted of iron pipes (Fig. 4b). According to the data shown in Fig. 4a and b, these conditions provided a more effective degradation, with complete pesticide degradation and a remarkable mineralization of 50 % (such mineralization is due not only to the carbon present in the pesticide paraquat but also to that in other organics present, as the pesticide employed is not analytical grade but rather commercial gramoxone – cf. Sect. 2.1). Thus it can be stated that the heterogeneous reaction is quite effective for PQ degradation and mineralization of organics.

It should be noted that, after some time the pH of water started to increase (Fig. 4c), being however more noticeable in the runs with longer pipe sections, in accordance with the explanations given before. One should take into account that mineralization is the highest in RUN #4, so more organics got oxidized and more CO_2 was formed.

From the analysis of Fig. 4d, it can be seen that the iron concentration in solution is the lowest for the experiment using the longer iron pipe length; on the other hand, for the same experiment, the total iron is the highest (Fig. 4e), which means that the iron that acts as catalyst is released from the

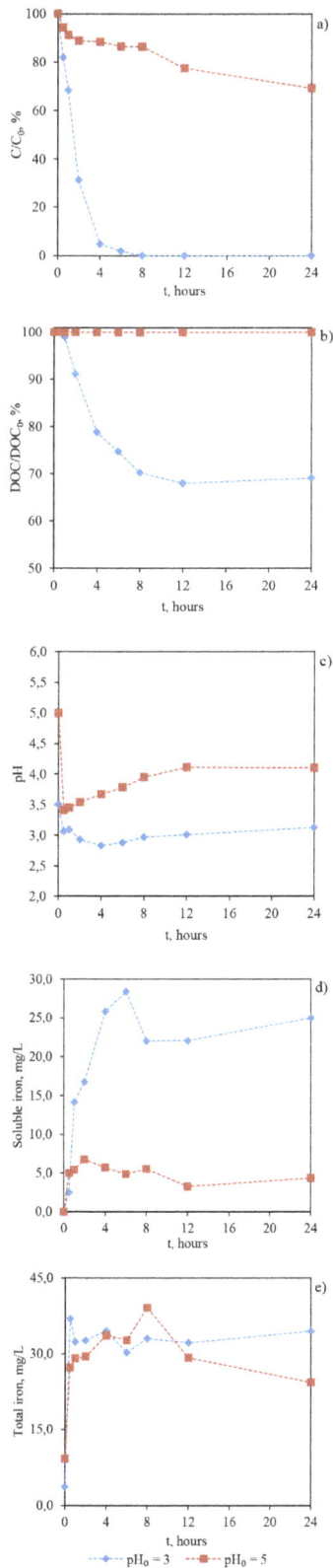

Figure 3. Effect of the initial pH in the PQ degradation (**a**), GMX mineralization (**b**), pH evolution (**c**), soluble iron concentration (**d**) and total iron concentration (**e**) along the time of reaction (Runs #1 and #2).

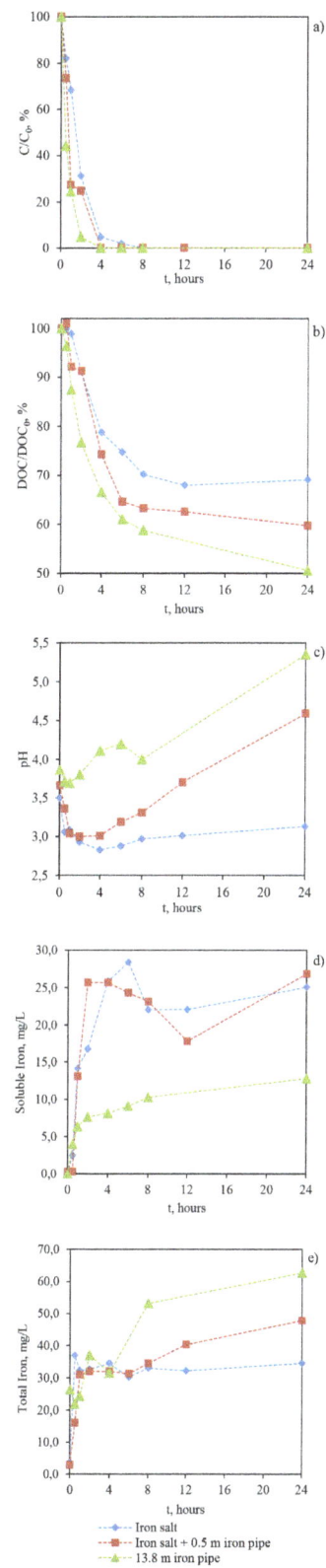

Figure 4. Effect of the use of iron pipes in the PQ degradation (**a**), GMX mineralization (**b**), pH evolution (**c**), soluble iron concentration (**d**) and total iron concentration (**e**) along the time of reaction (Runs #1, #3 and #4).

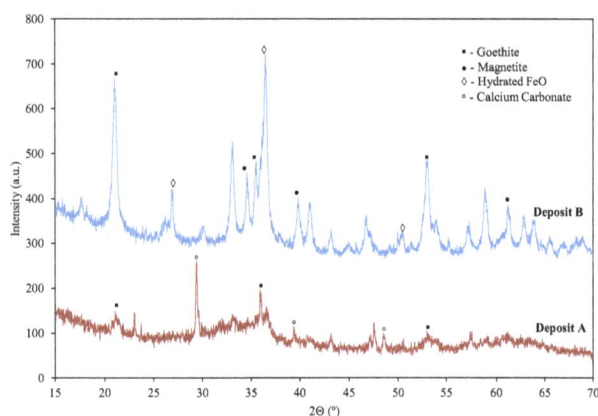

Figure 5. XRD patterns obtained for each deposit.

Table 3. Composition of the inorganic deposit samples used (determined by WDXRF).

Compounds	Deposit A	Deposit B
	Composition (wt. %)	
Fe_2O_3	73.35	81.51
SiO_2	3.23	2.71
CaO	5.65	0.73
SO_3	0.24	0.68
MnO	0.17	0.15
TiO_2	0.04	0.05
P_2O_5	2.15	–
Al_2O_3	–	0.44
Cr_2O_3	0.01	–
ZrO_2	–	0.01
ZnO	–	0.15
CuO	–	0.05
K_2O	–	0.02
NiO	–	0.01
BaO	0.24	–
MgO	0.14	–
SrO	0.05	–
ZnO	0.03	–
Cl	–	0.11
Water	14.70	13.38

pipes and remains in suspension in the solution, part of it being solubilized. In the case of the other two runs, dissolved iron is at a higher level, because $FeSO_4$ was added (Table 2).

Figure 4e shows that the longer is the iron pipes section more iron will be detected in solution as a consequence of the leaching phenomenon. The thickness of the iron pipe used in RUN #3 (2 % of pipes made of iron) was checked before and after the experiment. This measurement was done using ultrasound. The results showed that pipe wall thickness decreased 10 542 nm during the 48 h of reaction, meaning that pipe wall thickness decrease rate is approximately $1.92\,mm\,yr^{-1}$. In normal conditions, for the iron pipe, the thickness reduction along time would be around $160\,nm\,day^{-1}$. Using more aggressive conditions against pipes permanently, i.e. like the ones employed here for in situ decontamination ($pH = 3$ and $[H_2O_2] = 1.5 \times 10^{-2}\,M$), one can conclude that this pipe should be replaced only after 3–4 yr of use.

3.2.2 Loose deposits as catalysts

Two loose deposit samples (one obtained from a real drinking water distribution system – deposit A – and another obtained from the tower of the water distribution system – deposit B) were tested for their ability to catalyse the Fenton's reaction. These deposits were analysed by XRD and their characterization can be found in Table 3 and Fig. 5. Both deposits are mainly a mixture of iron oxyhydroxides; in the case of the deposit B, it has mostly goethite, magnetite and a hydrated phase of FeO; the sample named deposit A has in its composition mostly $CaCO_3$ and goethite.

The main minerals present in the samples were also determined by WDXRF. The results can be found in Table 3. It can be seen that both solids are quite complex, with numerous oxides in their composition, being however the iron-species the predominant, in agreement with the XRD data; for instance, for deposit A, it amounts to 73.35 wt. %, expressed as Fe_2O_3.

RUN #1 shows the degradation process in the presence of iron salt, while RUN #5 and RUN #6 present, respectively, the oxidation in the presence of the deposit A and the deposit B. The iron content was 73.4 % in deposit A and 81.5 % in deposit B (Table 3), which means that the iron concentration used was, respectively, $94\,mg\,Fe\,L^{-1}$ and $606\,mg\,Fe\,L^{-1}$ – Table 2. In the RUN #1, an iron concentration of $5.0 \times 10^{-4}\,M$ ($139\,mg\,Fe\,L^{-1}$) was provided by the addition of the $FeSO_4$.

Analysing Fig. 6a and b it can be said that the process is more effective when using $FeSO_4$; the experiment with deposit A yielded no paraquat degradation after 24 h, while the deposit B shows oxidation performance, being able to almost completely degrade paraquat. Significant difference in performances between deposits cannot be explained by the different doses used, but rather by their natures and composition. This evidences the possibility of using some pipe deposits as catalysts of the Fenton's process, which performance depends on the deposit used, in agreement with the results obtained previously and detailed below. However, this was now proved in a pilot-scale reactor.

It is of big interest to understand the evolution of the pH along time (Fig. 6c); as said before, the increase in the medium pH affects the availability of the Fe^{2+} to react with H_2O_2. Once the Fe^{2+} precipitates as Fe^{3+}, no more iron is available to react with the peroxide and thus no radicals are generated. Besides, upon increasing the medium pH, the peroxide is decomposed into oxygen and water. All these issues

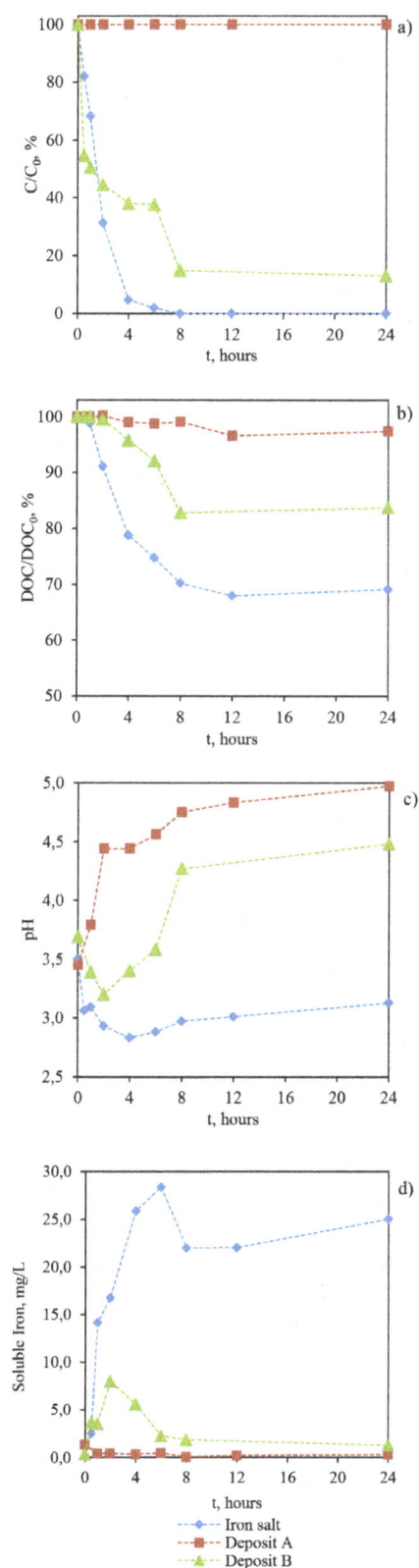

Figure 6. Effect of the loose deposits in the PQ degradation (**a**), GMX mineralization (**b**), pH evolution (**c**) and soluble iron concentration (**d**) along the time of reaction (Runs #1, #5 and #6).

Table 4. Way of H_2O_2 addition.

	Volume of H_2O_2, mL		
Time, h	Run #1	Run #7	Run #8
0	~ 160	32	15
2	–	32	15
4	–	32	30
6	–	32	30
8	–	32	70

are responsible for the decrease in the reaction performance. The increase in the pH along the reaction can be related also with the pH_{pzc} of each deposit, as reported previously (Oliveira et al., 2012). Loose deposits as sample A, rich in calcium carbonate, have higher pH_{pzc} values and thus are responsible for the higher pH in the medium and lower catalytic performance.

The above results are also in line with the dissolved iron (Fig. 6d), which is much lower for the experiment with the deposit A, where the final pH is higher.

The absence of catalytic activity of deposit A can be due to the fact of being mostly composed by $CaCO_3$ and goethite, which, according to Matta et al. (2007) and Oliveira et al. (2012), has a very low catalytic activity when compared with other iron minerals. Also the pH during experiment with deposit A (Fig. 6c) increased fast from the very beginning. It should be noted that in experiment with deposit B after pH raised to approximately 4.2 the degradation of the GMX and the mineralization also stopped.

3.3 Effect of the gradual addition of hydrogen peroxide in the oxidation process

The gradual addition of H_2O_2 was also tested because it is known to be a more effective way of oxidant use (Santos et al., 2011). Three experiments were performed: in the first one (RUN #1 – Table 2), 160 mL of oxidant were added at initial instant ($t = 0$ h); in the second the same amount of oxidant was used – RUN #7 – but divided in 5 doses: 32 mL of oxidant were added at 0, 2, 4, 6 and 8 h of reaction; in the third experiment (RUN #8), 15 mL were added at 0 and 2 h, and 30 mL were added after 4, 6 and 8 h of reaction (method described in Table 4). The experiments are compared in Fig. 7. All other experimental conditions were kept (cf. Table 2).

As can be seen in Fig. 7a and b, in the case of progressive addition of the oxidant (RUN #7 and RUN #8) the performance is similar to the one of RUN #1. However, after 24 h of reaction (Fig. 7b) much better mineralization degree was achieved as consequence of the decreased parallel and undesired reactions that are favoured by a higher H_2O_2 concentration – e.g., scavenging of radicals as shown in Reaction (4) (Laat and Le, 2006; Ramirez et al., 2007; Rodrigues et al., 2009). One should note also that for RUN #8 the final

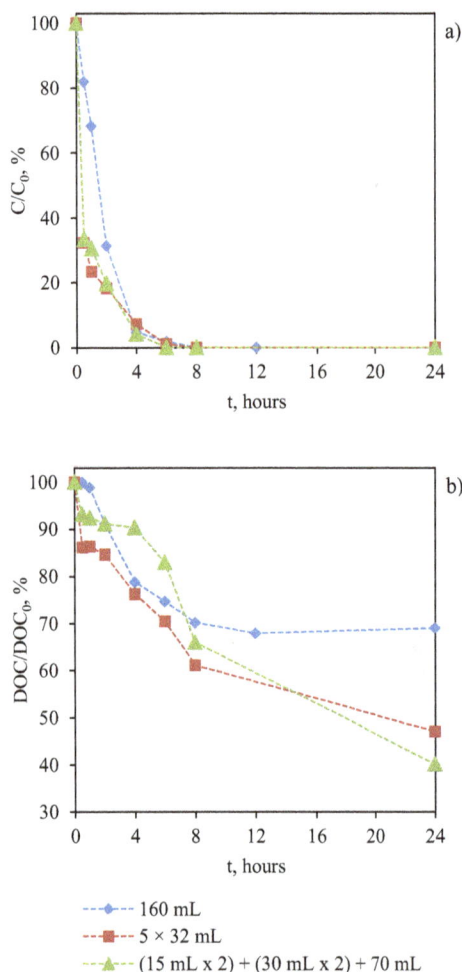

Figure 7. Effect of the gradual addition of H_2O_2 in the performance of PQ degradation (**a**) and mineralization (**b**) along the time of reaction (Runs #1, #7 and #8).

mineralization degree is even bigger that for the RUN #7, once the gradual addition in RUN #8 was made with increasing amounts of peroxide; a remarkable DOC reduction of 60 % (which is the best result among all the experiments) was reached in these conditions, which could probably be increased for higher reaction times.

$$H_2O_2 + HO^\bullet \rightarrow HO_2^\bullet + H_2O \qquad (R4)$$

From these experiments it can be concluded that the best paraquat degradation performance is achieved when gradual addition of H_2O_2 is used.

In spite of the promising results obtained in this study, some limitations were found. In fact, to be more representative of a real water network, the length of the loop should be increased, and the permanent circulation of fresh water should be included (flow-through situation, in opposition to a recirculation system). Other types of pipes with iron and/or other transition metals in their composition should also be

tested. This also applies to the loose deposits; deposits from different places and natures should be tested, considering the iron content and the content of other metals that can work as catalyst (iron, cobalt, nickel, copper, etc.).

It should be also remarked that the costs for the Fenton treatment are low, especially when compared with final disposal costs, being the more relevant those associated with the hydrogen peroxide consumption (Bigda, 1996). In addition, the operating costs are also reduced, once the process runs under moderate conditions of temperature and pressure. In the particular application studied in this work, the costs are further reduced because neither temperature control, neither the use of iron salt as catalyst is required. Other authors also claim that the Fenton's process is very cheap when compared to other AOP's such as photo-Fenton (Audenaert et al., 2011), ozonation or photocatalysis (e.g. photocatalytic oxidation with TiO_2 – Béltran, 2004); on the other hand, results of previous studies proved that the costs associated to AOP treatments are similar to the costs associated to well-established technologies of contaminants removal (Andreozzi et al., 1999). Of course, in an emergency situation like in the event of contamination of a water distribution system this aspect (cost) should not be the limiting issue.

4 Conclusions

It was found that Paraquat degradation can be done in the pilot loop, achieving similar results to those obtained in a lab scale reactor, i.e., homogeneous Fenton's reaction is an effective process in both scales for the pesticide degradation. Initial pH was proven to be a very important factor for Fenton reaction. Results showed complete paraquat degradation (nearly 100 % within 8 h) if initial pH was 3 and little degradation rate (30 % within 24 h) if initial pH was 5.

It was for the first time shown that distribution system pipes can work as a catalyst for Fenton reaction providing also complete paraquat degradation – the size of metallic pipes has a big influence in the oxidation process; the larger the pipe, the bigger is the contribution of the heterogeneous process. Once iron pipes work as catalysts, water decontamination can be done in-situ, using the appropriate operating conditions. Loose deposits can be used as catalysts, but special care must be taken to their composition.

The gradual addition of H_2O_2 showed to be the best option in the oxidation process, allowing reaching higher mineralization degrees, up to 60 % in only 24 h.

Summarizing, this work illustrated that it is possible to decontaminate water in a real water network by advanced Fenton oxidation. Besides, it is also possible to run the process efficiently using either the iron pipes or the loose deposits as the catalyst iron source.

Acknowledgements. This work was undertaken as part of the European Research Project SecurEau (http://www.secureau.eu/ – Contract no. 217976), supported by the European Commission within the 7th Framework Programme FP7SEC20071 – "Security; Increasing the security of Citizen; Water distribution surveillance". The authors are also grateful to the Portuguese Foundation for Science and Technology (FCT) for the financial support through the project PTDC/AAC-AMB/101687/2008. This work has partially been supported by the European Social Fund within the project "Support for the implementation of doctoral studies at Riga Technical University". Authors from RTU are also grateful to J. Rubulis, J. Neilands and S. Dejus for their support in the experiments.

References

Andreozzi, R., Caprio, V., Insola, A., and Marotta, R.: Advanced oxidation processes (AOP) for water purification and recovery, Catal. Today, 53, 51–59, 1999.

Audenaert W. T. M., Vermeersch Y., Van Hulle, S. W. H., Dejans, P., Dumouilin, A., and Nopens, I.: Application of a mechanistic UV/hydrogen peroxide model at full-scale: Sensitivity analysis, calibration and performance evaluation, Chem. Eng. J., 171, 113–126, 2011.

Bautitz, I. R. and Nogueira, R. F. L.: Degradation of tetracycline by photo-Fenton process – Solar irradiation and matrix effects, J. Photoch. Photobio. A, 187, 33–39, 2007.

Beltrán, F. J.: Ozone Reaction Kinetics for Water and Wastewater Systems, CRC Press, Florida, 2004.

Bigda, R. J.: Fenton's Chemistry: An effective advanced oxidation process, Environ. Technol., 6, 34–39, 1996.

De la Cruz, N., Giménez, J., Esplugas, S., Grandjean, D., De Alencastro, L. F., and Pulgarín, C.: Degradation of 32 emergent contaminants by UV and neutral photo-fenton in domestic wastewater effluent previously treated by activated sludge, Water Res., 46, 1947–1957, 2012.

Dhaouadi, A. and Adhoum, N.: Heterogeneous catalytic wet peroxide oxidation of paraquat in the presence of modified activated carbon, Appl. Catal. B-Environ., 97, 227–235, 2010.

Duarte, F., Maldonado-Hódar, F. J., Pérez-Cadenas, A. F., and Madeira, L. M.: Fenton-like degradation of azo-dye Orange II catalyzed by transition metals on carbon aerogels, Appl. Catal. B-Environ., 85, 139–147, 2009.

Hassan, D. H., Aziz, A. R. A., and Daud, W. M. A. W.: Oxidative mineralization of petroleum refinery effluent using Fenton-like process, Chem. Eng. Res. Des., 90, 298–307, 2012.

Herney-Ramirez, J., Vicente, M. A., and Madeira, L. M.: Heterogeneous photo-Fenton oxidation with pillared clay-based catalysts for wastewater treatment: a review, Appl. Catal. B-Environ., 98, 10–26, 2010.

Homem, V. and Santos, L.: Degradation and removal methods of antibiotics from aqueous matrices – A review, J. Environ. Manage., 92, 2304–2347, 2011.

Klamerth, N., Gernjak, W., Malato, S., Agüera, A., and Lendl, B.: Photo-Fenton decomposition of chlorfenvinphos: Determination of reaction pathway, Water Res., 43, 441–449, 2009.

Klamerth, N., Malato, S., Maldonado, M. I., Agüera, A., and Fernández-Alba, A. R.: Application of photo-Fenton as a ter-

tiary treatment of emerging contaminants in municipal wastewater, Environ. Sci. Technol., 44, 1792–1798, 2010.

Kralj, M. B., Trebse, P., and Franko, M.: Applications of bioanalytical techniques in evaluating advanced oxidation processes in pesticide degradation, TRAC-Trend. Anal. Chem., 26, 1020–1031, 2007.

Kuo, W. G.: Decolorizing dye wastewater with Fenton's reagent, Water Res., 26, 881–886, 1992.

Laat, J. D. and Le, T. G.: Effects of chloride ions on the iron (III)-catalyzed decomposition of hydrogen peroxide and on the efficiency of the Fenton-like oxidation process, Appl. Catal. B-Environ., 66, 137–146, 2006.

LVS EN 1484: Water analysis – Guidelines for the determination of total organic carbon (TOC) and dissolved organic carbon (DOC), International Organization for Standardization, 2000.

Ma, Y.-S., Sung, C.-F., and Lin, J.-G.: Degradation of carbofuran in aqueous solution by ultrasound and Fenton processes: Effect of system parameters and kinetic study, J. Hazard. Mater., 178, 320–325, 2010.

Matta, R., Hanna, K., and Chiron, S.: Fenton-like oxidation of 2,4,6-trinitrotoluene using different iron minerals, Sci. Total Environ., 385, 242–251, 2007.

Moctezuma, E., Leyva, E., Monreal, E., Villegas, N., and Infante, D.: Photocatalytic degradation of the herbicide "Paraquat", Chemosphere, 39, 511–517, 1999.

Navalon, S., Alvaro, M., and Garcia, H.: Heterogeneous Fenton catalysts based on clays, silicas and zeolites, Appl. Catal. B-Environ., 99, 1–26, 2010.

Neyens, E. and Baeyens, J.: A review of classic Fenton's peroxidation as an advanced oxidation technique, J. Hazard. Mater., 98, 33–50, 2003.

Oliveira, C., Santos, M. S. F., Maldonado-Hódar, F. J., Schaule, G., Alves, A., and Madeira, L. M.: Use of pipe deposits from water networks as novel catalysts in paraquat peroxidation, Chem. Eng. J., 210, 339–349, 2012.

Pera-Titus, M., García-Molina, V., Baños, M. A., Giménez, J., and Esplugas, S.: Degradation of chlorophenols by means of advanced oxidation processes: A general review., Appl. Catal. B-Environ., 47, 219–256, 2004.

Pérez, M., Torrades, F., Doménech, X., and Peral, J.: Fenton and photo-Fenton oxidation of textile effluents, Water Res., 36, 2703–2710, 2002.

Ramirez, J. H., Maldonado-Hódar, F. J., Pérez-Cadenas, A. F., Moreno-Castilla, C., Costa, C. A., and Madeira, L. M.: Azo-dye Orange II degradation by heterogeneous Fenton-like reaction using carbon-Fe catalysts, Appl. Catal. B-Environ., 75, 312–323, 2007.

Rodrigues, C., Madeira, L., and Boaventura, R.: Optimization of the azo dye Procion Red H-EXL degradation by Fenton's reagent using experimental design, J. Hazard. Mater., 164, 987–994, 2009.

Sanches, S., Crespo, M., and Pereira, V.: Drinking water treatment of priority pesticides using low pressure UV photolysis and advanced oxidation processes, Water Res., 44, 1809–1818, 2010.

Santos, M. S. F., Alves, A., and Madeira, L. M.: Paraquat removal from water by oxidation with Fenton's reagent, Chem. Eng. J., 175, 279–290, 2011.

Santos, M. S. F., Schaule, G., Alves, A., and Madeira, L. M.: Adsorption of paraquat herbicide on deposits from drinking water networks, Chem. Eng. J., 229, 324–333, 2013.

Venny, G. S. and Ng, H. K.: Current status and prospects of Fenton oxidation for the decontamination of persistent organic pollutants (POPs) in soils, Chem. Eng. J., 213, 295–317, 2012.

Walling, C.: Fenton's Reagent Revisited, Accounts Chem. Res., 8, 125–131, 1975.

23

Confirming anthropogenic influences on the major organic and inorganic constituents of rainwater in an urban area

K. Chon[1,2], Y. Kim[3], D. H. Bae[4], and J. Cho[5]

[1]Department of Civil and Environmental Engineering, Yonsei University, Yonsei-ro 50, Seoul 120-749, Korea
[2]Chemical Safety Division, National Institute of Agricultural Science, 166, Nongsaengmyeong-ro, Iseo-myeon, Wanju-gun, Jeollabuk-do, 565-851, Korea
[3]School of Environmental Science and Engineering, Gwangju Institute of Science and Technology (GIST), 123 Cheomdangwagi-ro, Buk-gu, Gwangju 500-712, Korea
[4]Department of Civil and Environmental Engineering, Sejong University, Neungdong-ro 209, Gwangjin-gu, Seoul 143-747, Korea
[5]School of Urban and Environmental Engineering, Ulsan Institute of Science and Technology (UNIST), UNIST-gil 50, Ulsan 689-798, Korea

Correspondence to: J. Cho (jaeweoncho@unist.ac.kr)

Abstract. Recently, rainwater composition affected by atmospheric pollutants has been the topic of intense study in East Asia because of its adverse environmental and human health effects. In the present study, the chemical composition and organic compounds of rainwater were investigated from June to December 2012 at Gwangju in Korea. The aim of this study is to determine the seasonal variation of rainwater chemical composition and to identify possible sources of inorganic and organic compounds. The volume-weighted mean of pH ranged from 3.83 to 8.90 with an average of 5.78. Of rainwater samples, 50 % had pH values below 5.6. The volume-weighted mean concentration (VWMC) of major ions followed the order $Cl^- > SO_4^{2-} > NH_4^+ > Na^+ > NO_3^- > Ca^{2+} > Mg^{2+} > K^+$. The VWMC of trace metals decreased in the order $Zn > Al > Fe > Mn > Pb > Cu > Ni > Cd > Cr$. The VWMCs of major ions and trace metals were higher in winter than in summer. The high enrichment factors indicate that Zn, Pb, Cu, and Cd originated predominantly from anthropogenic sources. Factor analysis (principal component analysis) indicates the influence of anthropogenic pollutants, sea salt, and crustal materials on the chemical compositions of rainwater. Benzoic acids, 1H-isoindole-1,3(2H)-dione, phthalic anhydride, benzene, acetic acids, 1,2-benzenedicarboxylic acids, benzonitrile, acetaldehyde, and acetamide were the most prominent pyrolysis fragments for rainwater organic compounds identified by pyrolysis gas chromatography/mass spectrometry (Py-GC/MS). The results indicate that anthropogenic sources are the most important factors affecting the organic composition of rainwater in an urban area.

1 Introduction

Rain is an efficient scavenging process for pollutants in the air and is becoming a source of pollution to the environment (Santos et al., 2011). The emission of SO_2 and NO_x from fossil fuel combustion and industrial processes has rapidly increased in East Asia due to its fast growing economy. These gases are converted into sulfuric and nitric acids before precipitating as acid rain (Lee et al., 2000; Báez et al., 2006). Consequently, there have been adverse environmental effects on aquatic, biological, and terrestrial systems (Bard, 1999; Başak and Alagha, 2004). Thus, the chemical composition of precipitation has been investigated all around the world during the last decade (Lara et al., 2001; Mouli et al., 2010; Santos et al., 2011).

Coal combustion, automobile exhaust, and industrial emissions represent the dominant anthropogenic sources of heavy metals in rainwater (Kaya and Tuncel, 1997; Hu and Balasubarmanian, 2003; Cheng et al., 2011). Heavy metals from precipitation accumulate in the biosphere and may cause adverse human health and environmental effects (Barrie et al., 1987; Báez et al., 2007). Thus, the studies of heavy metals in rainwater have increased in many countries (Pike and Moran, 2001; Al-Momani, 2003; Báez et al., 2007).

Dissolved organic carbon (DOC) is a major component of both continental (161 μM) and marine rain (23 μM) (Willey et al., 2000). DOC is a mixture of simple substances such as alkanes, carbohydrates, and fatty acids, and of complex polymeric molecules, such as aromatic, aliphatic, and carboxylic carbons (Muller et al., 2008). Rainwater DOC plays a significant role in the atmospheric carbon cycle, and the global rainwater flux of DOC is 430×10^{12} g C yr^{-1} (Willey et al., 2000). Sources of DOC compounds in rainwater are primary anthropogenic and biogenic emissions, and photochemical transformations of precursors (Klouda et al., 1996). While numerous studies have been conducted on the inorganic chemistry of rainwater, little attention has been paid to organic compounds. Kawamura et al. (1983) initiated the identification and distribution of organic compounds in rainwater. The role of organic compounds in atmospheric processes has gained much attention in the last few years, and knowledge of organic compounds has improved with various analytical methods (Kieber et al., 2002; Seitzinger et al., 2003). However, the organic compounds in the atmosphere are very complicated and chemical characterization of organic matter in precipitation requires further work (Santos et al., 2009).

Several studies have examined chemical composition of rainwater in Korea. Although many studies have been conducted on the seasonal variation of all the major ions or heavy metals in rainwater (Lee et al., 2000; Kang et al., 2004; Kim et al., 2012; Moon et al., 2012), there are few on the organic compounds and the sources and variation of DOC in precipitation (Yan and Kim, 2012).

In this study, rainwater samples were collected from June to December 2012 in an urban area of Gwangju, Korea. The objectives of this study are to investigate the seasonal variation of rainwater chemical composition and to identify possible sources of inorganic and organic compounds in rainwater.

Our framework for research design is as follows:

1. seasonal variation in chemical composition (ions and trace metals) of rainwater

2. enrichment factor analysis to evaluate the contribution of non-crustal sources

3. variation in chemical composition of initial and subsequent rainfall

4. factor analysis to investigate the influence of natural and anthropogenic sources

5. analysis of organic compounds in rainwater.

2 Materials and methods

2.1 Study site

Rainwater samples were collected at the Gwangju Institute of Science and Technology campus (35°13″ N, 126°50″ E) in Gwangju, Korea (Fig. 1), from June to December in 2012. Gwangju has an area of 501.34 km^2 and a population of around 1 480 000. The annual average temperature is 13.5 °C and the average precipitation is 1368 mm per year.

2.2 Sample collection and analysis

Rainwater samples were collected per event using a Teflon-coated collector designed to manually take samples with time. Samples from initial precipitation events were collected with care not to be mixed with later precipitation samples. Initial rainwater samples were collected from a single rain event or multiple rain events in a single day. We collected initial rainwater samples until a 2 L glass jar was filled during the beginning of the rain event. All the initial and later samples were transported to the laboratory and pH and electrical conductivity (EC) were measured. The remaining samples were filtered through 0.45 μm filters (mixed cellulose ester, Advantec, Japan) and then stored at 4 °C until further chemical analyses were performed within 1 week. pH and EC were measured using a pH meter and an EC meter, respectively (Orion 3-Star, Thermo Scientific, USA).

Major ions were quantified using a suppressed type ion chromatograph (DX-120, ICS-90, Dionex, Sunnyvale, CA, USA). An IonPac AS14 and an IonPac CS12A column (Dionex, Sunnyvale, CA, USA) were used for measurements of major anions (NO_3^-, SO_4^{2-}, and Cl^-) and cations (NH_4^+, Ca^{2+}, Mg^{2+}, Na^+, and K^+), respectively. The concentrations of trace metals were measured using inductively coupled plasma mass spectrometry (ICP-MS) (7500ce, Agilent,

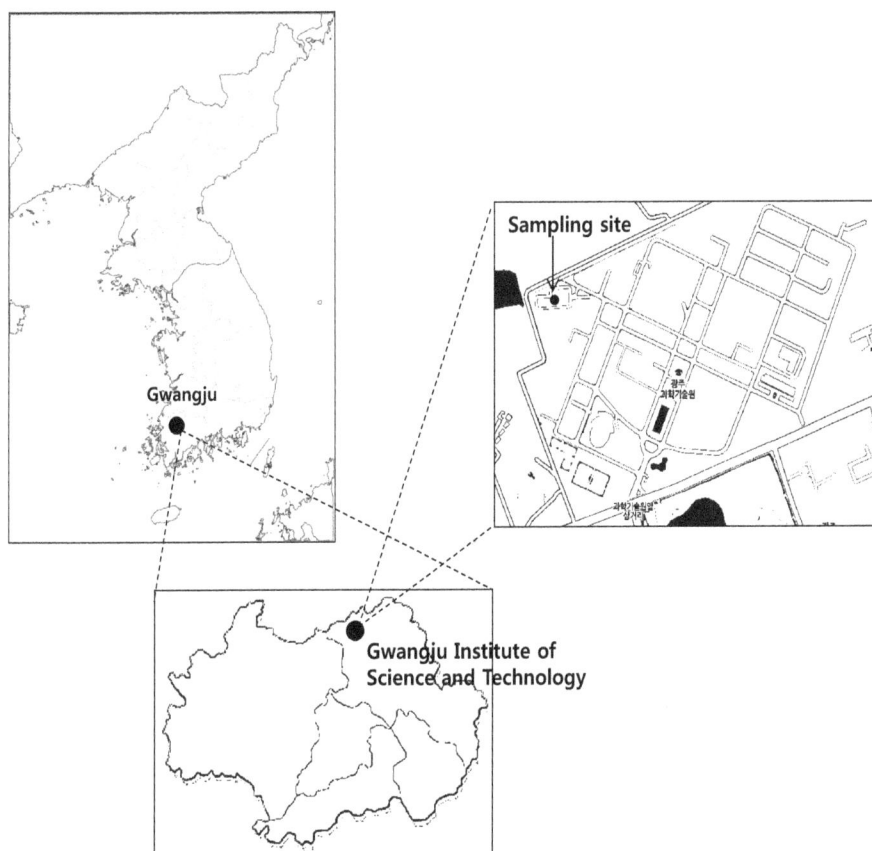

Figure 1. Location of the sampling site for this study.

Santa Clara, CA, USA). All the samples were acidified to a final nitric acid concentration of 2 % using a 70 % nitric acid solution. The levels of DOC and total nitrogen (TN) contents of rainwater samples were determined using a total organic carbon analyzer (TOC-V CPH, Shimadzu, Japan) equipped with a TN analyzer (TNM-1, Shimadzu, Japan). The UV absorbance of the samples at 254 nm (UV_{254}) was measured using a UV–vis spectrophotometer (UV-1601, Shimadzu, Japan). The specific UV absorbance (SUVA) value (an indicator of aromaticity) was calculated from the ratio of UV_{254} to DOC concentration. Rainwater samples were concentrated to a final DOC concentration of approximately $100\,\mathrm{mg\,C\,L^{-1}}$ with a rotary evaporator (Eyela, Japan). Approximately 20 mL of concentrated samples was freeze-dried using a freeze dryer (Ilshin, Korea) prior to pyrolysis gas chromatography/mass spectrometry (Py-GC/MS) analysis. Approximately 0.5–1.0 mg of freeze-dried sample powders were pyrolyzed at 590 °C for 5 s in the pyrofoil (Pyrofoil F590, Japan Analytical Industry, Tokyo, Japan) within a Curie point pyrolyzer (JCI-22, Japan Analytical Industry, Tokyo, Japan) coupled with an Agilent 7890A gas chromatograph coupled to a 5975C quadrupole mass spectrometer (ion source temperature 220 °C, scanning from 40 to 500 amu, electron energy 70 eV). Pyrolysis fragments were separated

by GC equipped with a DB-5MS (Agilent Technologies, USA) column and identified using a mass spectrometer. Helium was used as the carrier gas. The temperature program of the GC oven was initially maintained at 40 °C for 5 min, then increased at $7\,^{\circ}\mathrm{C\,min^{-1}}$ to a final temperature of 300 °C and remained there for 10 min, giving a total run time of 52.14 min. The interpretation of the pyrochromatograms was conducted according to the methods described in Bruchet et al. (1990).

3 Results and discussion

3.1 Variation of pH value and precipitation amount

The average monthly rainfall from June to December during 2012 was 195.4 mm and the highest amount of rainfall (473.5 mm) was observed in August 2012 (Fig. 2). Of the total rainfall, 79 % occurred during the summer (July–September). The percent frequency distribution of pH for the rainwater samples is presented in Fig. 3. The volume-weighted mean of pH was 5.78. The lowest pH value was observed on 11 July with a pH of 3.83, while the highest was on 22 August 2012 with a pH value of 8.90. Most pH values ranged between 5.0 and 5.5 (23.9 %), while about 2.6 %

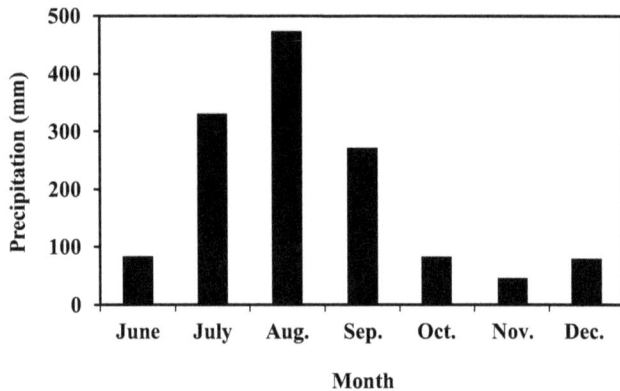

Figure 2. Monthly integrated precipitation at Gwangju during 2012.

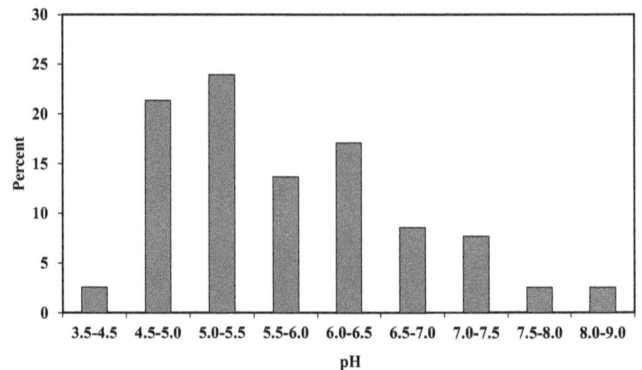

Figure 3. Distribution of pH in rainwater samples collected from Gwangju.

of pH values were in the range of 3.5–4.5. Of the samples, 50 % had pH values below 5.6, which is the value of unpolluted water equilibrated with atmospheric CO_2 (Charlson and Rodhe, 1982). On the other hand, approximately 12 % of the samples had a pH above 7.0, suggesting strong inputs of alkaline substances to rainwater in this area.

3.2 Ionic composition of rainwater

The volume-weighted mean concentration (VWMC), standard deviation of the VWMC (SDVWMC), and minimum (Min) and maximum (Max) concentrations of the rainwater chemical components are summarized in Table 1. VWMC was calculated by the following equation:

$$\mathrm{VWMC} = \sum_{i=1}^{n} Xi\,Pi \Big/ \sum_{i=1}^{n} Pi,$$

where Xi is the measured ion concentration, Pi is the precipitation amount, and n is the number of samples. The VWMC of major ions in rainwater follows the order $Cl^- > SO_4^{2-} > NH_4^+ > Na^+ > NO_3^- > Ca^{2+} > Mg^{2+} > K^+$. Among all the ions, Cl^- was the most abundant with an average of $123.5\,\mu eq\,L^{-1}$, accounting for 28.5 % of the ionic concentration. SO_4^{2-} was the second highest anion with an average of $91.9\,\mu eq\,L^{-1}$, accounting for 21.2 % of total ions. The contribution of NO_3 was 8.5 % (concentration of $36.8\,\mu eq\,L^{-1}$). The high value of SO_4^{2-} and highest of Cl^- can be attributed to emissions from fossil fuel combustion and typhoons and sea salts, respectively. As for cation species, NH_4^+ made the highest contribution at $63.0\,\mu eq\,L^{-1}$, accounting for 14.5 % of the total ions. Meanwhile, Na^+ was the second highest cation accounting for 13.5 % of the total ions. The concentration of Ca^{2+}, Mg^{2+}, and K^+ ions accounted for 8.5, 3.5, and 1.8 %, respectively. The high level of NH_4^+ in Gwangju was comparable to Seoul (with a mean concentration of $66.4\,\mu eq\,L^{-1}$; Lee et al., 2000). Wastes from agricultural and domestic activities as well as

the use of fertilizers are believed to be the main sources of the emission of gaseous ammonia (Dianwu and Anpu, 1994; Panyakapo and Onchang, 2008). Ca^{2+} may originate from wind-blown dust and calcareous soil, Mg^{2+} from sea salts and dusts in the atmosphere, while Na^+ from sea salts. EC of rainwater ranged from 0.87 to $169.00\,\mu S\,cm^{-1}$, with a VWM (volume-weighted mean) value of $27.82\,\mu S\,cm^{-1}$. The VWMs of UV and SUVA in rainwater were 0.0160 and 1.09, respectively. Organic matter in rainwaters exhibit relatively hydrophilic characteristics, with SUVA values less than 2 generally indicating a high fraction of hydrophilic non-humic matter with low UV absorbance.

The VWMC of major ions in rainwater at Gwangju were compared to those reported in other urban areas around the world (Table 2). The pH value measured in our study is lower than that in Tirupati, India, but higher than those reported for other sites. The concentration of Na^+ is comparable to that in Shanghai but lower than in Istanbul and higher than other areas. The Cl^- value is comparable to that in Istanbul and much higher than those reported for other sites. The high concentration of marine elements was likely due to typhoons during summer seasons. The value of NO_3^- is higher than that in southeastern Brazil and close to that in other sites. Regarding SO_4^{2-}, its concentration is higher than that in Seoul, Mexico, and Brazil and lower than in other areas. Shanghai shows the highest values of NO_3^- and SO_4^{2-}, indicating severe air pollution problems in China. Both of these ions were mainly derived from the high coal/fuel consumption and mobile sources. The concentrations of K^+, Ca^{2+}, and Mg^{2+} are higher than those in Seoul, Mexico, and Brazil and lower than in other sites. The concentrations of these ions are comparatively higher in Istanbul than in other areas. The value of NH_4^+ is comparable to that in Seoul and higher than that in India, Turkey, and Brazil.

3.3 Seasonal variation

Figure 4 displays the monthly variation of VWM of pH, conductivity, TOC, TN, SUVA, and UV of rainwater sam-

Table 1. Volume-weighted mean concentration (VWMC), standard deviation of the VWMC (SDVWMC), and minimum (Min) and maximum (Max) concentrations of chemical composition in rainwater collected from June to December during 2012.

$n = 113$		VWMC	SDVWMC	Min	Max
pH		5.78	1.66	3.83	8.90
Conductivity	$\mu S\,cm^{-1}$	27.82	11.13	0.87	169.00
TOC	$mg\,C\,L^{-1}$	1.49	0.53	0.15	6.90
UV	cm^{-1}	0.0160	0.0061	0.0006	0.0972
SUVA		1.09	0.30	0.03	2.42
TN	$mg\,N\,L^{-1}$	1.13	0.40	N.D.	8.62
Na^+	$mg\,L^{-1}$	1.4	0.7	0.0	19.65
	$\mu eq\,L^{-1}$	58.5	30.7	1.2	854.50
NH_4^+	$mg\,L^{-1}$	1.1	0.4	0.0	5.9
	$\mu eq\,L^{-1}$	63.0	22.1	1.6	327.1
K^+	$mg\,L^{-1}$	0.3	0.1	0.0	1.9
	$\mu eq\,L^{-1}$	7.7	4.1	0.3	53.0
Mg^{2+}	$mg\,L^{-1}$	0.2	0.1	N.D.	2.5
	$\mu eq\,L^{-1}$	15.3	7.5	N.D.	208.6
Ca^{2+}	$mg\,L^{-1}$	0.7	0.4	N.D.	6.7
	$\mu eq\,L^{-1}$	36.7	17.7	N.D.	337.1
NO_3^-	$mg\,L^{-1}$	2.28	0.9	N.D.	12.4
	$\mu eq\,L^{-1}$	36.8	14.8	N.D.	199.2
SO_4^{2-}	$mg\,L^{-1}$	4.4	2.1	0.0	19.5
	$\mu eq\,L^{-1}$	91.9	42.6	0.9	405.4
Cl^-	$mg\,L^{-1}$	4.4	3.0	0.0	39.3
	$\mu eq\,L^{-1}$	123.5	83.7	1.0	1106.4

N.D.: not detected, n: number of samples. WHO (World Health Organization) drinking water guidelines: pH: 6.5–8.5, NH_4^+: 1.5 mg L^{-1}, NO_3^-: 50 mg L^{-1}, SO_4^{2-}: 250 mg L^{-1}, Cl^-: 250 mg L^{-1}.

Table 2. Comparison between the VWMC of major ions ($\mu eq\,L^{-1}$) in precipitation in Gwangju and at other sites.

Location	Period	pH	Na^+	NH_4^+	K^+	Mg^{2+}	Ca^{2+}	NO_3^-	SO_4^{2-}	Cl^-	References
This study	June–Dec. 2012	5.78	58.5	63.0	7.7	15.3	36.7	36.8	91.9	123.5	
Seoul, Korea	May 1996–Apr. 1998	4.7	10.5	66.4	3.5	6.9	10.5	29.9	70.9	18.2	Lee et al. (2000)
Tirupati, India	July 2000–June 2001	6.78	33.08	20.37	33.89	50.51	150.66	40.84	127.96	33.91	Mouli et al. (2005)
Istanbul, Turkey	Oct. 2001–July 2002	4.81	75.2	12.8	57.4	99.6	285	33.4	115.2	124.8	Basak and Alagha (2004)
Mexico City, Mexico	May 2001–Oct. 2002	5.08	7	92.35	2.16	2.46	26.44	42.62	61.94	9.56	Baez et al. (2007)
Shanghai, China	2005	4.49	50.11	80.68	14.89	29.64	203.98	49.8	199.59	58.34	Huang et al. (2008)
Piracicaba, southeastern Brazil	Aug. 1997–July 1998	4.5	2.7	17.1	2.9	2.3	5.3	16.6	18.7	7	Lara et al. (2001)

ples. pH values measured during summer seasons (June–September) with greater rainfall were higher than those during the winter season (October–December). This is consistent with previous studies (Cerón et al., 2013). Lee et al. (2000) also mentioned that the amount of rainfall influences pH. Conductivity values increased from summer to winter seasons, in agreement with the seasonal trend of major ions. Lower conductivity during the summer was caused by the dilution effect of higher rainfall. The values of TOC show similar monthly trends to TN, increasing with decreasing rainfall amounts. Similar seasonal variations of TOC have been reported in other studies (Pan et al., 2010; Yan and Kim, 2012). DOC in the atmosphere is known to vary with the seasons due to biogenic emission form vegetation. Kieber et

al. (2002) reported relatively higher levels of DOC in rainwater in warmer seasons; in contrast, Yan and Kim (2012) indicated that it was due to the prevailing anthropogenic contributions (mostly fossil fuel burning) and low precipitation in the relatively dry winter season. They also found that DOC in precipitation over Seoul was mainly produced by incomplete combustion of fossil fuel. The values of TN display similar monthly trends with TOC. They were higher in spring, fall, and winter, which are relatively dryer than summer. As expected, this is consistent with the monthly variation of NO_3^- and NH_4^+. The value of UV_{254} measured during the winter season (November–December) is higher than in the summer season (July–September). SUVA values higher than 1.4 were measured in September, November, and December, but in the

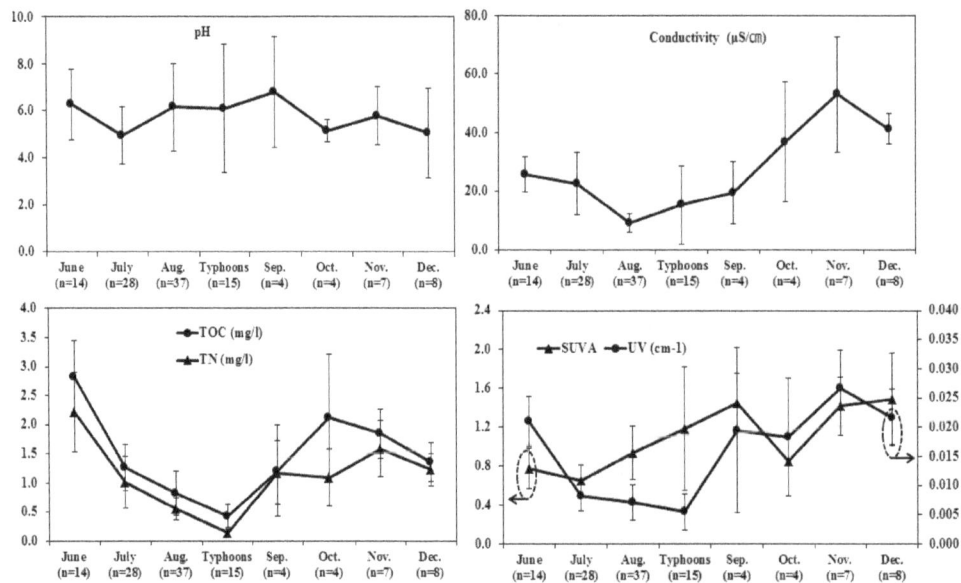

Figure 4. Monthly variation of the volume-weighted mean value of pH, conductivity, TOC, TN, SUVA, and UV absorbance in rainwater samples collected in Gwangju during 2012.

Table 3. Correlation matrix between ions in rainwater samples. The Pearson correlation coefficient and the P values are shown. Bold numbers presents $R^2 > 0.6$.

Variables	Na^+	NH_4^+	K^+	Mg^{2+}	Ca^{2+}	NO_3^-	Cl^-
NH_4^+	0.19						
	0.04						
K^+	0.50	0.50					
	0.00	0.00					
Mg^{2+}	**0.99**	0.23	0.53				
	0.00	0.01	0.00				
Ca^{2+}	0.41	0.55	**0.67**	0.48			
	0.00	0.00	0.00	0.00			
NO_3^-	0.39	**0.86**	0.59	0.44	**0.78**		
	0.00	0.00	0.00	0.00	0.00		
Cl^-	**1.00**	0.19	0.49	**0.99**	0.39	0.38	
	0.00	0.04	0.00	0.00	0.00	0.00	
SO_4^{2-}	0.29	**0.82**	**0.61**	0.32	**0.66**	**0.81**	0.29
	0.00	0.00	0.00	0.00	0.00	0.00	0.00

other months they were lower than 1.0. The monthly VWMC of major ions in rainwater samples are depicted in Fig. 5. The mean concentration of major ions in samples during summer (July–September), typhoons, and winter (November–December) are illustrated in Fig. 6. More than half of the total rainfall occurs in summer, while in the winter, precipitation is less than 10 % of the total in Korea (Lee et al., 2000). Thus, seasonal variations of ionic concentration in rainwater were mainly affected by precipitation patterns and monsoonal winds (Lee et al., 2000). The major ions had relatively higher concentrations in the winter, which is a pattern opposite to the trend in rainfall. The ionic concentrations showed

decreasing trends with increasing rainfall amounts, suggesting a dilution effect of rainwater on precipitation chemistry. The lowest ion concentrations were observed during typhoon periods, except for Na^+ and Cl^- which are typical marine components.

3.4 Correlation analysis

To investigate and identify potential correlation between major ions in rainwater samples, the Pearson correlation analysis was applied (Table 3). A strong correlation was found between Na^+ and Mg^{2+} ($R^2 = 0.99$) and between Na^+ and Cl^- ($R^2 = 1.00$), suggesting a marine source. The high cor-

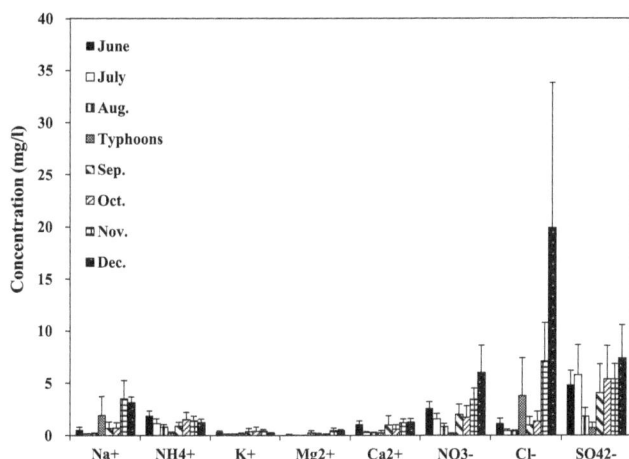

Figure 5. Monthly variation of VWMC of major ions in rainwater samples during 2012.

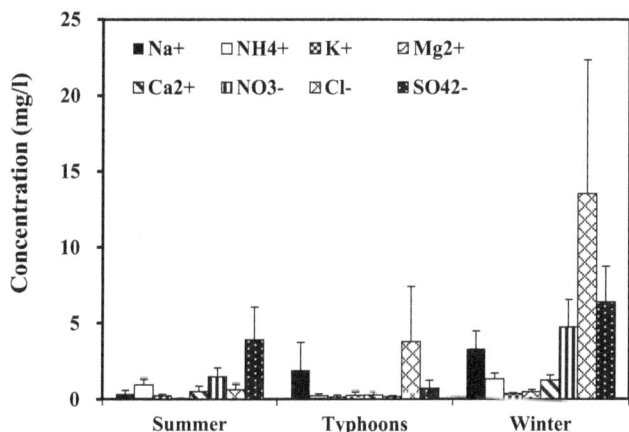

Figure 6. Mean concentration of major ions in rainwater samples during summer (July–September), typhoons, and winter (November–December).

Table 4. Concentrations of trace metals in rainwater samples (ppb).

Trace metals	VWMC	SDVWMC	Min	Max
Zn	18.78	14.08	N.D.	567.20
Al	12.99	6.64	N.D.	104.60
Fe	11.00	6.55	N.D.	74.51
Mn	4.58	2.89	N.D.	33.45
Pb	3.10	1.77	N.D.	25.01
Cu	1.69	0.89	N.D.	14.70
Ni	0.28	0.18	N.D.	8.04
Cd	0.09	0.05	N.D.	1.14
Cr	0.06	0.03	N.D.	0.47

WHO drinking water guidelines: Zn: $3\,mg\,L^{-1}$, Al: $0.\,mg\,L^{-1}$, Fe: $0.3\,mg\,L^{-1}$, Mn: $0.5\,mg\,L^{-1}$, Pb: $0.01\,mg\,L^{-1}$, Cu: $2\,mg\,L^{-1}$, Ni: $0.07\,mg\,L^{-1}$, Cd: $0.03\,mg\,L^{-1}$, Cr: $0.05\,mg\,L^{-1}$.

ity of their behavior in precipitation. Correlations were also found between NO_3^- and Ca^{2+} ($R^2 = 0.78$), and between SO_4^{2-} and Ca^{2+} ($R^2 = 0.66$), suggesting that $Ca(NO_3)_2$ and $CaSO_4$ are formed from the neutralization reactions of acidity in rainwater (Zhao et al., 2013). Moderate correlation was found between K^+ and Ca^{2+}, indicating a common origin in marine aerosol. K^+ and SO_4^{2-} were somewhat correlated ($R^2 = 0.61$), suggesting the occurrence of reactions between the acid H_2SO_4 and the alkaline compounds carried into the atmosphere by windblown sea salts/dusts (Huang et al., 2008).

3.5 Trace metals

The VWMCs of trace metals in rainwater samples are presented in Table 4. The concentration of trace metals decreased in the order $Zn > Al > Fe > Mn > Pb > Cu > Ni > Cd > Cr$. Table 5 provides the comparison of trace metal concentrations in precipitation from this study with values obtained from other rural and urban regions of the world. This order of element abundance is similar to those obtained from precipitation in Chuncheon, Suwon, Mexico, and New Castle in the USA. While Al and Fe were much higher in Ankara, the authors indicate that this is due to its location in the middle of the semi-arid Anatolia Plateau. In general, the concentrations of trace metals in Gwangju are lower than those in other countries, due to the lack of industry in the city. With respect to Al, Fe, Ni, Cd, and Cr, concentrations in this study were the lowest of all the other countries. Most of the industries in Korea are located in and near Seoul and the Gyeonggi Province, but Gwangju is a city of education, culture, universities and limited industrial activity. With respect to trace metals, values measured in Ankara, Turkey, were the highest of all of the other countries with the exception of Zn.

relation between Mg^{2+} and Cl^- ($R^2 = 0.99$) also corresponds to salts of marine origin. Relatively high correlations between NO_3^- and NH_4^+ ($R^2 = 0.86$), and between SO_4^{2-} and NH_4^+ ($R^2 = 0.82$), may reveal co-occurrence of NH_4^+ with SO_4^{2-} and NO_3^- in rainwater. This was probably due to dissolution of secondary inorganic aerosols (($NH_4)_2SO_4$ and NH_4NO_3) in rainwater (Panyakapo and Onchang, 2008). Among the compounds of ammonium, ammonium sulfate is known to predominate in the atmosphere (Seinfeld, 1986). However, our study showed slightly higher correlation between NH_4^+ and NO_3^- ($R^2 = 0.86$) than between NH_4^+ and SO_4^{2-} ($R^2 = 0.82$), indicating that NH_4NO_3 dominates over ($NH_4)_2SO_4$. The dominance of NH_4NO_3 has also been reported previously (Saxena et al., 1996). The high correlation between SO_4^{2-} and NO_3^- suggests a common source origin, due to the co-emission of precursors (SO_2 and NO_x) from the same sources, such as fossil fuel burning, and the similar-

Table 5. Comparison between the VWMC of trace metals in precipitation in Gwangju and at other locations.

Location	Zn	Al	Fe	Mn	Pb	Cu	Ni	Cd	Cr	References
This study	18.78	12.99	11.00	4.58	3.10	1.69	0.28	0.09	0.06	
Chuncheon, Korea	9.9	13.85		3.23	1.51	1.73	0.52	0.07		Kim et al. (2012)
Suwon, Korea	13.06				9.98	6.08	0.66	0.2	0.43	Jung et al. (2007)
Mexico		50.7		9.64	2.48		3.37	0.41	0.52	Baez et al. (2007)
Ankara, Turkey	0.03	980	750		19.1	6.1	4.1	9.5	3	Kaya and Tuncel (1997)
Singapore	7.23	18.44	23.91	2.78	3.37	5.58	3.86	0.33	1.62	Hu and Balasubarmanian (2003)
Jordan	6.52	382	92	2.11	2.57	3.08	2.62	0.42	0.77	Al-Momani et al. (2003)
New Castle, NH, USA	25.5	24.4	22.8		1.31	1.33	0.85	0.24	0.14	Pike and Moran (2001)

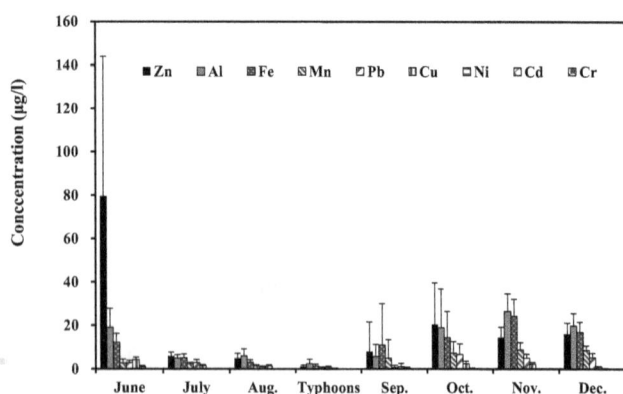

Figure 7. Monthly variation of VWMC of trace metals in rainwater samples during 2012.

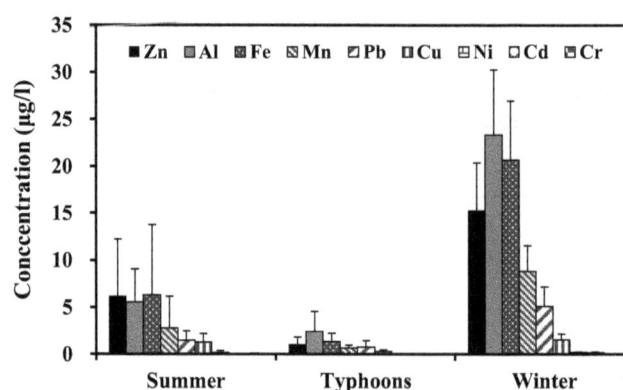

Figure 8. Mean concentration of trace metals in rainwater samples during summer (July–September), typhoons, and winter (November–December).

3.6 Variations in trace metal levels in rainwater

Monthly variations of the VWMC of trace metals in rainwater are shown in Fig. 7. As expected, the lowest concentrations of trace metals were found during typhoon periods (July–September), caused by more frequent rain events and the consequent dilution effect. The samples collected in winter had higher concentrations of trace metals than the samples in summer (Fig. 8). Similar seasonal variations have been reported in other studies (Cheng et al., 2011; Kim et al., 2012).

3.7 Correlation analysis

Table 6 lists the matrix correlation between ions and trace metals (correlation coefficients greater than 0.5 are marked in bold letters). Moderate correlation was found among the trace metals and NH_4^+, NO_3^-, and SO_4^{2-}, with the exception of Cr and Zn, suggesting the anthropogenic origin of the species. These correlations were also observed in previous studies (Garcia et al., 2006; Jung et al., 2011). Significant correlations between SO_4^{2-} and trace metals in rainwater were observed in Clarke and Radojevic (1987).

3.8 Enrichment factor analysis

An enrichment factor (EF) has been used to evaluate the contribution of non-crustal sources (Kaya and Tuncel, 1997; Kim et al., 2012). The EF values were calculated using the following relation:

$$EF = (C_X/C_{Al} \text{ sample})/(C_X/C_{Al} \text{ crust}),$$

where (C_X / C_{Al} sample) is the ratio of the concentration of an element X and Al in the rainwater sample and (C_X / C_{Al} crust) is the same ratio in crustal material, adapted from Wedepohl (1995).

EFs from 1 to 10 suggest crustal origin; EFs from 10 to 100 suggest moderate anthropogenic enrichment; and EFs > 100 indicate anthropogenic origin. Fig. 9 presents the bars of the EF of the trace elements. Fe and Cr with EF values of 1–10 were significantly enriched by soil. Mn and Ni with EF values in the range of 10–100 were moderately enriched by anthropogenic sources, while Zn, Pb, Cu, and Cd with EF values exceeding 100 were highly enriched by human activities. Ni and Mn are mainly emitted from manufacture of ferroalloys and from oil-fired furnaces (Báez et al., 2007). The primary sources of Zn, Pb, Cu, and Cd are industrial and traffic activities such as metal smelting and fuel combustion (Al-Momani, 2003; Cheng et al., 2011). The variation of monthly

Table 6. Correlation coefficients and P value between ions and trace metals in rainwater 1 samples from Gwangju ($n = 113$).

	Na^+	NH_4^+	K^+	Mg^{2+}	Ca^{2+}	NO_3^-	Cl^-	SO_4^{2-}
Al	0.32	**0.60**	0.44	0.36	**0.51**	**0.66**	0.33	**0.59**
	0.00	0.00	0.00	0.00	0.00	0.00	0.00	0.00
Cr	0.17	0.41	0.19	0.20	0.30	0.41	0.18	0.42
	0.07	0.00	0.04	0.03	0.00	0.00	0.06	0.00
Mn	0.36	0.35	0.29	0.40	**0.52**	**0.57**	0.36	0.38
	0.00	0.00	0.00	0.00	0.00	0.00	0.00	0.00
Fe	0.36	**0.53**	0.39	0.41	0.49	**0.64**	0.37	**0.53**
	0.00	0.00	0.00	0.00	0.00	0.00	0.00	0.00
Ni	0.02	**0.53**	0.32	0.05	0.23	0.39	0.03	0.37
	0.82	0.00	0.00	0.63	0.01	0.00	0.79	0.00
Cu	0.05	**0.67**	0.35	0.07	0.35	**0.58**	0.06	**0.58**
	0.59	0.00	0.00	0.43	0.00	0.00	0.54	0.00
Zn	0.08	0.24	0.32	0.12	0.33	0.22	0.08	0.20
	0.39	0.01	0.00	0.19	0.00	0.02	0.38	0.03
Cd	0.31	**0.59**	0.42	0.36	**0.50**	**0.61**	0.33	**0.64**
	0.00	0.00	0.00	0.00	0.00	0.00	0.00	0.00
Pb	**0.50**	**0.56**	0.44	**0.54**	0.39	**0.63**	**0.51**	**0.63**
	0.00	0.00	0.00	0.00	0.00	0.00	0.00	0.00

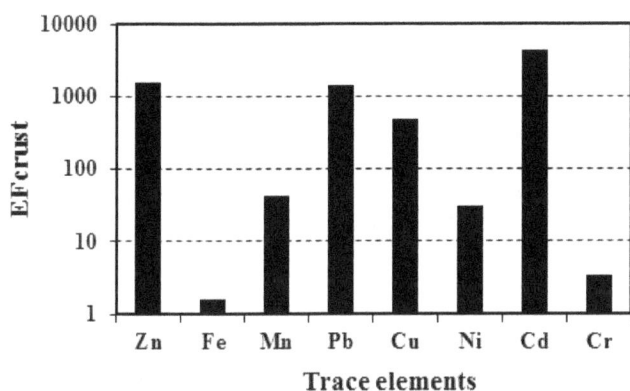

Figure 9. Average EFs of trace metals in rainwater in the Gwangju area during 2012.

Table 7. Mean values of TOC, TN, UV, SUVA, and pH and conductivity in initial ($n = 40$) and subsequent ($n = 73$) rainfall during 2012.

	Initial rainfall		Subsequent rainfall	
	Mean	SD	Mean	SD
TOC (mg $C\,L^{-1}$)	2.18	1.41	1.11	0.68
TN (mg $N\,L^{-1}$)	1.95	1.19	0.79	0.47
UV (cm^{-1})	0.0244	0.0135	0.0111	0.0073
SUVA	1.19	0.39	1.06	0.33
pH	5.97	0.52	5.55	0.64
Conductivity ($\mu S\,cm^{-1}$)	38.17	25.67	25.72	19.07

SD: standard deviation.

EFs of trace metals in rainwater is depicted in Fig. 10. The monthly EFs of trace metals were relatively high during summer (from June to September). Cd and Cr were not detected during the typhoon periods or in September. The highest EF values of Fe and Mn, Pb, Cu, Cd, and Cr, and Zn and Ni were found during September, July, and June, respectively.

3.9 Comparison of chemical components in initial and subsequent rainfall

A comparison of the mean concentrations of major ions and trace metals in initial ($n = 40$) and subsequent ($n = 73$) rainfall samples is presented in Fig. 11. Greater amounts of major ions and trace metals appeared in initial rainfall samples than in subsequent samples as large amount of pollutants were dissolved in less water. Mean values of TOC, TN, UV,

SUVA, pH, and conductivity in initial ($n = 40$) and subsequent ($n = 73$) samples are shown in Table 7. Again, higher mean concentrations of TOC, TN, UV, SUVA, pH, and conductivity were observed in initial rainfall, with the values of TOC, TN, and UV in initial rainfall being twice as high as in subsequent rainfall.

3.10 Factor analysis

Factor analysis (principal component analysis) has been widely applied in previous studies to investigate the influence of different sources on the chemical components in precipitation (Báez et al., 2006; Garcia et al., 2006; Panyakapo and Onchang, 2008). A varimax-rotated principal component analysis (PCA) was performed using Minitab version 16 for Windows. Table 8 presents the result of the factor analy-

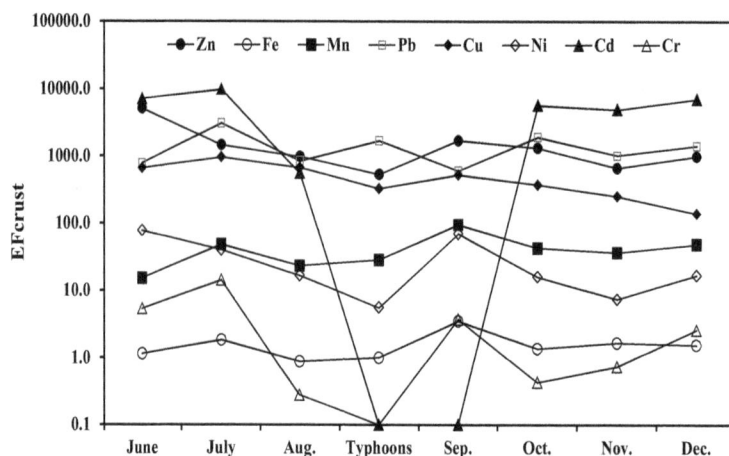

Figure 10. Monthly variation of EFs of trace metals in rainwater in the Gwangju area during 2012.

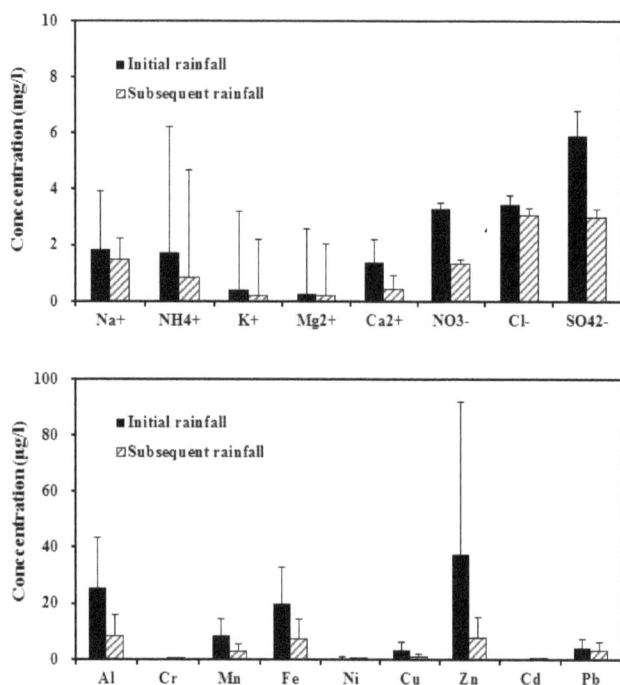

Figure 11. Comparison of mean concentration of major ions and trace metals in initial ($n = 40$) and subsequent ($n = 73$) rainfall during 2012.

Table 8. Factor analysis of chemical constituents in rainwater. Bold numbers are significant at > 0.5.

Variable	Factor 1	Factor 2	Factor 3
Na^+	0.12	**0.97**	0.10
NH_4^+	**0.88**	0.03	0.27
K^+	**0.61**	0.46	0.11
Mg^{2+}	0.15	**0.96**	0.15
Ca^{2+}	**0.67**	0.39	0.19
NO_3^-	**0.85**	0.29	0.28
Cl^-	0.11	**0.96**	0.12
SO_4^{2-}	**0.85**	0.18	0.23
Al	0.51	0.20	**0.71**
Cr	0.14	0.08	**0.79**
Fe	0.41	0.28	**0.76**
Ni	0.44	−0.18	**0.50**
Zn	0.04	0.02	**0.62**
Cd	0.47	0.24	**0.63**
Pb	0.44	0.42	**0.59**
Mn	0.36	0.35	**0.50**
Cu	**0.75**	−0.14	0.41
% total variance	28.5	22.4	22.2

sis that identified three factors that explained approximately 73.1 % of the total variance. Factor 1 explained 27.1 % of the total variance with high loadings for SO_4^{2-}, NO_3^-, and NH_4^+, and moderate loadings for K^+, Ca^{2+}, and Cu. This factor indicates marine sources for K^+ and Ca^{2+}, with an anthropogenic source for Cu, as is indicated by EF. Cu is a typical metal contaminant emitted from industrial processes (Wong et al., 2003). The loadings of SO_4^{2-}, NO_3^-, and NH_4^+ suggest that these ions come from anthropogenic sources associated with precursor gases such as SO_2 and NO_x. The co-occurrence of NH_4^+ with NO_3^- and SO_4^{2-} are caused by dissolution of aerosols and secondary pollutants containing $(NH_4)_2SO_4$ and NH_4NO_3 in rainwater. Factor 2 accounted for 22.4 % of the total variance with high loading for Na^+, Mg^{2+}, and Cl^-, indicating the influence of sea spray. Factor 3, which explained 22.2 % of the total variance, proposed moderate loadings of most of the trace metals. This factor indicates the contribution of anthropogenic sources and significant crustal contributions for Al, Cr, and Fe, as indicated by EF. Overall, the anthropogenic pollutants, sea salts, and crustal materials were the main sources of the chemical species in rainwater in Gwangju.

Table 9. Dominant pyrolysis fragments for rainwater samples as identified by Py-GC/MS.

Pyrolysis fragments

Ranking	2 June	8 June	18 June	19 June	29 June	30 June	12 July	18 July (Typhoon Khanun)	6 Aug.	10–15 Aug.
1	Benzoic acid, 2,4-dichloro-(PHA)	Benzoic acid, 2,6-dichloro-(PHA)	Benzoic acid, 2,6-dichloro-(PHA)	Benzoic acid, 2,6-dichloro-(PHA)	Benzoic acid, 2,6-dichloro-(PHA)	Benzoic acid, 2,6-dichloro-(PHA)	Acetic acid (PS,AS)	Acetamide, 2-fluoro-(PS,AS)	1H-Isoindole-1,3(2H)-dione (Pr)	1H-Isoindole-1,3(2H)-dione (Pr)
2	1H-Isoindole-1,3(2H)-dione (Pr)	Benzoic acid, 4-chloro-(PHA)	Acetic acid (PS,AS)	Benzoic acid, 4-chloro-(PHA)	Acetic acid (PS,AS)	Phthalic anhydride (PHA)	1H-Isoindole-1,3(2H)-dione (Pr)	1,3-Benzenedicarbonitrile (Pr,PHA)	Phthalic anhydride (PHA)	Phthalic anhydride (PHA)
3	2,5-Pyrrolidinedione (Pr)	2,4-Dichlorobenzamide (PHA)	Benzenecarboxylic acid (PHA)	2,4-Dichlorobenzamide (PHA)	Benzoic acid, 4-chloro-(PHA)	1H-Isoindole-1,3(2H)-dione (Pr)	Furan, 2-methyl-(PS)	1H-Isoindole-1,3(2H)-dione (Pr)	Furan, 2-methyl-(PS)	Furan, 3-methyl-(PS)
4	Benzenecarboxylic acid (PHA)	Benzenecarboxylic acid (PHA)	Benzonitrile (Pr)	Benzene, 1,4-dichloro-(PHA)	Benzonitrile (Pr)	Acetic acid (PS,AS)	Benzonitrile (Pr)	1,2-Benzenedicarboxylic acid (PHA)	2,3-Butanedione (PS)	Acetonitrile, dichloro-(Pr)
5	Benzene, 1,4-dichloro-(PHA)	Benzonitrile (Pr)	Benzene (PHA)	1H-Isoindole-1,3(2H)-dione (Pr)	Benzene (PHA)	Benzoic acid, 2-chloro-(PHA)	Phthalic anhydride (PHA)	Benzene (PHA)	Benzenecarboxylic acid (PHA)	Benzoic acid, 2-chloro-(PHA)
6	Benzoic acid, 4-chloro-(PHA)	Benzene, 1,4-dichloro-(PHA)	Benzoic acid, 4-chloro-(PHA)	Acetic acid (PS,AS)	1H-Isoindole-1,3(2H)-dione (Pr)	Dodecanoic acid (Lp)	2(5H)-Furanone, 3,5,5-trimethyl-(PS)	Benzene (PHA)	Benzonitrile (Pr)	
7	Acetamide (PS,AS)	2,5-Pyrrolidinedione (Pr)	n-Hexadecanoic acid (Lp)	Benzenecarboxylic acid (PHA)	Benzenecarboxylic acid (PHA)	Furan, 2-methyl-(Pr)	n-Hexadecanoic acid (Lp)	Acetic acid (PS,AS)		
8	Benzonitrile (PR)	n-Hexadecanoic acid (Lp)	1H-Isoindole-1,3(2H)-dione (Pr)	Benzoic acid, 2,4-dichloro-(PHA)	Phthalic anhydride (PHA)	2,5-Pyrrolidinedione (PR)	1,2-Benzenedicarbonitrile (Pr, PHA)			
9	Phenol (PR,PHA)	1,4-Benzenedicarbonitrile (Pr,PHA)	2,5-Pyrrolidinedione (Pr)	Benzonitrile (Pr)	2,5-Pyrrolidinedione (Pr)	Benzene, 1-chloro-2-ethoxy-(PHA)	Furan, 2-methyl-(PS)			
10	2(5H)-Furanone, 3,5,5-trimethyl-(PS)	Octadecanoic acid (Lp)	Acetamide (PS,AS)	1,2-Benzenedicarbonitrile (Pr, PHA)	Furan, 2-methyl-(PS)	Tetradecanoic acid (Lp)	Formic acid, ethenyl ester (Lp)			

Ranking	23–24 Aug.	27–28 Aug. (Typhoon Bolaven)	30 Aug. (Typhoon Tenbin)	12 Sep.	22 Oct.	5–11 Nov.	13–26 Nov.	12 Dec.
1	1H-Isoindole-1,3(2H)-dione (PR)	Benzene (PHA)	Acetic acid (PS, AS)	o-Cyanobenzoic acid (PHA)	1,2-Benzenedicarboxylic acid (PHA)	Benzonitrile (Pr)	Benzonitrile (Pr)	Acetaldehyde (PS)
2	Phthalic anhydride (PHA)	Acetic acid (PS, AS)	Phthalic anhydride (PHA)	Phthalic anhydride (PHA)	1H-Isoindole-1,3(2H)-dione (Pr)	Phthalic anhydride (PHA)	1H-Isoindole-1,3(2H)-dione (Pr)	Acetaldehyde (PS)
3	Benzoic acid, 4-(1-methylethyl)-(PHA)	Phenol (Pr,PHA)	Benzonitrile (Pr)	1,2-Benzenedicarboxylic acid (PHA)	Acetic acid (PS, AS)	Acetic acid (PS, AS)	Acetic acid (PS, AS)	1H-Isoindole-1,3(2H)-dione (Pr)
4	Acetic acid (PS,AS)	Benzene (PHA)	Benzene (PHA)	Benzonitrile (PHA)	Benzonitrile (Pr)		Benzene (PHA)	Acetic acid (PS, AS)
5	Benzoic acid, 4-chloro-(PHA)	Styrene (Pr)	2-Pentenoic acid, 4-methyl-(Pr)				Phthalic anhydride (PHA)	
6	Benzoic acid, 4-methyl-(PHA)	α-D-Glucopyranose, 4-O-β-D-galactopyranosyl-(PS)	Styrene (Pr)				Furan, 2,5-dimethyl-(PS)	
7	Benzonitrile (Pr)	2-Dodecenal, (E)-(Lp)	Pyridine (Pr)					
8	Benzenecarboxylic acid (PHA)	2H-Pyran, 3,4-dihydro-4-methyl-(PS)	Benzenecarboxylic acid (PHA)				Benzonitrile, 4-[2-(methylsulfonylethenyl)]-(Pr)	Benzenecarboxylic acid (PHA)
9	Furan, 2-methyl-(PS)	Benzene, 1-methyl-4-(1-methylethenyl)-(PHA)	Benzene, 1-methyl-4-(1-methylethenyl)-(PHA)					1,2-Benzenedicarbonitrile (Pr, PHA)
10	Benzoic acid, 4-(1-methylethyl)-(PHA)	Benzaldehyde (Lg)						

Pr: protein; PS: polysaccharides; AS: amino sugars; PHA: polyhydroxy aromatics; Lg: lignins; Lp: lipids.

3.11 Organic compounds in rainwater

Table 9 lists dominant pyrolysis fragments for rainwater organic compounds collected from June to December 2012 as identified by Py-GC/MS based on peak area percent of pyrochromatograms. Each sample could be fractionated and categorized into polysaccharides (PS), amino sugars (AS), proteins (PR), polyhydroxy aromatics (PHA), lignins (LG), and lipids (Lp) (Leenher and Croue, 2003). Benzoic acids were the most abundant compound during June, which originates from anthropogenic sources such as fossil fuel combustion. Kawamura et al. (1986) identified benzoic acids in used engine oil and motor exhaust. Benzoic acids have been found in rainfall in Los Angeles and Norway (Lunde et al., 1977; Kawamura and Kaplan, 1983). 2-Fluoro-acetamide and acetic acids were the most abundant compounds in July and September. Acetamide was one of the major pyrolysis organic compounds with precipitation in Königstein, Germany (Laniewski et al., 1998), which is a source of plasticizer and industrial solvents and normally found near industrial areas (Cho and Shin, 2013). Acetic acids originate from primary natural sources such as emissions from microbial activity and vegetation and from anthropogenic sources (biomass burning and traffic exhaust) (Avery et al., 1991). It has been found in rainwaters from both developed and remote areas (Galloway et al., 1982; Keene and Galloway, 1984). 1H-Isoindole-1,3(2H)-dione, phthalic anhydride, benzene, and acetic acids were the most abundant species during August. 1H-Isoindole-1,3(2H)-dione was one of the major pyrolysis organic compounds from precipitation in Königstein, Germany (Laniewski et al., 1998), and was found to be from tire-derived aggregates. Phthalic anhydride originates from agricultural crop burning and open burning of scrap tires (Lemieux et al., 2004). Benzene is an important aromatic compound from gasoline, automobile exhaust, and the urban atmosphere (Sigsby et al., 1987). Benzene was generally found as the major compound of volatile organic carbons in rainwater from Yokohama, Japan (Sato et al., 2006). 1,2-Benzenedicarboxylic acids were the most abundant compound in October and have been identified to come from particulate automobile exhaust emissions, cigarette combustion, degradation of plastics, and products of lignin-type material metabolized by microorganisms (Johnstone and Plimmer, 1959; Simoneit, 1985; Kawamura and Kaplan, 1987). Benzonitrile was the most abundant species during November, which has been found in accidental vehicle fires (Estrellan and Lino, 2010) and biomass combustion in improved stoves in rural China (Wang et al., 2009). Acetaldehyde was the most abundant compound in December, probably derived from primary incomplete combustion of fossil fuels and secondary photochemical reactions of hydrocarbons emitted from anthropogenic sources (Sakugawa et al., 1993). Acetaldehyde was also detected in Los Angeles (Kawamura et al., 2001) and Spain (Peña et al., 2002). Among fatty acids, n-hexadecanoic acid (C_{16}), octadecanoic acid (C_{18}), and tetradecanoic acid (C_{14}) were found in the top 10 organic compounds during June and July. This is in agreement with previous work (Kawamura and Kaplan, 1986), where lower molecular fatty acids (C_{12}–C_{19}) are major compounds while higher molecular weight fatty acids (C_{20}–C_{30}) are minor. A major source of fatty acids ($C < 20$) is known to come from cooking and biomass combustion in urban areas (Rogge et al., 1991; Xu et al., 2008). All results in the present study indicate that anthropogenic sources were significant contributors to the organic compounds present in rainwater.

4 Conclusions

Rainwater samples collected from Gwangju, Korea, during June–December 2012 were analyzed to determine the chemical composition and organic compounds present in rainwater. Even though our study period was quite short compared to other studies, we tried to present a valuable data analysis of the chemical composition and organic compounds in rainwater between summer and winter, as there are two distinct seasons with rain in Korea. The VWM of pH was 5.78 (ranging from 3.83 to 8.90) and acid rain (pH < 5.6) frequency was 50 %. The VWMC of major ions and trace metals followed the order $Cl^- > SO_4^{2-} > NH_4^+ > Na^+ > NO_3^- > Ca^{2+} > Mg^{2+} > K^+$ and $Zn > Al > Fe > Mn > Pb > Cu > Ni > Cd > Cr$. The higher VWM values of TOC, TN, UV, SUVA, pH, conductivity, major ions, and trace metals appeared in the initial rainfall events rather than in subsequent rainfall events. The VWMCs of major ions and trace metals were relatively lower in summer due to the dilution effect. The lowest VWMCs of ions (except for Na^+ and Cl^-) and trace metals were observed during typhoons. Based on EF values, Zn, Pb, Cu, Ni, Cd, and Mn were emitted mainly from anthropogenic sources whereas Fe and Cr originated from crustal sources. The factor analysis (principal component analysis) revealed that the anthropogenic pollutants, sea salts, and crustal materials were the main sources of ionic species and trace metals in rainwaters. The most abundant pyrolysis fragments for rainwater organic compounds were benzoic acids, 1H-isoindole-1,3(2H)-dione, phthalic anhydride, benzene, acetic acids, 1,2-benzenedicarboxylic acids, benzonitrile, acetaldehyde, and acetamide, indicating that anthropogenic pollutants are significant contributors to rainwater organic compounds.

Acknowledgements. This research was supported by the National Research Foundation of Korea (NRF) grant funded by the Korean government (MSIP) (no. 2011-0030040 (ERC) and no. NRF-2015R1A5A7037825).

References

Al-Momani, I. F.: Trace elements in atmospheric precipitation at Northern Jordan measured by ICP-MS: acidity and possible sources, Atmos. Environ., 37, 4507–4515, 2003.

Avery Jr., G. B., Willey, J. D., and Wilson, C. A.: Formic and acetic acids in coastal North Carolina rainwater, Environ. Sci. Technol., 25, 1875–1880, 1991.

Báez, A. P., Belmont, R. D., García, R. M., Torres, M. C. B., and Padilla, H. G.: Rainwater chemical composition at two sites in Central Mexico, Atmos. Res., 80, 67–85, 2006.

Báez, A. P., Belmont, R. D., García, R. M., Padilla, H. G., and Torres, M. C. B.: Chemical composition of rainwater collected at a southwest site of Mexico City, Mexico, Atmos. Res., 86, 61–75, 2007.

Bard, S. M.: Global transport of anthropogenic contaminants and the consequences for the arctic marine ecosystem, Mar. Pollut. Bull., 38, 356–379, 1999.

Barrie, L. A., Lindberg, S. E., Chan, W. H., Ross, H. B., Arimoto, R., and Church, T. M.: On the concentration of trace metals in precipitation, Atmos. Environ., 21, 1133–1135, 1987.

Başak, B. and Alagha, O.: The chemical composition of rainwater over Büyükçekmece Lake, Istanbul, Atmos. Res, 71, 275–288, 2004.

Bruchet, A., Rousseau, C., and Mallevialle, J.: Pyrolysis-GC/MS for investigating high-molecular weight THM precursors and other refractory organics, J. Am. Water Works Ass., 82, 66–74, 1990.

Cerón, R., Cerón, J., Carballo, C., Aguilar, C., Montalvo, C., Benítez, J., Villareal, Y., and Gómez, M.: Chemical composition, fluxes and seasonal variation of acid deposition in Carmen Island, Campeche, Mexico, J. Environ. Prot., 4, 50–56, 2013.

Charlson, R. J. and Rodhe, H.: Factors controlling the acidity of natural rainwater, Nature, 295, 683–685, 1982.

Cheng, M. C., You, C. F., Lin, F. J., Huang, K. F., and Chung, C. H.: Sources of Cu, Zn, Cd and Pb in rainwater at a subtropical islet offshore northern Taiwan, Atmos. Environ., 45, 1919–1928, 2011.

Cho, Y. H. and Shin, H. S.: Determination of trace levels of acetamide, propanamide, and butyramide in surface and drinking water using gas chromatography–mass spectrometry after derivatization with 9-xanthydrol, Anal. Chim. Acta, 787, 111–117, 2013.

Clarke, A. G. and Radojevic, M.: Oxidation of SO_2 in rainwater and its role in acid rain chemistry, Atmos. Environ., 21, 1115–1123, 1987.

Dianwu, Z. and Anpu, W.: Estimation of anthropogenic ammonia emissions in Asia, Atmos. Environ, 28, 689–694, 1994.

Estrellan, C. R. and Iino, F.: Toxic emissions from open burning, Chemosphere, 80, 193–207, 2010.

Galloway, J. N., Likens, G. E., Keene, W. C., and Miller, J. M.: The composition of precipitation in remote areas of the world, J. Geophys. Res.-Oceans (1978–2012), 87, 8771–8786, 1982.

Garcia, R., del Torres Ma, C., Padilla, H., Belmont, R., Azpra, E., Arcega-Cabrera, F., and Báez, A.: Measurement of chemical elements in rain from Rancho Viejo, a rural wooded area in the State of Mexico, Mexico, Atmos. Environ., 40, 6088–6100, 2006.

Hu, G. P. and Balasubramanian, R.: Wet deposition of trace metals in Singapore, Water Air Soil Pollut., 144, 285–300, 2003.

Huang, K., Zhuang, G., Xu, C., Wang, Y., and Tang, A.: The chemistry of the severe acidic precipitation in Shanghai, China, Atmos. Res, 89, 149–160, 2008.

Johnstone, R. A. W. and Plimmer, J. R.: The Chemical Constituents of Tobacco and Tobacco Smoke, Chem. Rev., 59, 885–936, 1959.

Jung, G. B., Kim, M. K., Lee, J. S., Kim, W. I., Kim, G. Y., and Ko, B. G.: Wet Deposition of Heavy Metals during Farming Season in Taean, Korea, Korean J. Environ. Agric., 30, 153–159, 2011.

Kang, G. G., Collett, J. L., Shin, D. Y., Fujita, S. I., and Kim, H. K.: Comparison of the chemical composition of precipitation on the western and eastern coasts of Korea, Water Air Soil Pollut., 151, 11–34, 2004.

Kawamura, K. and Kaplan, I. R., Organic compounds in the rainwater of Los Angeles, Environ. Sci. Technol., 17, 497–501, 1983.

Kawamura, K. and Kaplan, I. R.: Biogenic and anthropogenic organic compounds in rain and snow samples collected in southern California, Atmos. Environ. (1967), 20, 115–124, 1986.

Kawamura, K. and Kaplan, I. R.: Motor exhaust emissions as a primary source for dicarboxylic acids in Los Angeles ambient air, Environ. Sci. Technol., 21, 105–110, 1987.

Kawamura, K., Steinberg, S., Ng, L., and Kaplan, I. R.: Wet deposition of low molecular weight mono-and di-carboxylic acids, aldehydes and inorganic species in Los Angeles, Atmos. Environ., 35, 3917–3926, 2001.

Kaya, G. and Tuncel, G.: Trace element and major ion composition of wet and dry deposition in Ankara, Turkey, Atmos. Environ., 31, 3985–3998, 1997.

Keene, W. C. and Galloway, J. N.: Organic acidity in precipitation of North America, Atmos. Environ. (1967), 18, 2491–2497, 1984.

Kieber, R. J., Peake, B., Willey, J. D., and Avery, G. B.: Dissolved organic carbon and organic acids in coastal New Zealand rainwater, Atmos. Environ., 36, 3557–3563, 2002.

Kim, J. E., Han, Y. J., Kim, P. R., and Holsen, T. M.: Factors influencing atmospheric wet deposition of trace elements in rural Korea, Atmos. Res., 116, 185 194, 2012.

Klouda, G. A, Lewis, C. W., Rasmussen, R. A., Rhoderık, G. C, Sams, R. L., Stevens, R. K., Currie, L. A., Donahue, D. J., Jull, A. J. T., and Seila, R. L.: Radiocarbon measurements of atmospheric volatile organic compound: quantifying the biogenic contribution, Environ. Sci. Technol., 30, 1098–1105, 1996.

Laniewski, K., Borén, H., Grimvall, A., and Ekelund, M.: Pyrolysis–gas chromatography of chloroorganic compounds in precipitation, J. Chromatogr. A, 826, 201–210, 1998.

Lara, L. B. L. S., Artaxo, P., Martinelli, L. A., Victoria, R. L., Camargo, P. B., Krusche, A., and Ballester, M. V.: Chemical composition of rainwater and anthropogenic influences in the Piracicaba River Basin, Southeast Brazil, Atmos. Environ., 35, 4937–4945, 2001.

Lee, B. K., Hong, S. H., and Lee, D. S.: Chemical composition of precipitation and wet deposition of major ions on the Korean peninsula, Atmos. Environ., 34, 563–575, 2000.

Leenher, J. A. and Croue, J. P.: Characterizing aquatic dissolved organic matter, Environ. Sci. Technol., 37, 18–26, 2003.

Lemieux, P. M., Lutes, C. C., and Santoianni, D. A.: Emissions of organic air toxics from open burning: a comprehensive review, Prog. Energ. Combust., 30, 1–32, 2004.

Lunde, G., Gether, J., Gjùs, N., and Stùbet Lande, M. B.: Organic micropollutants in precipitation in Norway, Atmos. Environ., 11, 1007–1014, 1977.

Moon, S. H., Lee, J. Y., Lee, B. J., Park, K. H., and Jo, Y. J.: Quality of harvested rainwater in artificial recharge site on Jeju volcanic island, Korea, J. Hydrol., 414, 268–277, 2012.

Mouli, P., Mohan, S., and Reddy, S.: Rainwater chemistry at a regional representative urban site: influence of terrestrial sources on ionic composition, Atmos. Environ., 39, 999–1008, 2005.

Muller, C. L., Baker, A., Hutchinson, R., Fairchild, I. J., and Kidd, C.: Analysis of rainwater dissolved organic carbon compounds using fluorescence spectrophotometry, Atmos. Environ., 42, 8036–8045, 2008.

Pan, Y., Wang, Y., Xin, J., Tang, G., Song, T., Wang, Y. H., Li, X. R., Wu, F. K.: Study on dissolved organic carbon in precipitation in Northern China, Atmos. Environ., 44, 2350–2357, 2010.

Panyakapo, M. and Onchang, R.: A four-year investigation on wet deposition in western Thailand, J. Environ. Sci. (China), 20, 441–448, 2008.

Peña, R. M., García, S., Herrero, C., Losada, M., Vázquez, A., and Lucas, T.: Organic acids and aldehydes in rainwater in a northwest region of Spain, Atmos. Environ., 36, 5277–5288, 2002.

Pike, S. M. and Moran, S. B.: Trace elements in aerosol and precipitation at New Castle, NH, USA, Atmos. Environ., 35, 3361–3366, 2001.

Rogge, W. F., Hildemann, L. M., Mazurek, M. A., Cass, G. R., and Simoneit, B. R. T.: Sources of fine organic aerosol: 1. Charbroilers and meat cooking operations, Environ. Sci. Technol., 25, 1112–1125, 1991.

Sakugawa, H., Kaplan, R., and Shepard, L.: Measurements of H_2O_2, aldehydes and organic acids in Los Angeles rainwater: their sources and deposition rates, Atmos. Environ. B, 27, 203–219, 1993.

Santos, P. S., Otero, M., Duarte, R. M., and Duarte, A. C.: Spectroscopic characterization of dissolved organic matter isolated from rainwater, Chemosphere, 74, 1053–1061, 2009.

Santos, P. S, Otero, M., Santos, E. B., and Duarte, A. C.: Chemical composition of rainwater at a coastal town on the southwest of Europe: What changes in 20 years?, Sci. Total. Environ, 409, 3548–3553, 2011.

Sato, E., Matsumoto, K., Okochi, H., and Igawa, M.: Scavenging effect of precipitation on volatile organic compounds in ambient atmosphere, Bull. Chem. Soc. Jpn., 79, 1231–1233, 2006.

Saxena, A., Kulshrestha, U. C., Kumar, N., Kumari, K. M., and Srivastava, S. S.: Characterization of precipitation at Agra, Atmos. Environ., 30, 3405–3412, 1996.

Seinfeld, J. H.: Atmospheric Chemistry and Physics of Air Pollution, Wiley, New York, 1986.

Seitzinger, S. P., Styles, R. M., Lauck, R., and Mazurek, M. A.: Atmospheric pressure mass spectrometry: a new analytical chemical characterization method for dissolved organic matter in rainwater, Environ. Sci. Technol., 37, 131–137, 2003.

Sigsby, J. E., Tejada, S., Ray, W., Lang, J. M., and Duncan, J. W.: Volatile organic compound emissions from 46 in-use passenger cars, Environ. Sci. Technol., 21, 466–475, 1987.

Simoneit, B. R. T.: Application of Molecular Marker Analysis to Vehicular Exhaust for Source Reconciliations, Int. J. Environ. Anal. Chem., 22, 203–233, 1985.

Wang, S., Wei, W., Du, L., Li, G., and Hao, J.: Characteristics of gaseous pollutants from biofuel-stoves in rural China, Atmos. Environ., 43, 4148–4154, 2009.

Wedepohl, H. K.: The composition of the continental crust, Geochim. Cosmochim. Acta, 59, 1217–1232, 1995.

Willey, J. D., Kieber, R. J., Eyman, M. S., and Avery Jr., G. B.: Rainwater dissolved organic carbon: concentrations and global flux, Global Biogeochem. Cy., 14, 139–148, 2000.

Wong, C. S. C., Li, X. D., Zhang, G., Qi, S. H., and Peng, X. Z.: Atmospheric depositions of heavy metals in the Pearl River Delta, China, Atmos. Environ., 37, 767–776, 2003.

Xu, T., Song, Z., Liu, J., Wang, C., Wei, J., and Chen, H.: Organic composition in the dry season rainwater of Guangzhou, China, Environ. Geochem. Health, 30, 53–65, 2008.

Yan, G. and Kim, G.: Dissolved organic carbon in the precipitation of Seoul, Korea: Implications for global wet depositional flux of fossil-fuel derived organic carbon, Atmos. Environ., 59, 117–124, 2012.

Zhao, M., Li, L., Liu, Z., Chen, B., Huang, J., Cai, J., and Deng, S.: Chemical Composition and Sources of Rainwater Collected at a Semi-Rural Site in Ya'an, Southwestern China, Atmos. Clim. Sci., 3, 486–496, 2013.

24

Immobilized photocatalyst on stainless steel woven meshes assuring efficient light distribution in a solar reactor

A. S. El-Kalliny[1,2,3], **S. F. Ahmed**[1], **L. C. Rietveld**[2], **and P. W. Appel**[1]

[1]Product and Process Engineering, ChemE, Delft University of Technology, Julianalaan 136, 2628 BL Delft, the Netherlands
[2]Sanitary Engineering, Faculty of Civil Engineering and Geosciences, Delft University of Technology, Stevinweg 1, 2628 CN Delft, the Netherlands
[3]Department of Water Pollution Research, National Research Centre, Dokki, 12622 Giza, Egypt

Correspondence to: A. S. El-Kalliny (kalliny78@hotmail.com)

Abstract. An immobilized TiO_2 photocatalyst with a high specific surface area was prepared on stainless steel woven meshes in order to be used packed in layers for water purification. Immobilization of such a complex shape needs a special coating technique. For this purpose, dip coating and electrophoretic deposition (EPD) techniques were used. The EPD technique gave the TiO_2 coating films a better homogeneity and adhesion, fewer cracks, and a higher $^{\bullet}OH$ formation than the dip coating technique. The woven mesh structure packed in layers guaranteed an efficient light-penetration in water treatment reactor. A simple equation model was used to describe the distribution of light through the mesh layers in the presence of absorbing medium (e.g., colored water with humic acids). Maximum three or four coated meshes were enough to harvest the solar UV light from 300 nm to 400 nm with a high penetration efficiency. The separation distance between the mesh layers played an important role in the efficiency of solar light penetration through the coated mesh layers, especially in case of colored water contaminated with high concentrations of humic acid.

1 Introduction

The use of photocatalytic oxidation to remove organic pollutants from drinking water has attracted considerable interests in recent years (Malato et al., 2009). In particular, the use of TiO_2 photocatalysts for water purification has been of great interest because TiO_2 is stable, harmless, inexpensive, and can potentially be activated by solar radiation (Malato et al., 2009). The application of TiO_2 in suspension (e.g., TiO_2-P25 Degussa) is effective in capturing sun light, because suspended TiO_2 powders have a high specific surface area in the range from $50\,m^2\,g^{-1}$ to higher than $300\,m^2\,g^{-1}$ (Balasubramanian et al., 2004a; Malato et al., 2002; Sunada et al., 2003; Gumy et al., 2006; Thiruvenkatachari et al., 2008), which in turn helps in avoiding mass transfer limitations, resulting in a high photocatalytic activity (Mehrotra et al., 2003). How-

ever, a light transport limitation appears with a high catalyst loading. Besides, it is difficult to separate the small TiO_2 particles from water after treatment (Thiruvenkatachari et al., 2008; McCullagh et al., 2011; Feitz et al., 2000). To overcome this, the catalyst particles can be immobilized on a surface. Yet this may lower the oxidation potential per volume of water compared to the slurry system, due to mass transfer limitation and light transport limitation caused by (i) a lower catalyst surface-to-volume ratio, (ii) the presence of substrate that absorbs light and worsens its distribution, and (iii) a lack of movement of particles (Feitz et al., 2000; Dijkstra et al., 2001).

There have been many attempts to immobilize TiO_2 photocatalyst over different structures of supports along with increasing the surface/volume ratio at the same time, which consequently enhances the photocatalytic oxidation efficiency. However, the surface area can only be efficient if it allows efficient absorption of light. There are different kinds of materials that have been used as a support to fix TiO$_2$, such as glass and borosilicate glass (Fujishima et al., 2008; Yang et al., 2004; Parra et al., 2004; Zhang et al., 2012; Ziegmann et al., 2006), cellulose fibres (Goetz et al., 2009), and stainless steel (Yanagida et al., 2006; Chen and Dionysiou, 2006a, b, 2007). Among the different supports, stainless steel is an excellent substrate material for many reasons. First, it keeps its structural integrity under the high temperature required for calcination of the TiO$_2$ films whereas quartz glass, for example, softens and deforms. Second, it is not susceptible to attack during the coating process (e.g., sodium ions diffused from the soda-lime glass into the TiO$_2$ film and decreased the photocatalytic activity, Nam et al., 2004). Third, stainless steel can be used in the electrochemical process whereas quartz and ceramics cannot be used because of their dielectric properties. At last, it can be easily used in complex shapes and has excellent mechanical properties (Balasubramanian et al., 2004b). The structures of substrates so far used do not allow an even light distribution in a fixed bed reactor and therefore the photocatalytic efficiency is much lower compared to that of a slurry reactor. Therefore, the challenge of designing an efficient photocatalytic reactor is in using a suitable catalyst structure to optimize both the area covered by photocatalytic particles and the light distribution. The design criteria of such a reactor should be (i) good vertical mixing, (ii) high surface area per unit reactor volume, and (iii) no shadowing effect.

A stainless steel woven mesh catalyst substrate is applied in the field of air purification as it shows a higher mechanical strength and a lower pressure drop than typical ceramic monolith honeycomb catalysts (Ismail et al., 2007). Also, Shang et al. (2003) used a stainless steel webnet coated by TiO$_2$ for the degradation of gaseous formaldehyde in a cylindrical flowing photo-reactor. This webnet structure showed a good activity due to its large surface area, good ventilation for gases passing, and good utilization of UV light. In the water purification field, the Ti/TiO$_2$ mesh structure was prepared by an anodization method and the photocatalytic efficiency was studied by several researchers (F. B. Li et al., 2002; X. Z. Li et al., 2000, 2002; Yoon et al., 2009). However, the anodization process requires a high energy input and moreover Ti mesh is expensive, which makes this technique not feasible for large-scale applications. On the other hand, although a mixture of anatase and rutile (e.g., the TiO$_2$ crystallites of Evonik (Degussa) P25 contain a combination of anatase ($\sim 80\,\%$) and rutile, $\sim 20\,\%$) gives a high photocatalytic activity due to the intimate contact between the two phases, enhances the separation of photo-generated electrons and holes, and results in a reduced recombination (Pelaez et al., 2012), yet, it is difficult to control the anatase-to-rutile ratio in the anodization process to have the highest photocatalytic activity.

In this paper, stainless steel woven meshes were selected to be a substrate for immobilization of TiO$_2$ particles. An immobilized photocatalyst was prepared by using the sol gel dip coating technique and electrophoretic deposition (EPD) technique. These immobilization techniques are suitable for coating substrates of complex shape such as woven meshes (Chen and Dionysiou, 2006c; Yanagida et al., 2005). For example, EPD was used for immobilizing TiO$_2$ over different types of substrates such as carbon and SiC fiber (Boccaccini et al., 2004), metal fiber or plate (Boccaccini et al., 2001; Sakamoto et al., 1998), and stainless steel meshes (Yanagida et al., 2005; Ghorbani, 2008). However, using stainless steel meshes coated with TiO$_2$ packed in layers in the water purification field has not been done so far. The production of $^{\bullet}$OH radicals was detected by a photoluminescence (PL) technique in order to evaluate the performance of the coated meshes. Based on the design criteria of packed bed solar reactor, solar light transmission through the woven mesh layers in the presence of light absorbing medium such as humic acid (HA) was studied in order to determine the optimum number of mesh layers and the separation distance between them.

2 Experimental methods

2.1 Experimental approach

Immobilization of the TiO$_2$ photocatalyst over stainless steel woven meshes was done by using sol gel dip coating and EPD techniques in order to reach the best coating films with such a complex shape of the substrate. PL technique was used in order to evaluate the photocatalytic activity of the TiO$_2$ films. Besides, the thickness, the roughness, and the microscopic images of TiO$_2$ films were investigated to optimize the best coating films. Then the distribution of the solar light through the catalyst meshes structure in the presence of light non-absorbing or light absorbing medium was detected in order to optimize the number of mesh layers and the separation distance between them. The transmitted light through the mesh layers was measured and compared with the calculated values via a developed equation model to investigate its validity. The developed equation model was used for the determination of the fraction of the captured light by meshes, available for the production of $^{\bullet}$OH radicals on the catalyst surface, with different values of dimensionless parameter representative for the separation distance between meshes and the extinction coefficient and concentration of the medium.

2.2 Preparation of the immobilized TiO$_2$ catalyst

2.2.1 Treatment of the substrate

Uncoated 304 stainless steel meshes (wire diameter 0.355 mm, aperture diameter 0.915 mm, and open area 52 %) and flat plates (thickness 2 mm), used as substrates, were firstly rinsed with deionized water. After the stainless steel was dried in air, ethanol (96 %, VWR International) and methyl ethyl ketone (99.8 %, Fisher Scientific) were used for further cleaning of the substrates. The clean substrates were dried at 125 °C for 24 h (Chen and Dionysiou, 2006c). The flat plate substrates were used only for characterization of TiO_2 films.

2.2.2　Sol gel dip coating method

The TiO_2 was immobilized onto discs of stainless steel meshes and flat plates of 5.7 cm diameter by the dip coating technique horizontally for 1 min. Besides, discs of stainless steel meshes of 16.5 cm diameter were used in light distribution experiments. The commercial sols O500 and O510 (TIPI Technology) were used for coating. Besides, a modified commercial sol was used by adding 10 g TiO_2 P-25 Degussa powder per 1 L of sol O500 or O510.

2.2.3　Electrophoretic deposition method

The TiO_2 was immobilized onto the stainless steel substrate meshes or flat plate discs of 5.5 cm diameter by the EPD technique using the vertical coating setup (Fig. 1a). In order to test the up-scalability, the horizontal coating setup (Fig. 1b) was used for coating of large-scale substrates of dimensions 32 cm × 19 cm. The orientation of the electrodes whether vertically or horizontally did not affect the coating process, where the mesh with the large size requires the horizontal setup to be handled more easily. The commercial sol O500 (TIPI Technology) modified with suspension AERODISP W740X (40 % P-25, Evonik Industries AG) was mixed by 5 % v/v ratio and used as electrolyte. Figure 1 shows a schematic view of the EPD cells. The EPD process is a combination of two steps: (1) the motion of charged particles in a stable suspension under sufficient electric field (electrophoresis) and (2) the coagulation and deposition of the particles on the electrode (Ghorbani, 2008; Sarkar and Nicholson, 1996). The suspension of TiO_2 was mixed with O500 sol gel and formed a stable TiO_2 suspension sol gel. TiO_2 particles are negatively charged in the presence of the anionic surfactant which is used to stabilize the TiO_2 suspension. TiO_2 particles moved towards the anode by the action of the DC current (6 V, 2.5 A) during 1 min deposition time and then neutralized as they touch the depositing electrode or deposit and became static (Sarkar and Nicholson, 1996; Grillon et al., 1992).

2.2.4　Treatment of the coating films

The coating films were dried in order to adhere to the surface of the stainless steel substrate. The drying was done in a furnace by increasing the temperature at a ramp rate of 15 °C min^{-1} up to 100 °C; this temperature was held for 15 min. The temperature of the furnace was subsequently increased at a ramp rate of 15 °C min^{-1} to the final calcina-

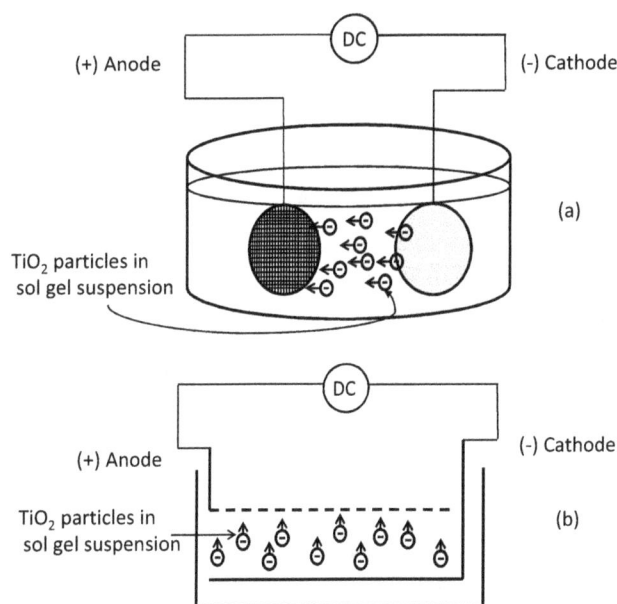

Figure 1. A schematic view of EPD cells. (a) A vertical coating setup for the small mesh discs. (b) A horizontal coating setup for the large-scale meshes.

tion temperature (200–600 °C) and was held at this value for 15 min. The optimum final drying temperature and the number of performed dip coatings were determined as they affect strongly the adhesion of the catalyst film over the stainless steel substrate and consequently the photocatalytic efficiency (Chen and Dionysiou, 2006c). The optimum drying temperature determined when using the dip coating technique was also used in the EPD technique. Finally, the films were cooled at ambient conditions and then rinsed by deionized water to remove any non-attached particles.

2.3　Characterization of the TiO_2 coatings

Surface thickness and roughness of the TiO_2 film were measured with the Veeco Dektak 8 Stylus Profiler fitted with a 12.5 μm tip. Reflection of the immobilized TiO_2 film was detected by using the integrated sphere of UV/VIS/NIR spectrophotometer, model Lambda 90, Perkin Elmer (300–400 nm). The homogeneity of the coating layer was investigated by the optical stereo microscope "Stereo-Discovery V8" from Carl Zeiss. The adhesion of the TiO_2 film was investigated by using a tape casting method similar to the ASTM (ASTM D3359) test. The cross-hatch in the coating TiO_2 film was cut with a razor blade (single-edge #9, VWR Scientific, Media, PA, USA), followed by applying the tape (3M Scotch Tape #810) to the cross-hatch area and pulling the tape back off.

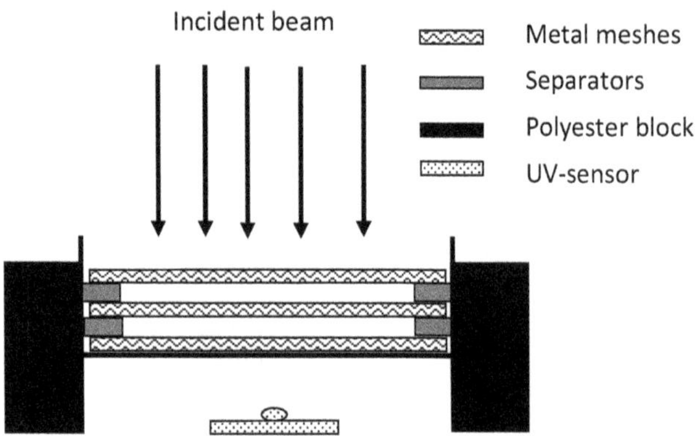

Figure 2. A schematic view of the experimental setup for light distribution measurements.

2.4   Measurements of transmitted light through the mesh structure

The apparatus designed to measure the light distribution in one or more stainless steel metal meshes with immobilized catalysts is shown in Fig. 2. The intensity of the transmitted UV solar light (300–400 nm) passing through the metal mesh, immersed in deionized water was measured by a XenoCal UV-sensor (Atlas) with a resolution of $0.1\,\mathrm{Wm}^{-2}$. The uncoated or coated discs of metal meshes (diameter 16.5 cm) were placed in a Pyrex glass jar with an internal diameter of 18.5 cm and separated with silicone rubber rings with a height of 0.5 cm. The XenoCal UV-sensor was placed underneath the Pyrex glass vessel that was supported by a black polyester block. The Pyrex glass jar with deionized water absorbed 23 % of the intensity of the UV light in the range between 300 nm and 400 nm. The percentage of the transmitted light through the mesh layer is the ratio of the intensity of light measured by the $\mathrm{UV}_{300-400}$-sensor through the mesh, water, and the glass vessel (I) to that measured through water and the glass vessel (I_o).

2.5   Theory and model equations

To calculate the light transmission through the woven meshes and the water media considering that a system in which UV radiation with an initial intensity I_o (Wm^{-2}) falls on stacked meshes immersed in water and it was assumed that the equal distance between the meshes is such that the light transmitted through a mesh is fully homogenized before it enters the next mesh. The meshes were considered to be infinitely thin. Further, it was assumed that back and forth reflection of the light by the meshes can be neglected. Then, the intensity of the light I_n (Wm^{-2}) on the top of n meshes (n is an integer) is

$$I_n = I_\mathrm{o}(1-r)^{n-1} \cdot e^{-\varepsilon cnd}, \tag{1}$$

where r (dimensionless) is the fraction of light reduced by mesh. For uncoated meshes with an open area of 52 %,

$r = 0.48$. ε ($\mathrm{L\,mg}^{-1}\,\mathrm{cm}^{-1}$) and c ($\mathrm{mg\,L}^{-1}$) are the extinction coefficient and the concentration of the medium, respectively, and d (cm) is the separation distance between the meshes.

The XenoCal UV-sensor measures the intensity of the light as an integrated value over the wavelength range of 300–400 nm. Therefore, in order to compare theoretical values with experimental values, the theoretical values must also be integrated values over the same wavelength interval. The fraction of the light absorbed at a wavelength range of 300–400 nm is given by (Goldstein et al., 2007)

$$\propto = \frac{\int_{300}^{400} I(\lambda)\,A\,\mathrm{d}\lambda}{\int_{300}^{400} (\lambda)\,\mathrm{d}\lambda}, \tag{2}$$

where $I(\lambda)$ is the normalized photonic flux (arbitrary units) at wavelength λ and $A(\lambda) = 1 - T(\lambda)$ is the absorbance. In order to calculate the fraction of the absorbed light $\propto$ of HA, the emission spectrum of the light $I(\lambda)$ and the transmission spectrum of the HA $T(\lambda)$ were measured over the 300–400 nm range every 0.5 nm and 2 nm, respectively (see Fig. 3). Then the integration was done numerically by approximated intervals of 2 nm width. The values of $\propto_{300-400}$, $T_{300-400}$, and $\varepsilon_{\mathrm{HA},300-400\,\mathrm{nm}}$ are equal to 0.24, 0.76, and $0.028\,\mathrm{L\,mg}^{-1}\,\mathrm{cm}^{-1}$, respectively. The extinction coefficient $\varepsilon_{\mathrm{HA},300-400\,\mathrm{nm}}$ was used to estimate the transmission through HA layers between coated mesh layers.

The value of the captured light by meshes is inversely proportional to the path length of the light through the colored medium layers (nd) in cm, the extinction coefficient (ε) in $\mathrm{L\,mg}^{-1}\,\mathrm{cm}^{-1}$, and the concentration (c) in $\mathrm{mg\,L}^{-1}$ of the colored medium. Thus, it is essential to optimize these three factors to get as much captured light as possible.

The fraction of the captured light for a system with n meshes (η_n) immersed in a light absorbing medium was determined by

$$\eta_n = \sum_{k=1}^{n} (1-r)^{k-1} \cdot e^{-k\beta} \cdot r. \tag{3}$$

2.6   Evaluation of photocatalytic activity

Photocatalytic activity measurements were carried out by following up the •OH formation. Hydroxyl radicals (•OH) were detected by PL using terephthalic acid (TA) from Sigma Aldrich with 98 % purity as a probe molecule. TA readily reacts with •OH to produce a highly fluorescent product: 2-hydroxyterephthalic acid (2-OHTA) (Yu and Wang, 2010). The PL signal intensity of 2-OHTA at 425 nm is proportional to the amount of •OH produced in water. The photocatalytic activity experiments were done with meshes in 250 mL Pyrex glass beakers (internal diameter 7 cm) using a magnetic stirrer (IKA, RT15) with an equal stirring rate at 550 rpm (rotation per minute). The temperature was controlled to be

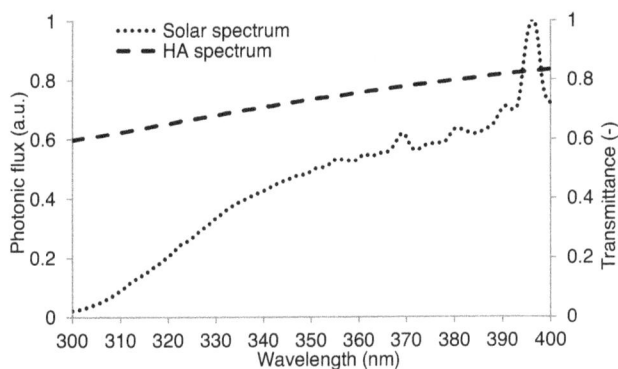

Figure 3. Transmission spectrum of HA was measured by Hach Lange DR 5000 spectrophotometer and the normalized emission spectrum of artificial solar light was measured by Black C-50 spectrometer, a product of StellarNet Inc.

$32 \pm 1\,°C$ by letting the water flow in a water bath through a recirculation cooler (Julabo, FL300). The entire system was placed inside the chamber of the SUNTEST XXL+ (Atlas) which consists of 3 xenon lamps irradiating UV solar light (300–400 nm) with an intensity of $40\,Wm^{-2}$. The solar spectrum of these xenon lamps was measured by a Black C-50 spectrometer, a product of StellarNet. The coated stainless steel meshes with a diameter of 5.7 cm were separated with silicone rubber rings of a diameter of 0.2 cm. The deionized water used in the preparation of TA (5×10^{-4} M dissolved in 2×10^{-3} M NaOH aqueous solution) was aerated before the reaction to let the dissolved oxygen concentration reach about $8\,mg\,L^{-1}$. For each experiment, 100 mL of this TA solution was stirred for 30 min with the immobilized TiO_2 catalyst in the dark to take the water bath temperature and to ensure equilibrium adsorption.

2.7 Analyses

PL spectra of generated 2-OHTA due to the reaction of TA with the ·OH radicals were measured by a Photoluminescence Spectrophotometer PIT Quanta Master Model QM-1. After light irradiation for 45 min, the reaction solution was used to measure the increase of the PL intensity at 425 nm excited by 315 nm light.

The concentration of the working solution of HA in all light experiments was approximately $10\,mg\,L^{-1}$. It was prepared from a $1000\,mg\,L^{-1}$ stock solution of HA sodium salt (Sigma Aldrich) which was prepared by dissolving it in deionized water and filtering it through a $0.45\,\mu m$ syringe-driven filter unit (Millex) to remove suspended solids ($0.03\,g\,L^{-1}$). The concentration was monitored with a Hach Lange DR 5000 spectrophotometer by measuring the UV absorbance at 254 nm (UV_{254}), which represents the aromatic moieties (Yigit and Inan, 2009). The UV_{254} calibration curve with five concentration levels of HA is shown as a straight line with R^2 equal to 0.991. The extinction

coefficient of HA determined by Lambert Beer's law is $\varepsilon_{HA,254\,nm} = 0.0717\,L\,mg^{-1}\,cm^{-1}$.

3 Results and discussion

3.1 Performance and characterization of coated meshes

3.1.1 Sol gel dip coating

Figure 4 shows the PL spectra of 2-OHTA formed by the reaction of TA with ·OH radicals after the irradiation of the different coated meshes with UV solar light ($40\,Wm^{-2}$) for 45 min. The blank line represents the TA solution with meshes coated with O510-P25 in the dark. No PL signal was observed in the absence of UV light irradiation, which indicates that there was no formation of 2-OHTA and consequently no formation of ·OH radicals. Coated films with plain sol gels O500 and O510 showed some photocatalytic activity as they contain nano TiO_2 particles, while those coated with sol gel modified by P25 Degussa powder gave the highest PL peak at 425 nm. The loading of P25 had considerably increased the photocatalytic activity of the plain sol gel, which is consistent with what was found by Chen and Dionysiou (2006a, b, 2007). The presence of TiO_2-P25 particles increased the photocatalytic activity of the coating film as P25 consists of an anatase/rutile phase mixture ratio of approximately 80/20 (Collins-Martínez et al., 2007). The mixture of these two phases is beneficial in reducing the recombination of photo-generated electrons and holes (Yu et al., 2007a, b).

Figure 5a shows the effect of the final drying temperature of one dip coated woven meshes (with O500-P25 and O510-P25 sol gels) on the maximum PL intensity at 425 nm after 45 min irradiation with solar light in the presence of basic TA solution. The drying of the coating film is necessary for dehydration, which increases the adhesion forces between the TiO_2-TiO_2 particles and TiO_2-stainless steel surface (Chen and Dionysiou, 2006b). The final drying temperature of 300 °C showed for both films the highest PL intensity at 300 °C. This indicates that at this final temperature the TiO_2 particles showed the optimum adhesion between each other and the stainless steel surface. As the final drying temperature of the film was increased above 300 °C, cracks were formed and TiO_2 particles were detached during the washing step, leading to a decrease of ·OH formation. The formation of cracks is probably due to two reasons: first, the difference of expansion of the stainless steel and the film and second, the increasing contraction of TiO_2-TiO_2 particles by further dehydration. Nevertheless, choosing the smallest wire diameter possible (0.355 mm) preserved the minimum cracks formation, as by further decreasing of the wire diameter the curvature increases and the possibility of cracks formation increases.

Figure 5b shows the effect of the number of dip coatings on the photocatalytic activity of the coating films. Three dip

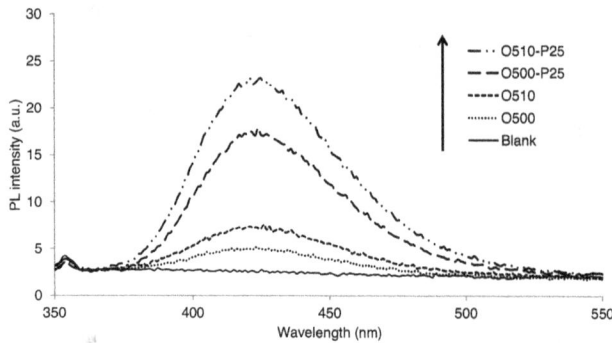

Figure 4. PL spectra of 2-OHTA formed by 45 min irradiation of different coated meshes. Initial TA concentration is 5×10^{-4} M; excitation is at 315 nm; final drying temperature is 300 °C.

coatings of O500-P25 exhibited the maximum PL intensity at 425 nm when it was irradiated for 45 min with solar light in the presence of a basic TA solution. While the low photocatalytic activity performance for the films coated with O510-P25 sol gel with more than one dip coating was due to its high viscosity which increased the possibility of cracks formation.

Figure 6a shows the microscopic image of the coating films of O500-P25 with different numbers of dip coatings. In the first image the texture of the stainless steel substrate appears clearly as being coated by O500 sol without TiO_2-P25. The density of the TiO_2 agglomerates is increased by increasing the number of dip coatings of O500-P25. The area covered by the agglomerates per unit surface area of the substrate increased with an increase in the number of dip coatings until three dip coatings. There was no significant increase in the density of TiO_2 agglomerates observed by increasing the number of dip coatings from three to five. This explains the increase of the ·OH formation from one dip coating to three dip coatings and no further improvement in the photocatalytic activity after three dip coatings.

Figure 6b shows the microscopic image of the coated stainless steel woven mesh with O500-P25 sol gel with three dip coatings at 300 °C. The commercial O500 sol acts as a binder for the TiO_2 P-25 Degussa particles agglomerates (Chen and Dionysiou, 2006a). The coating TiO_2 film shows a good coverage of the complex woven mesh shape. However, by dipping the mesh in a sol gel and then lifting it, a meniscus surface of the sol was formed and then solidified by heating. As a consequence, this meniscus surface formed film wings. These wings may get detached by attrition causing a damage of the coating film over the mesh wires.

Figure 6c shows the effect of the number of dip coatings with O500-P25 at 300 °C on the average film thickness and roughness. There is a more or less linear relationship between the average film thickness and the number of dip coatings. This is in agreement with what was observed by Chen and Dionysiou (2006c). The average film thickness is approxi-

mately $2.2\,\mu m$ per dip coating. It was found that the photocatalytic activity of a TiO_2 thin film reached its maximum at a film thickness of ca. 140 nm (Eufinger et al., 2008) and of 360–430 nm (Nam et al., 2004). The difference between the two studies is due to the difference in the thin film preparation techniques and/or the experimental setup for testing their photocatalytic activity. Tada and Tanaka (1997) proposed that there is a limited diffusion length of the charge carriers and a cumulative light adsorption with increasing film thickness. They found that the critical film thickness is ca. 100–150 nm, from which they deduced a charge carrier diffusion length of about 300 nm. Eufinger et al. (2008) deduced that most of the incoming light is absorbed in the first few 100 nm of the film. Therefore, it can be concluded that the increase in photocatalytic efficiency with an increasing number of dip coatings is not due to the increase in film thickness. Figure 6c shows an increase in average roughness from ca. $1\,\mu m$ to $2\,\mu m$ by increasing the number of dip coatings from 1 to 3 Afterwards, no further increase in roughness was observed. This indicates that new TiO_2 agglomerates were deposited on the surface of the film by each dip coating until three dip coatings and the active sites on the TiO_2 film surface was increased. The conclusion is that the average film roughness is an important parameter for the photocatalytic activity of thick TiO_2 films as the critical average film thickness (ca. $6.4\,\mu m$) gave the highest average film roughness (ca. $2\,\mu m$).

Since adhesion for immobilized TiO_2 films is an important property in water treatment, the cross-hatch tape test was performed on the coating films over stainless steel flat plate to evaluate the strength of adhesion. According to ASTM 3359, the result of the adhesion test has been qualitatively divided into 6 grades, from 5 to 0, where grade 5 represents the highest level of adhesion and 0 the lowest level of adhesion. The O500-P25 coating film with three dip coatings at 300 °C showed the best adhesion. About 75 % of the squares of the lattice were not detached after pulling off the tape, corresponding to ASTM Class 2B.

3.1.2 Electrophoretic deposition

Figure 7a and b shows the microscopic views of the coated stainless steel woven mesh with O500-P25 sol gel by EPD using DC current (6 V, 2.5 A) during 1 min deposition time. The coating TiO_2 film with the EPD method shows better homogeneity than that coating by dip coating method especially with large dimensions. This would be due to the applied electric field which enabled a uniform distribution of TiO_2 particles on the mesh surface. Less film wings were observed with EPD method because of the action of the applied electric field which forced the TiO_2 particles to attach with mesh wires and prevented the formation of a meniscus between the mesh wires. Figure 7c shows the area covered by the TiO_2 agglomerates per unit surface area of the substrate

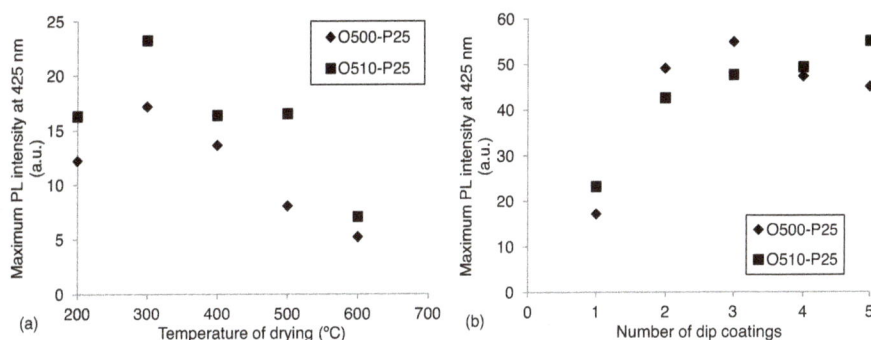

Figure 5. Maximum PL intensity at 425 nm for woven meshes coated with O500-P25 and O510-P25 sol solutions as a function of **(a)** final drying temperature (one dip coating) and **(b)** number of dip coatings (final drying at $T = 300\,°C$).

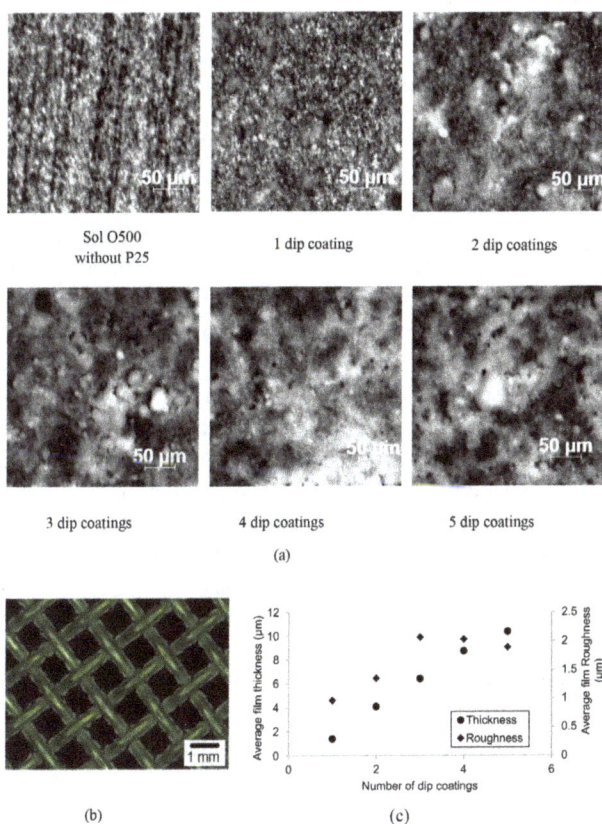

Figure 6. (a) Microscopic images of coating films showing increasing of TiO$_2$ agglomerates by increasing of number of dip coatings. **(b)** Microscopic image for woven mesh coated by three dip coatings on O500-P25 sol gel and dried at final temperature 300 °C. **(c)** Average surface thickness and roughness of O500-P25 TiO$_2$ film over stainless steel flat plate with different number of dip coatings at 300 °C measured by Veeco Dektak 8 Stylus Profiler.

higher than that obtained with the sol gel dip coating technique.

The TiO$_2$ coating film with EPD shows a little bit better adhesion than that coating by dip coating technique three

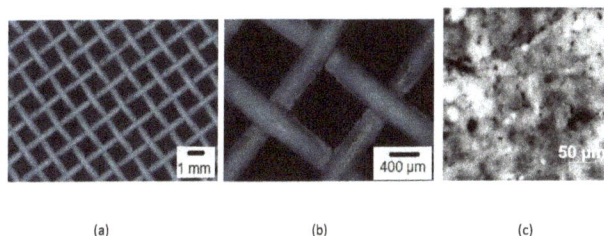

Figure 7. Microscopic image of the stainless steel woven mesh coated with EPD method using O-500 sol gel modified with TiO$_2$ suspension: **(a)** low magnification; **(b)** high magnification; **(c)** microscopic image of coating film with O500-P25 sol gel by EPD using DC current (6 V, 2.5 A) during 1 min deposition time.

times using the O500-P25 sol gel. The cross-hatch tape test was performed on the coating film by the EPD technique over stainless steel flat plate. About 80–85 % of the squares of the lattice were not detached after pulling off the tape, corresponding to ASTM Class 2B. On the other hand, EPD technique produced a thicker TiO$_2$ film ranging between $8.95\,\mu m$ and $9.29\,\mu m$ with a higher average roughness of $2.2\,\mu m$ compared to that obtained by dip coating technique three times using the O500-P25 sol gel. This indicates that more TiO$_2$ agglomerates were deposited on the surface of the substrate introducing more active sites for the catalyst and explaining the higher maximum PL intensity obtained by the coating films by EPD technique compared with that obtained by the dip coating technique (see Fig. 7).

Figure 8 shows the formation of 2-OHTA as a PL intensity at 425 nm for the small single woven disc meshes (diameter 5.5 cm) and the large single woven sheet (19 cm × 32 cm) coated by EPD. The area-to-volume ratio for the large mesh was higher than that for the small mesh. Thus, in order to adjust the PL intensity results from the large mesh, they were divided by a factor 2.15. This factor is the result of dividing the area-to-volume ratio of the small mesh ($23.75 \times 10^{-4}\,m^2\,0.1\,L^{-1}$) by the area-to-volume ratio of the large mesh ($510 \times 10^{-4}\,m^2\,1\,L^{-1}$). The results show a linear relationship between the formation of 2-OHTA and the time.

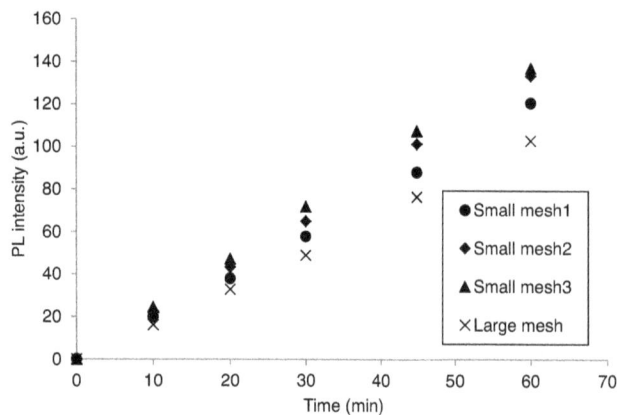

Figure 8. Maximum PL intensity at 425 nm as a function of time for three single woven disc meshes and single woven sheet (19 cm × 32 cm), coated by EPD, after corrections for area-to-volume ratio.

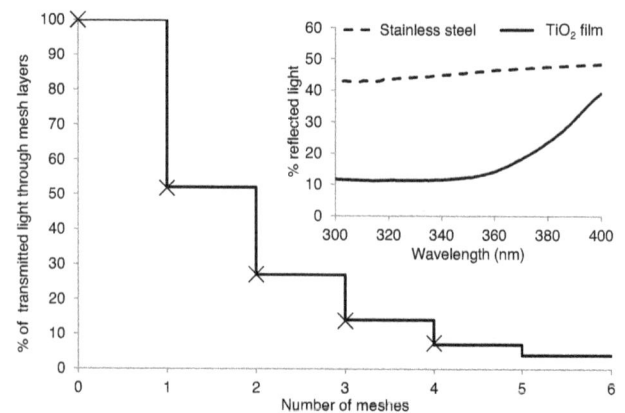

Figure 9. Light transmission through uncoated woven meshes (for which $r = 0.48$) in a non-absorbing medium (distilled water). The solid line represents model calculations according to Eq. (1); (×) measured intensities with 1 cm mesh separation. Inset: reflection of stainless steel flat plate and TiO$_2$ film measured with the integrated sphere spectrometer.

The PL intensity results for the large sheet mesh were slightly lower, but of the same order of magnitude, than those for the small disc meshes. This means that the amount of ˙OH radicals with the large mesh was close to the amount that was produced with the small meshes.

The conclusion was, therefore, that coating of stainless steel woven meshes by the EPD is up-scalable and gives a higher ˙OH radical production, a lower formation of cracks, and a higher adhesion than the dip coating technique. On the other hand, a comparison between the mesh structure and the flat plate in the fixed-bed reactor on the photodegradation of HA is given in detail by El-Kalliny (2013).

3.2 Light distribution in a structure of packed meshes

3.2.1 Light transmission through the mesh structure

Figure 9 shows a comparison between the measured and calculated percentages of the UV light transmitted through uncoated mesh layers immersed in a non-absorbing medium (distilled water). It shows the stepwise decrease of the light intensity at the passage of each mesh according to Eq. (1), while the intensity between the meshes remains constant (solid line). The top ends of the vertical lines represent the light intensity falling on the mesh and the bottom ends represent the light intensity transmitted through the mesh. For a non-absorbing medium, ε is near to be zero and the right term in Eq. (1) becomes equal to 1. Thus, the reduction of light is due to the mesh layers only, which is represented by the left term in Eq. (1). Using the experimental setup for light distribution measurements described in Sect. 2.3, the light transmission through different stacks of meshes was measured for separation distances of 1 cm. The calculated and the experimental measurements of the intensities of UV solar light in the range of 300–400 nm after passing (n) meshes coincide with each other, which validates the model Eq. (1). This indi-

cates that Eq. (1) can be applied to investigate the transmitted light profile not only for meshes immersed in non-absorbing medium but also for those immersed in absorbing medium.

The inset in Fig. 9 shows the reflection of light ($\lambda = 300$–400 nm) for stainless steel and TiO$_2$ film measured by the integrated sphere spectrometer. The largest contribution of reflected light will be experienced by the second mesh: from this mesh the light is reflected upwards and backwards from the down side of the first mesh. The calculation shows that the largest contribution of reflected light is an extra 2 % using stainless steel (average reflection of 40 %) and 0.2 % using TiO$_2$ coated meshes (average reflection of 13 %). This shows that the assumption not to consider back and forth reflection by the meshes in the calculations was justified.

Figure 10 shows the measurements of the percentage of transmitted light through coated woven meshes immersed in a non-absorbing medium (distilled water) at two different separation distances. For coated meshes with an open area of 55 %, $r = 0.45$ due to a slight increase in the mesh wire thickness with coating. The measurement values were more or less the same, indicating that the drop of the light transmission was due to the meshes only and that there was no effect of the separation distances between the mesh layers. In addition, the mesh was like a diffuser which intends to break up the light beams without introducing shadows below it. Thus, the system in practice behaves as an ideal stacked mesh system. Shifting or turning one mesh over the next, there was not an observable influence on the intensity readings, indicating that the transmitted light below the mesh was distributed homogeneously and the homogeneity of the coating film over the mesh was good enough to distribute the light. Practically, the orientation of the meshes was fixed randomly which gave a good opportunity to subject the mesh wires to the light

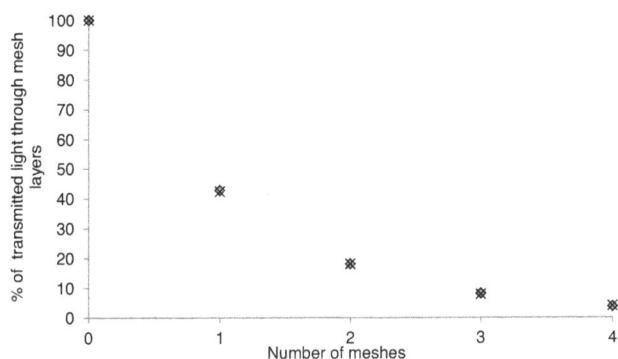

Figure 10. Light transmission through coated woven meshes ($r = 0.45$) in a non-absorbing medium (distilled water). Measured intensities: (♦) 0.2 cm mesh separation; (×) 1 cm mesh separation.

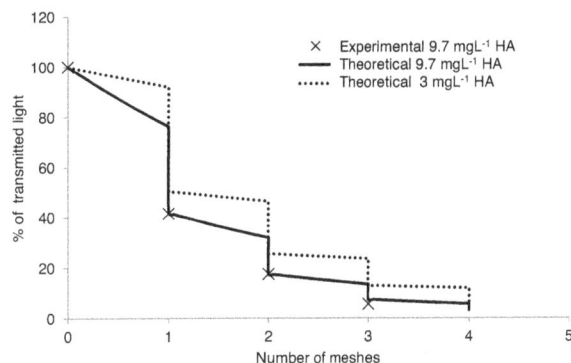

Figure 11. Experimental measurements and theoretical calculations for the percentage of the transmitted UV solar light (300–400 nm) through coated woven meshes ($r = 0.45$) with separation distance 1 cm in the presence of different concentrations of HA.

radiation and prevent the shadowing effect. In conclusion, Eq. (1) can be used to calculate the percentage of transmitted light through coated meshes.

3.2.2 Effect of a light absorbing medium

The experimental setup for light distribution measurements described in Sect. 2.4 was used for measuring the light transmission through different stacks of meshes for separation distances of 1 cm immersed in a light absorbing medium ($9.7 \, mg \, L^{-1}$ HA, which is comparable to the concentration of HA in surface water).

Figure 11 shows a comparison between the measured and calculated percentages of the UV light transmitted through the coated mesh layers and immersed in a light absorbing medium. The curves for the light transmission through the colored water layers follow Lambert Beer's law with an exponential decrease in transmission. Thus the transmitted light through meshes and HA of $3 \, mg \, L^{-1}$ is higher than the one transmitted through meshes and HA of $9.7 \, mg \, L^{-1}$. The measured transmission values of the UV solar light in the wavelength range of 300–400 nm with HA of $9.7 \, mg \, L^{-1}$ coincide with the calculated values. This indicates that Eq. model (1) which describes the transmission of light through mesh layers and Eq. (2) which calculates the fraction of the light absorbed by HA layers can be used to describe the transmission light profile inside the packed bed reactor. The transmission through the coated mesh structure was strongly affected by the presence of a light absorbing medium or colored water due to the absorption of light in the range of 300–400 nm. The lower the transmission values through HA solution, the lower the interaction of light with the catalyst surface was. It was calculated that a maximum of 3.6 % of the incident light irradiated the coated fifth mesh with HA concentration of $3 \, mg \, L^{-1}$. Thus, three, or maximum four meshes, were sufficient to harvest more than 96 % of the irradiated solar light. Figure 11 also shows that as the HA concentration decreased the percentage of the transmitted light through the

HA layers increased and consequently the fraction of light available for the production of $^{\bullet}OH$ radicals increases. Similarly, as the separation distance between mesh layers decreased the percentage of the transmitted light through the HA layers increased (see Eq. 1). The photocatalytic oxidation efficiency will be the highest when the fraction of the captured light by the meshes available for the production of $^{\bullet}OH$ radicals is optimized. This can be done by introducing the dimensionless parameter $\beta = \varepsilon \cdot C \cdot d$ in order to be applicable for various light absorbing mediums with different extinction coefficients and for different concentrations.

The dimensionless parameter β approached 0 for very clean water and approximately 1 for highly contaminated water with, for example, HA concentration of $20 \, mg \, L^{-1}$ and a maximum separation distance between the meshes of 2 cm, assuming that there is no shadowing effect that influences the light transmission, where the shadowing effect is not present with a separation distance larger than 0.2 cm (see Fig. 10). This realistic range of β is due to that the fine suspended particles were not taken into consideration and the absorbance of light was only due to the dissolved organic materials of HA. Figure 12 shows the decrease in the fraction of the captured light as a function of β for systems with different number of meshes. It shows that as β increased, less number of meshes is required. Three meshes are enough for capturing the light for $\beta > 0.4$ and two meshes are enough for $\beta > 0.7$ as there is no contribution for the third mesh. This means that with a very low concentration of colored water contaminated with micro-pollutants, a multi mesh system would be used. The intercept of the curves with y axis represent the fraction of the captured light by meshes in the presence of non-absorbing medium.

The separation distance should be low as much as possible in order to decrease the effect of absorption of light with HA layers and decrease the value of β. The lower separation distance provides a good vertical mixing and a higher captured number of photons per unit reactor volume. Practically, the

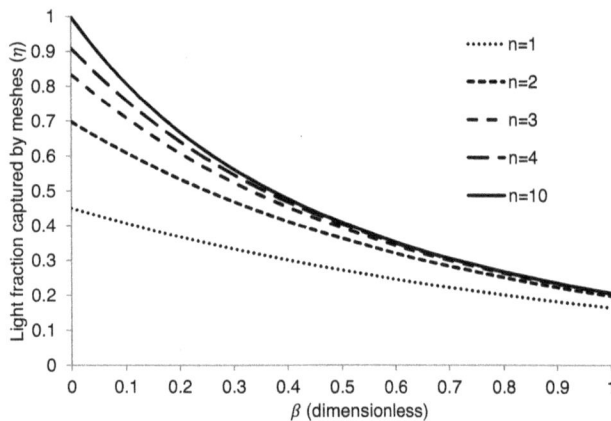

Figure 12. The fraction of captured light by the coated meshes ($r = 0.45$) as a function of the dimensionless β parameter in a light absorbing medium.

separation distance can be decreased as much as possible to be 0.2 cm in order to avoid any attachment of the mesh layers because the meshes were not flattened perfectly. On one hand, as the open area of the mesh increases, the fraction of light reduced by mesh r decreases and more meshes should be used in order to capture the light as much as possible. On the other hand, as the mesh wire diameter decreases, the surface area increases and the shadowing effect is minimized.

4 Conclusions

- The average film roughness is an important parameter for the photocatalytic activity of thick TiO_2 films as the critical average film thickness (ca. $6.4\,\mu m$) gave the highest average film roughness (ca. $2\,\mu m$) in a dip coating technique.

- A promising immobilized photocatalyst was prepared by EPD technique on stainless steel woven meshes using a stable TiO_2 suspension in sol gel O500 as electrolyte at $300\,^\circ C$ dryness temperature. This technique shows a higher $^\bullet OH$ formation, a lower cracks formation, a better homogeneity and adhesion than dip coating technique. Coating with EPD is up-scalable for large stainless steel woven meshes.

- A stainless steel woven wire mesh was a good photocatalyst substrate due to its large surface area and the possibility of the light to be effectively distributed through it.

- The equation model (1) can be used to describe the distribution of the light through the catalyst woven mesh structure not only immersed in non-absorbing medium, but also in a light absorbing medium.

- The dimensionless parameter $\beta = \varepsilon \cdot C \cdot d$ can be used to optimize the fraction of the captured light by meshes

in a colored medium in order to be applicable for various light absorbing mediums with different extinction coefficients and for different concentrations.

- Three mesh layers or maximum four were enough to harvest more than 96 % of the light in the presence of colored water with HA.

- A 0.2 cm separation distance was an optimum separation distance between mesh layers to have as much as possible illuminated light on top of the last mesh layer at the bottom of the reactor. This distance also meets the design criteria of having a high vertical mixing and a high captured number of photons per unit reactor volume.

Acknowledgements. This research work has been carried out within the framework of INNOWATOR project, financed by Agentschap NL.

References

Balasubramanian, G., Dionysiou, D. D., Suidan, M. T., Baudin, I., and Laîné, J. M.: Evaluating the activities of immobilized TiO_2 powder films for the photocatalytic degradation of organic contaminants in water, Appl. Catal. B-Environ., 47, 73–84, 2004a.

Balasubramanian, G., Dionysiou, D. D., and Suidan, M. T.: Titanium dioxide coatings on stainless steel, in: Dekker Encyclopedia of Nanoscience and Nanotechnology, Marcel Dekker, edited by: Schwarz, J. A. and Contescu, C. I., ISBN: 978-0-8247-5055-8, Vol. 6 (Chapter 311), 2004b.

Boccaccini, A. R., Schindler, U., and Krüger, H.-G.: Ceramic coatings on carbon and metallic fibres by electrophoretic deposition, Mater. Lett., 51, 225–230, 2001.

Boccaccini, A. R., Karapappas, P., and Marijuan, J. M.: TiO_2 coatings on silicon carbide and carbon fibre substrates by electrophoretic deposition, J. Mater. Sci., 39, 851–859, 2004.

Chen, Y. and Dionysiou, D. D.: TiO_2 photocatalytic films on stainless steel: The role of degussa P-25 in modified sol–gel methods, Appl. Catal. B-Environ., 62, 255–264, 2006a.

Chen, Y. and Dionysiou, D. D.: Effect of calcination temperature on the photocatalytic activity and adhesion of TiO_2 films prepared by the P-25 powder-modified sol–gel method, J. Mol. Catal. A-Chem., 244, 73–82, 2006b.

Chen, Y. and Dionysiou, D. D. : Correlation of structural properties and film thickness to photocatalytic activity of thick TiO_2 films coated on stainless steel, Appl. Catal. B-Environ., 69, 24–33, 2006c.

Chen, Y. and Dionysiou, D. D.: A comparative study on physicochemical properties and photocatalytic behavior of macroporous TiO_2-P25 composite films and macroporous TiO_2 films coated on stainless steel substrate, Appl. Catal. A-Gen., 317, 129–137, 2007.

Collins-Martínez, V., Ortiz, A. L., and Elguézabal, A. A.: Influence of the anatase/rutile ratio on the TiO_2 photocatalytic activity for the photodegradation of light hydrocarbons, Int. J. Chem. React. Eng., 5, A92, 2007.

Dijkstra, M. F. J., Buwalda, H., de Jong, A. W. F., Michorius, A., Winkelman, J. G. M., and Beenackers, A. A. C. M.: Experimental comparison of three reactor designs for photocatalytic water purification, Chem. Eng. Sci., 56, 547–555, 2001.

El-Kalliny, A. S.: Photocatalytic Oxidation in Drinking Water Treatment Using Hypochlorite and Titanium Dioxide, Ph.D. thesis, Delft University of Technology, Delft, the Netherlands, 85–100, 2013.

Eufinger, K., Poelman, D., Poelman, H., De Gryse, R., and Marin, G. B.: TiO$_2$ thin films for photocatalytic applications in Thin Solid Films: Process and Applications, ISBN: 978-81-7895-314-4, 189–227, 2008.

Feitz, A. J., Boyden, B. H., and Waite, T. D.: Evaluation of two solar pilot scale fixed-bed photocatalytic reactors, Water Res., 34, 3927–3932, 2000.

Fujishima, A., Zhang, X., and Tryk, D. A.: TiO$_2$ photocatalysis and related surface phenomena, Surf. Sci. Rep., 63, 515–582, 2008.

Ghorbani, M.: Electrophoretic deposition of titanium dioxide nanopowders films in isopropanol as a solvent, Int. J. Mod. Phys. B, 22, 2989–2994, 2008.

Goetz, V., Cambon, J. P., Sacco, D., and Plantard, G.: Modeling aqueous heterogeneous photocatalytic degradation of organic pollutants with immobilized TiO$_2$, Chem. Eng. Process., 48, 532–537, 2009.

Goldstein, S., Aschengrau, D., Diamant Y., and Rabani, J.: Photolysis of aqueous H$_2$O$_2$: quantum yield and applications for polychromatic UV actinometry in photoreactors, Environ. Sci. Technol., 41, 7486–7490, 2007.

Grillon, F., Fayeulle, D., and Jeandin, M.: Quantitative image analysis of electrophoretic coating, J. Mater. Sci. Letters, 11, 272–275, 1992.

Gumy, D., Rincon, A. G., Hajdu, R., and Pulgarin, C.: Solar photocatalysis for detoxification and disinfection of water: Different types of suspended and fixed TiO$_2$ catalysts study, Sol. Energy, 80, 1376–1381, 2006.

Ismail, K. N., Hamid, K. H. K., Kadir, S. A. S. A., Musa, M., and Savory, R. M.: Woven stainless steel wire mesh supported catalyst for NO$_X$ reduction in municipal solid waste flue (MSW) gas: synthesis and characterization, The Malaysian Journal of Analytical Sciences, 11, 246–254, 2007.

Li, F. B., Li, X. Z., Kang, Y. H., and Li, X. J.: An innovative Ti/TiO$_2$ mesh photoelectrode for methyl orange photoelectrocatalytic degradation, J. Environ. Sci. Heal. A, 37, 623-640, 2002.

Li, X. Z., Liu, H. L., and Yue, P. T.: Photoelectrocatalytic oxidation of rose bengal in aqueous solution using a Ti/TiO$_2$ mesh electrode, Environ. Sci. Technol., 34, 4401–4406, 2000.

Li, X. Z., Liu, H. L., Li, F. B., and Mak, C. L.: Photoelectrocatalytic oxidation of rhodamine B in aqueous solution using Ti/TiO$_2$ mesh photoelectrodes, J. Environ. Sci. Heal. A, 37, 55–69, 2002.

Malato, S., Blanco, J., Vidal, A., and Richter, C.: Review: Photocatalysis with solar energy at a pilot-plant scale: An overview, Appl. Catal. B-Environ., 37, 1–15, 2002.

Malato, S., Fernandez-Ibanez, P., Maldonado, M. I., Blanco, J., and Gernjak, W.: Decontamination and disinfection of water by solar photocatalysis: Recent overview and trends, Catal. Today, 147, 1–59, 2009.

McCullagh, C., Skillen, N., Adams, M., and Robertson, P. K. J.: Photocatalytic reactors for environmental remediation: A review, J. Chem. Technol. Biot., 86, 1002–1017, 2011.

Mehrotra, K., Yablonsky, G. S., and Ray, A. K.: Kinetic studies of photocatalytic degradation in a TiO$_2$ slurry system: Distinguishing working regimes and determining rate dependences, Industrial & Engineering Chemical Research, 42, 2273–2281, 2003.

Nam, H.-J., Amemiya, T., Murabayashi, M., and Itoh, K.: Photocatalytic activity of sol-gel TiO$_2$ thin films on various kinds of glass substrates: the effects of Na$^+$ and primary particle size, J. Phys. Chem. B, 108, 8254–8259, 2004.

Parra, S., Stanca, S. E., Guasaquillo, I., and Thampi, K. R.: Photocatalytic degradation of atrazine using suspended and supported TiO$_2$, Appl. Catal. B-Environ., 51, 107–116, 2004.

Pelaez, M., Nolan, N. T., Pillai, S. C., Seery, M. K., Falaras, P., Kontos, A. G., Dunlop, P. S. M., Hamilton, J. W. J., Byrne, J. A., O'Shea, K., Entezari, M. H., and Dionysiou, D. D.: A review on the visible light active titanium dioxide photocatalysts for environmental applications, Appl. Catal. B-Environ., 125, 331–349, 2012.

Sakamoto, R., Nishimori, H., Tatsumisago, M., and Minami, T.: Preparation of titania thick films by electrophoretic sol-gel deposition using hydrothermally treated particles, Journal of Ceramic Society of Japan, 106, 1034–1036, 1998.

Sarkar P. and Nicholson, P. S.: Electrophoretic deposition (EPD): Mechanisms, kinetics, and application to ceramics, J. Am. Ceram. Soc., 79, 1987–2002, 1996.

Shang, J., Li, W., and Zhu, Y.: Structure and photocatalytic characteristics of TiO$_2$ film photocatalyst coated on stainless steel webnet, J. Mol. Catal. A-Chem., 202, 187–195, 2003.

Sunada, K., Watanabe, T., and Hashimoto, K.: Studies on photokilling bacteria on TiO$_2$ thin film, J. Photoch. Photobio. A, 156, 227–233, 2003.

Tada, H. and Tanaka, M.: Dependence of TiO$_2$ photocatalytic activity upon its film thickness, Langmuir, 13, 360–364, 1997.

Thiruvenkatachari, R., Saravanamuth, V., and Moon, I. S.: A review on UV/TiO$_2$ photocatalytic oxidation process, Korean J. Chem. Eng., 25, 64–72, 2008.

Yanagida, S., Nakajima, A., Kameshima, Y., Yoshida, N., Watanabe, T., and Okada, K.: Preparation of a crack-free rough titania coating on stainless steel mesh by electrophoretic deposition, Mater. Res. Bull., 40, 1335–1344, 2005.

Yanagida, S., Nakajima, A., Kameshima, Y., and Okada, K.: Effect of applying voltage on photocatalytic destruction of 1,4-dioxane in aqueous system, Catal. Commun., 7, 1042–1046, 2006.

Yang, H., Zhu, S., and Pan, N.: Studying the mechanisms of titanium dioxide as ultraviolet-blocking additive for films and fabrics by an improved scheme, J. Appl. Polym. Sci., 92, 3201–3210, 2004.

Yigit, Z. and Inan, H.: A study of the photocatalytic oxidation of humic acid on anatase and mixed-phase anatase–rutile TiO$_2$ nanoparticles, Water Air Soil Poll., 9, 237–243, 2009.

Yoon, J., Shim, E., and Joo, H.: Photocatalytic reduction of hexavalent chromium (Cr(VI)) using rotating TiO$_2$ mesh, Korean J. Chem. Eng., 26, 1296–1300, 2009.

Yu, J., Su, Y., and Cheng, B.: Template-free fabrication and enhanced photocatalytic activity of hierarchical macro-/mesoporous titania, Adv. Funct. Mater., 17, 1984–1990, 2007a.

Yu, J., Zhang, L., Cheng, B., and Su, Y.: Hydrothermal preparation and photocatalytic activity of hierarchically sponge-like macro-/mesoporous titania, J. Phys. Chem. C, 111, 10582–10589, 2007b.

Yu, J. and Wang, B.: Effect of calcination temperature on morphology and photoelectrochemical properties of anodized titanium dioxide nanotube arrays, Appl. Catal. B-Environ., 94, 295–302, 2010.

Zhang, W., Li, Y., Wu, Q., and Hu, H.: Removal of endocrine-disrupting compounds, estrogenic activity, and escherichia coliform from secondary effluents in a TiO_2-coated photocatalytic reactor, Environ. Eng. Sci., 29, 195–201, 2012.

Ziegmann, M., Doll, T., and Frimmel, F. H.: Matrix effects on the photocatalytical degradation of dichloroacetic acid and atrazine in water, Acta Hydroch. Hydrob., 34, 146–154, 2006.

25

Household water treatment and safe storage – effectiveness and economics

Stefanie M. L. Stubbé[1], Alida Pelgrim-Adams[2], Gabor L. Szántó[2], and Doris van Halem[1]

[1]TU Delft, Civil Engineering, Sanitary Department Stevinweg 1 (Building 23),
2628 CN Delft, the Netherlands
[2]PRACTICA Foundation, Geulwg 16, 3356 LB, Papendrecht, the Netherlands

Correspondence to: Stefanie Stubbé (s.m.l.stubbe@student.tudelft.nl)

Abstract. Household Water Treatment and safe Storage (HWTS) systems aim to provide safe drinking water in an affordable manner to users where safe piped water supply is either not feasible or not reliable. In this study the effectiveness, economic parameters and costs of three selected HWTS systems were identified. The selected systems are SODIS, ceramic filter and biosand filter. These options were selected based on their accessibility, affordability and available scientific data. Data was obtained through peer-reviewed literature, reports, webpages and informal sources. The findings show a wide dispersion for log removal of effectiveness of the HWTS systems. For bacteria (*E. coli*), log removals of 1–9 (SODIS), 0.5–7.2 (ceramic) and 0–3 (biosand) were reported. In the case of viruses (mostly echovirus and bacteriophages), log removals of 0–4.3 (SODIS), 0.09–2.4 (ceramic) and 0–7 (biosand) were found. The dispersions of log removals for both bacteria and viruses range from non-protective to highly protective according to WHO performance targets. The reported costs of HWTS systems show a wide range as well. The price per cubic meter water is found to be EUR 0–8 (SODIS), EUR 0.37–6.4 (ceramic) and EUR 0.08–12.3 (biosand). The retail prices found are: negligible (SODIS), USD 1.9–30 (ceramic) and USD 7–100 (biosand). No relationship was observed between removal efficiency and economics of the three systems.

1 Introduction

In many parts of the world, people do not have access to safe drinking water, this is especially true in rural areas of developing countries (Unicef et al., 2012). Conventional piped water delivery and similar centralized systems are not feasible for rural and peri-urban communities in the near future, implying that they are left with the responsibility (and need) to collect, treat and store their own water (Brown Sobsey and Loomis, 2008). Where groundwater is inaccessible or contaminated, these users depend on household water treatment (HWTS) systems for safe drinking water (Sobsey et al., 2008). These HWTS systems have the goal to provide safe drinking water in an affordable and sustainable manner (Duke et al., 2006) while being simple and easy to manage by their users (Heinsbroek and Peters, 2014). As such, these systems are crucial in reducing occurrence of diarrheal and other debilitating illnesses (Meierhofer and Landolt, 2009; Stauber et al., 2009). Efficiency in providing safe water differs per method. To indicate removal efficiency, the WHO produced guidelines (WHO, 2004) to define default performance targets to indicate a certain removal efficiency for different pathogens as "interim", "protective" or "highly protective" (see Fig. 1).

When looking at the economics of HWTS systems, it is common practise to look at the price per produced m^3 water (NWP, 2010). Generally, this is calculated by dividing the investment and operational costs over the produced water during the lifetime of the technology (NWP, 2010).

$$\text{Price}\left[\text{Euro}\,\text{m}^{-3}\right] = \frac{\text{Investment} + \text{operational costs}}{\text{Produced water}} \quad (1)$$

The objective of this paper is to give an overview of the potential effectiveness according to the WHO performance tar-

Target	Log$_{10}$ reduction required: **Bacteria**	Log$_{10}$ reduction required: **Viruses**	Log$_{10}$ reduction required: **Protozoa**
Highly protective	≥ 4	≥ 5	≥ 4
Protective	≥ 2	≥ 3	≥ 2
*Interim**	Achieves "protective" target for two classes of pathogens and results in health gains		

Summary of performance requirements for small-scale and household drinking-water treatment, based on reference pathogens *Campylobacter jejuni*, *Cryptosporidium* and rotavirus (see Appendix 1).

Figure 1. WHO guidelines on default performance targets of HWTS systems (WHO, 2004).

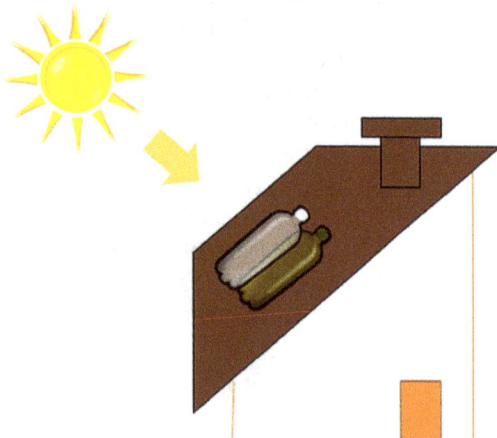

Figure 2. Schematic illustration of SODIS.

gets and the costs paid by the user of three HWTS systems: SODIS, ceramic filters and biosand filters. These three systems offer the most optimal combination of (i) accessibility and affordability even for the poorest, (ii) most widespread in use in low-income settlements and (iii) a considerable body of literature exists for these three systems.

1.1 SODIS

SODIS is based on the principle of disinfection by solar radiation (see Fig. 2). The procedure is straightforward; an unscratched and uncoloured PET or glass bottle is filled with water and exposed to direct sunlight for a minimum of 6 h (Heinsbroek and Peters, 2014). Water with low oxygen and high turbidity levels has to be pre-treated (Acra et al., 1990; Meierhofer and Landolt, 2009).

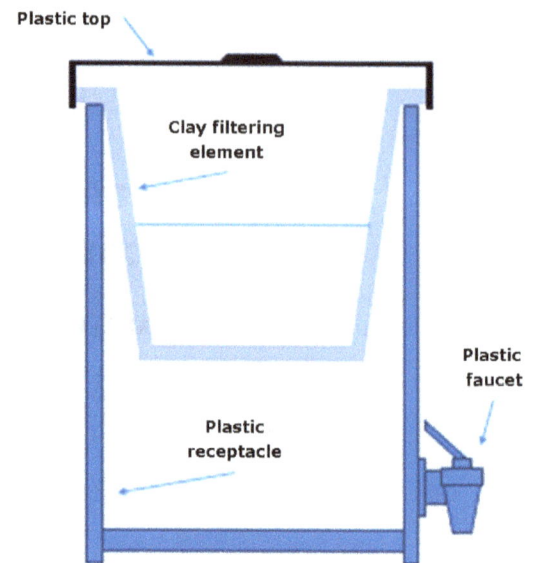

Figure 3. Schematic illustration of ceramic Pot filter (Van Halem et al., 2007).

1.1.1 Ceramic filters

The ceramic filter is based on the following principle: a porous media of fired clay that retains microbes by size exclusion and high tortuous properties (it traps microbes in the sharp bends; Hunter, 2009; Sobsey et al., 2008; van der Laan et al., 2014). Many variations of ceramic filters exist; e.g. pot filters or "water purifier" (see Fig. 3; Akvopedia, 2014b; Potters for Peace, 2014a), candle filters (Sobsey et al., 2008) and Tulip siphon filter (Basic Water Needs, 2014; Tulipfilter, 2013). Periodic scrubbing and rinsing is necessary to remove impurities (Sobsey et al., 2008).

1.2 Biosand filter

Biosand filters consist of a concrete or plastic frame filled with crushed rock (sand) filter media of 0.15–0.35 mm particle (Murphy et al., 2010b; see Fig. 4). Two filter mechanisms govern the removal principle of biosand filters: physical removal of organic matter and turbidity (Sobsey et al., 2008) and biological removal of colloidal particles and harmful pathogens in the so-called Schmutzdeke (Duke et al., 2006; Hunter, 2009; Weber-Shirk and Dick, 1997). The filter can be cleaned manually by removing the top few centimetres of sand and disposing the overlying water (Sobsey et al., 2008).

2 Effectiveness

In this section an in-depth description is given of the removal mechanisms of each of the selected HWTS systems and the corresponding removal efficiency. Both lab and field studies are used to give an overview of the reported effectiveness.

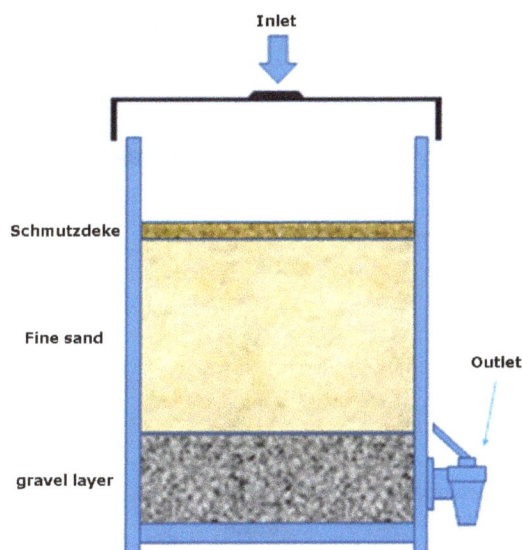

Figure 4. Schematic illustration of biosand filter (not on scale).

Insufficient data was found on the log removal of protozoa, so this pathogen is excluded from this study.

2.1 SODIS

The inactivation mechanisms of the solar radiation is based on direct UVB absorption (damaging the pathogenic DNA), optical inactivation (via reactive oxygen species) and thermal inactivation (denaturation; Reed, 2004). A synergy between optical inactivation and thermal inactivation was signalled at temperatures between 40–50 °C (Reed, 2004; Wegelin et al., 1994). Several parameters were suggested to enhance the SODIS treatment: black background surface (to reflect sunlight; Martín-Domínguez et al., 2005; Wegelin and Sommer, 1998), unscratched container material (diminish scattering; Wegelin and De Stoop, 1999), added photosensitizers (increase production of oxygen reactive species; Chilvers et al., 1999) and glass bottles (Duffy et al., 2004). Critics are focused on the potential leaching of plasticizers into the treated water (Reed, 2004). However, Wegelin et al. (2001) has shown that this is only the case at the outer surface of the bottles and not in the treated water.

In Figs. 5 and 6, a summary is given of the found removal efficiencies of SODIS for bacteria (*E. coli*) and viruses (mostly echovirus and bacteriophage) respectively. The majority of the research conditions lies between 40–65 °C and 4–6 h. For bacteria and viruses the log removal was between 1–9 and 0–4.3 respectively (Acra et al., 1990; Akvopedia, 2013; Dejung et al., 2007; Fujioka and Yoneyama, 2002; Heaselgrave et al., 2006; Joyce et al., 1996; Lonnen et al., 2005; Martín-Domínguez et al., 2005; McGuigan et al., 2012; Meyer and Reed, 2004; Sodis.ch, 2011). The majority of the results are centred around 2.5–5 log removal for bacteria and 1–4 log removal for viruses respectively.

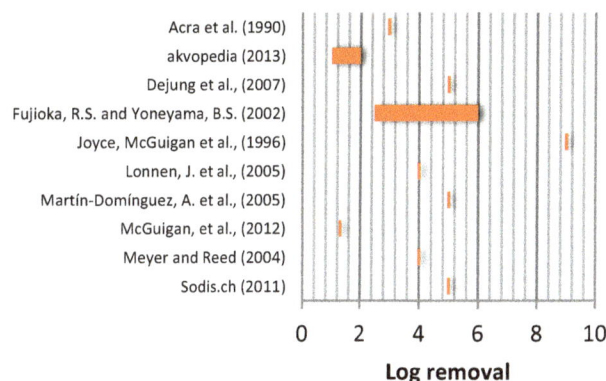

Figure 5. Log removal of SODIS for bacteria.

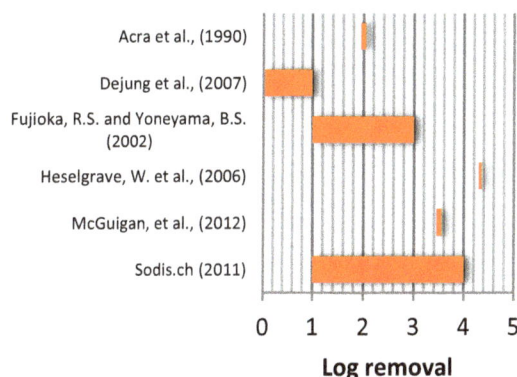

Figure 6. Log removal of SODIS for viruses.

2.2 Ceramic filters

By means of meta-regression, Hunter (2009) concluded that compared to other interventions (chlorine, SODIS, biosand filter and combined coagulant-chlorine), the ceramic filter shows the highest effectiveness on the long term. Most filters are manufactured by adding colloidal silver to increase efficiency. Silver inactivates bacteria and other pathogens through three mechanisms: reaction with thiol (in structural groups and functional proteins), structural changes in cell membrane and reaction with nucleic acids (Russell et al., 1994) . There are different ways to impregnate silver in the filter: dipping, painting, pulse injections and fire-in (Oyanedel-Craver and Smith, 2007; Ren and Smith, 2013). Van der Laan et al. (2014) and Oyanedel-Craver and Smith (2007) did not find a significant difference in removal efficiency for different silver application methods. On the contrary, the storage time in the receptacle of a silver-impregnated filter was found to be an important parameter in the bacterial removal efficiency; lengthy contact time in the receptacle led to higher removal efficiencies (van der Laan et al., 2014). Neither does an addition of iron appear to increase the removal efficiency in their research (Brown et al., 2008). Concerns exist about the virus removal of ceramic fil-

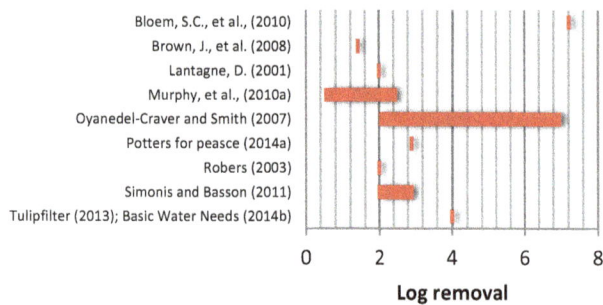

Figure 7. Log removal of ceramic filters for bacteria.

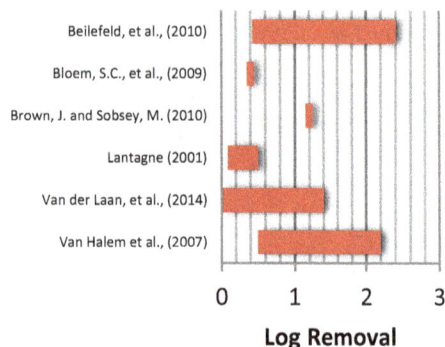

Figure 8. Log removal of ceramic filters for viruses.

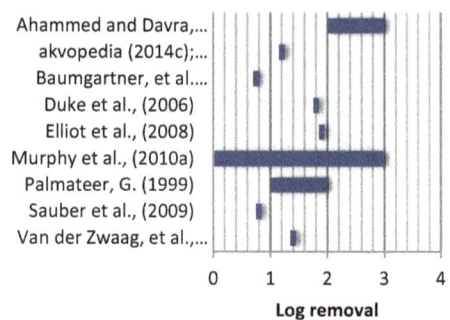

Figure 9. Log removal of biosand filters for bacteria.

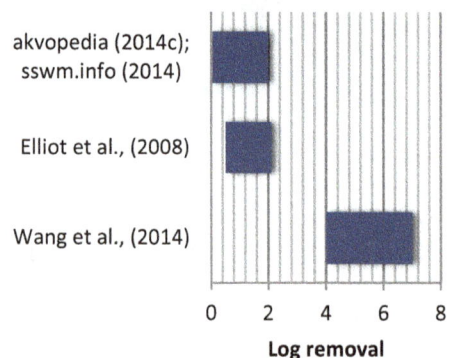

Figure 10. Log removal of biosand filters for viruses.

ters, since reported removal efficiencies do not reach WHO guidelines (Murphy et al., 2010a; van der Laan et al., 2014), and show high distribution (Bielefeldt et al., 2010). No critical parameter was yet identified to improve the virus removal efficiency (van der Laan et al., 2014).

In Figs. 7 and 8, a summary is given of the found removal efficiencies of ceramic filters for bacteria (*E. coli*) and viruses (mostly echovirus and bacteriophage) respectively. The log removal of ceramic filters for bacteria and viruses were between 0.5–7 and 0.09–2.4 respectively (Basic Water Needs, 2014; Bielefeldt et al., 2010; Bloem et al., 2009; Brown et al., 2008; Brown and Sobsey, 2010; Lantagne, 2001; Murphy et al., 2010a; Oyanedel-Craver and Smith, 2007; Potters for Peace, 2014; Roberts, 2003; Simonis and Basson, 2011; Tulipfilter, 2013; van der Laan et al., 2014; Van Halem et al., 2007). For removal of bacteria, most information sources report log removals between 1.3 and 4. The log removals of viruses are considerably lower with most information sources reporting log removals between 0.4 and 1.4.

2.3 Biosand filters

The removal mechanism of the biosand filter is based on the slow sand filtration principle and depends on the daily volume loaded to the filter (Elliott et al., 2008). The optimal volume is investigated to be equal to or smaller than the pore volume (Elliott et al., 2011). When larger charge volumes are exposed to the filter, a decrease in removal efficiency is found

(Baumgartner et al., 2007). Although this HWTS system is designed for intermitted use, continuous use of the biosand filter has higher removal efficiencies (Young-Rojanschi and Madramootoo, 2014). Introduction of iron oxide in the sand layer shows improved levels of pathogen removal and is especially beneficial after cleaning or in the ripening period (Ahammed and Davra, 2011). It is suggested that the Schmutzdeke contributes to the virus attenuation by the production of microbial exo-products (proteolytic enzymes) or grazing bacteria on viruses (Elliott et al., 2011; Huisman et al., 1974). Concerns exist about the lack of guidelines for the post-treatment of the removed Schmutzdeke during maintenance since this contains opportunistic pathogens and therefore poses an health risk to consumers (Hwang et al., 2014).

Figures 9 and 10 provide a summary of the reported removal efficiencies of biosand filters. Overall, the reported log removals of biosand filters for bacteria (*E. coli*) and viruses (mostly echovirus and bacteriophage) are between 0–3 and 0–7, respectively (Ahammed and Davra, 2011; Akvopedia, 2014c; Baumgartner et al., 2007; Duke et al., 2006; Elliott et al., 2008; Murphy et al., 2010a; Palmateer, 1999; Sswm.info, 2014; Stauber et al., 2009; Vanderzwaag et al., 2009; Wang et al., 2014). The log removal of bacteria is centred on 0.4–2; while the distribution of log removals reported for viruses is widely scattered.

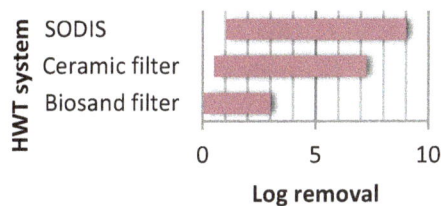

Figure 11. Overview of the overall range of found log removals in Sects. 2.1–2.3 for bacteria.

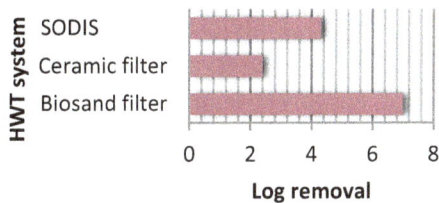

Figure 12. Overview of the overall range of found log removals in Sects. 2.1–2.3 for viruses.

2.4 Overview of effectiveness

Figures 5–10 show that the removal efficiency of HWTS systems differs per pathogen type and per study. The removal efficiencies found in the reviewed articles, are not always compatible with the target performance of the WHO (see Fig. 1), which corresponds with the results of previous studies such as Murphy et al. (2010a) and van der Laan et al. (2014). The difference between highest and lowest reported efficiencies of each HWTS system is what makes the difference between safe or unsafe water produced with this particular HWTS system. Hence, the question arises whether certain removal efficiency can be guaranteed for the HWTS systems.

In Figs. 11 and 12, the total range of lowest to highest reported log removal reported is shown per HWTS system. It can be seen that SODIS have the highest reported efficiency for bacteria removal (9 log removal) whereas, biosand filter report the highest reported efficiency for virus removal (7 log removal). Biosand filters show the lowest (zero) removal efficiency for bacteria whereas for virus removal, all HWTS systems have been reported with a zero log removal in one or more studies.

In the past, numerous studies questioned the effectiveness of HWTS systems. Various field test results indicated that HWTS systems may not always improve and sometimes even worsen the pathogenic state of the water (Murphy et al., 2010a). The lack of blinding and considerable heterogeneity in the results of HWTS systems show signs of concerns (Hunter, 2009). Moreover, it is reported that the research method can have a big impact on the reported efficiency (van der Laan et al., 2014). The reported removal efficiency also depends on the indicator pathogen used, as shown by Palmateer (1999) and Elliott et al. (2008). Quality tests are not

yet globally standardized (Rayner et al., 2013), so that a fair comparison between data sets is challenging.

2.5 Human factors

Operating conditions can also reduce the effectiveness of HWTS systems (Baumgartner et al., 2007). The effectiveness of HWTS systems does not only depend on technology, but also on human factors. When the HWTS system is not operated properly, exposure to pathogens can remain high. For example, it is common that people use the storage container of the device to collect dirty untreated water to feed the HWTS system, reducing the effectiveness of the device (Murphy et al., 2010a). Other reasons why in practice the effectiveness of HWTS systems is reduced: (i) only part of the used water is treated (Sobsey et al., 2008), as the water production of HWTS systems can be reduced in time due to clogging (ii) replacement-purchases are unfeasible (Brown et al., 2009; Hunter, 2009; Meierhofer and Landolt, 2009), (iii) the water is only intermittently treated (Sobsey et al., 2008), (iv) limited guidance to determine whether pre-treatment is necessary (Sobsey et al., 2008), (v) usage of the device simply stopped (Hunter, 2009), or (vi) selling it to a friend or relative (Brown et al., 2009). For ceramic filters, the rate of participation reduction is estimated at 2 % per month (Brown et al., 2009). The found diversity in effectiveness prompts that sufficient training and continued monitoring is needed to increase and sustain proper HWTS device management. Preferably, this could be done by a well-embedded local agent in order to increase acceptability (Meierhofer, 2006). Understanding the human factors that influence the real effectiveness of the HWTS systems is crucial for widespread adoption and sustained usage (Sobsey et al., 2008).

3 Economical evaluation

In this section, the parameters that determine the purchase price of a HWTS system (see Fig. 12) and the reported prices for the three selected HWTS systems are discussed (see Table 1).

3.1 Economic Parameters

The price of HWTS systems depends strongly on project area (Potters for Peace, 2014b; H. Jansen, personal communication, 2014). This can be explained by the fact that the price of HWTS systems is determined by (at least) four parameters (see Fig. 12). The first cost-parameter for HWTS systems is the production costs (or investment costs) including, materials (plastic, sand, ceramic), labour and basic tools (Basic Water Needs, 2014; Potters for Peace, 2014). These costs depend on the type of HWTS system and the region of production. Factories in China and India are frequently used, due to lower labour costs. The second parameter is distribution (Stuurman

Table 1. Overview of the price and retail price of HWTS systems.

Technology	Adjustment	Total Water Production (m^3 unit^{-1})[b]	Price (USUSD m^{-3})[a]	Retail Price (USUSD)[a] (without subsidies)	Reference
SODIS	–	0.243	0.53–2.03 new		NWP (2010)
			0–0.27 used	0	NWP (2010)
			3–8		Akvopedia (2013)
		0.292		0	CAWST (2012)
Ceramic	General	43.8	0.3–0.61		NWP (2010)
				5–15	Oyanedel-Craver and Smith (2007)
				8–10	Brown et al. (2009); Sobsey et al. (2008)
	Tulip Siphon Filter		0.4–0.44	5.69–7.32	NWP (2010)
			1.14–5.2	26.02	Tulipfilter (2013)
		7	2–5		Basic Water Needs (2014)
	Pot Filter		1		Akvopedia (2014b)
				9.76–17.89	CAWST (2012)
		36.5–54.75		7.5	Roberts (2003)
				5.4–28	NWP (2010)
	Water4life	70	0.46		Akvopedia (2014a)
				1.87–20.33	NWP (2010)
	Candle Filter			15–30	CAWST (2012)
	Potters for Peace			15–25	CAWST (2012); Potters for Peace (2014c)
				12	S. Chan (personal communication, 2014)
Biosand	Concrete	219	0.07–0.15	6.99–22.76	NWP (2010)
			10	12–40	Akvopedia (2014c); Sswm.info (2014)
				25–100	Sobsey et al. (2008)
				7–28	Ahammed and Davra (2011)
		262.8–788.4		12–30	CAWST (2012)
	Plastic	87.6–262.8		75	CAWST (2012)
	Iron oxide filter			15–36	Ahammed and Davra (2011)

[a] Conversion used where necessary 1.23 EUR/USD (Bloomberg, 2014). [b] Conversion used where necessary; daily production (m^3) × 365 (days yr^{-1}) × unit life-time.

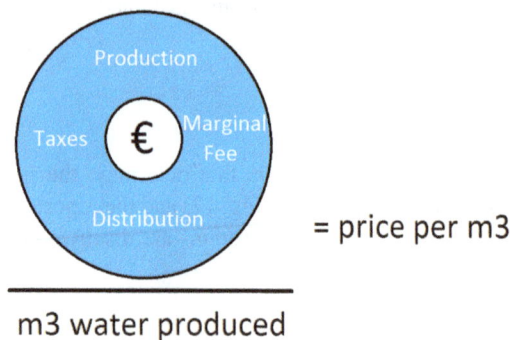

Figure 13. Economic Parameters that determine the costs of HWTS systems.

et al., 2010). Transport costs between production and project area depend on quantity and weight. In-land and over-sea transportation can differ considerably in total cost. For example, getting new ceramic filters to Ethiopia from India is more economic over-sea than over land (Basic Water Needs, 2014). This parameter is estimated to be the most dominant (Basic Water Needs, 2014). However, high variability in both

manufacturing and transportation costs translate into a severe limitation in data regarding the relation between costs in logistics and retail price. Local production factories are established to diminish distribution costs and enhance local economy (Brown, 2007). The third parameter is taxes. Depending on the country, HWTS systems need to be imported and import fees are involved. A possible fourth parameter is the (local) distributor's fee that is required to maintain his business. A retailer (of spare parts) of HWTS systems in Ethiopia, for example, can only remain in business if earnings are sufficiently attractive (Basic Water Needs, 2014). Depending on the developed supply chain, a (local) distributor will organize or co-organize distribution and sales in the project area.

3.2 Costs of HWTS systems

Only a limited number of peer-reviewed articles mention costs of HWTS systems, and in general only retail prices were mentioned. Retail price depends on the four parameters mentioned in the previous section and is the price eventually paid by the user. The retail price could be converted to the price per m^3, when the potential volume of water that can be treated with one filter is known (see Fig. 13). This potential

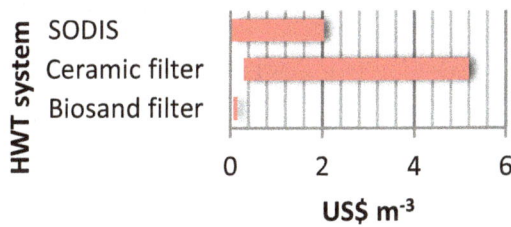

Figure 14. Price ranges (USD m^{-3}) for the three selected HWTS systems (outlier of biosand filter is neglected).

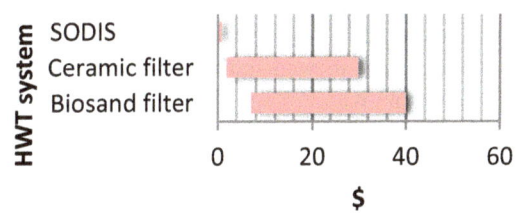

Figure 15. Ranges for the retail price of the three selected HWTS systems (outlier of SODIS is neglected).

volume is dependent on the lifetime of the filter, the flowrate, the sustainability of the system, etc. Reliable and sufficient information about this total potential volume is often lacking, therefore the retail price is mentioned separately from the price per m^3 water produced (see Table 1). Also, subsidies are not included in the calculations as it would be an unfair comparison with unsubsidized HWTS systems. However, it is noted that subsidies compromise a realistic and omnipresent part of emerging water sectors in developing countries. In practically emerging water sectors, subsidies play a crucial and – at present – decisive role for NGOs to arrange donor projects where consumers pay a reduced and affordable price for water treatment devices (Stuurman et al., 2010).

In Table 1, a summary is given of the price per m^3 and retail price per HWTS system. SODIS has a retail price of USD 0, since old PET bottles can be used. When new bottles are used, only a small investment is necessary (NWP, 2010). The costs per m^3 are related with the retail price. The outlier of USD 3–8 m^{-3} by Akvopedia (2013) is assumed to be an error, since it does not correspond with the numbers in the rest of the text of the same reference. For ceramic filters, the range of retail prices is between USD 1.9–30, with most of the references mentioning prices of around USD 15. The differences in price can be explained by the parameters elaborated in the previous section. The price per m^3 ranges between USD 0.3 and USD 5.2, which depends on the estimation of the potential amount of water that can be treated with the filter. For biosand filters, large ranges are found in the price per m^3 and retail price: 0.06–8.13 and USD 7–100 respectively (conversion 1.23 EUR/USD used where necessary following; Bloomberg, 2014). The outlier in price per m^3 of USD 10 is unreliable, since no argumentation is given in the reference (Akvopedia, 2014c). The outlier in retail price of USD 100 for concrete is also stated without further explanation, (Sobsey et al., 2008).

Overall, it is found that the biosand filter has the lowest price per m^3 produced what can be explained by its long life time, low maintenance costs and sustainable flowrate. Biosand filters do have the highest retail price (even when the highest outlier is neglected). Although SODIS is also a cheap technology, it requires a (small) investment when new (glass) bottles are used and it only produces little amount of

water per bottle. SODIS does have the cheapest retail price. It is shown that ceramic filters have the biggest range of price per m^3 water with the highest numbers. Ceramic filters are prone to breakage and the flowrate can decrease over time due to clogging. By far most independent research exists on ceramic filters compared to the other HWTS systems. In Figs. 14 and 15, an overview of the price ranges is given, neglecting the outliers mentioned above.

3.3 Constraints to economic evaluation

Most cost estimations of HWTS systems are found on websites of coordinating NGOs or device suppliers. Because the information is practice-oriented, the reliability of this information is likely to be fluctuating. More direct information from local producers turned out to be necessary. For example, Resource Development International in Cambodia reveals a standing quotation of a ceramic filter for USD 12 (RDI, 2014), which is in line with the prices in other sources. Since the price of HWTS systems does not only depend on the four parameters mentioned before, but is also fluctuating in time and susceptible to exchange rates. The price of the HWTS system today is therefore different from the price indicated for 2007 (PRACTICA Foundation, 2014). This study does not include these changes. The prices mentioned in table 1 are considered to be valid for the year of the respective reference.

4 Conclusions

In this study the removal efficiencies and economics of three selected Household Water Treatment and safe Storage (HWTS) systems were compared: SODIS, ceramic filters and biosand filters. Overall, no direct relationship between HWTS system's removal efficiency and economics was observed. This article aimed to be a guide through the currently available HWTS, however, it may be concluded that insufficient reliable information is available for a straightforward recommendation for the most effective and affordable HWTS.

For SODIS, low retail prices and intermediate prices per m^3 were observed with a range of removal efficiencies for bacteria (*E. coli*) from "non-protective" to "highly protective" and for viruses (mostly *echovirus* and *bacteriophages*) from "non-protective" to "protective" according to WHO targets. Ceramic filters showed intermediate retail prices and high prices per m^3 with a range of removal efficiencies for bacteria from "non-protective" to "highly protective" and for viruses "non-protective". Biosand filters had high retail prices and low prices per m^3 with a range of removal efficiencies for bacteria from "non-protective" to "protective" and for viruses from "non-protective" to "highly protective". The reported log removal should be viewed with some precaution, as parameters like indicator pathogen, research method and human factors are of influence. Also, most studies are short-time (around 26 weeks) and designed poorly unblinded, which could have given biased results (Hunter, 2009; Sobsey et al., 2008). The costs of HWTS were based on four parameters: production, distribution, taxes and marginal fees. They influence the price paid by the consumer besides other factors (interest, inflation). Additionally, the produced volume of water, or lifetime of the HWTS determines the actual price per m^3.

Acknowledgements. The authors would like to thank Jens Groot (Basic Water Needs), Herman Jansen (PRACTICA Foundation), Sreymeh Chan (RDI), Kaira Wagoner (Potters for Peace) and Laura Schuelert (CAWST) for their contribution to this research.

References

Acra, A., Jurdi, M., Mu'allem, H., Karahagopian, Y., and Raffoul, Z.: Water disinfection by solar radiation: assessment and application, International Development Research Centre, 1990.

Ahammed, M. M. and Davra, K.: Performance evaluation of biosand filter modified with iron oxide-coated sand for household treatment of drinking water, Desalination, 276, 287–293, 2001.

Akvopediam: UV treatment/ Solar disinfection. Retrieved 26th November, 2014, available at: http://akvopedia.org/wiki/Sodis (last access: 5 April 2016), 2013.

Akvopedia: Ceramic Candle filter, available at: http://akvopedia.org/wiki/Ceramic_candle_filter, last access: 26 November 2014a.

Akvopedia: Ceramic pot filter, available at: http://akvopedia.org/wiki/Ceramic_pot_filter, last access: 7 March 2014b.

Akvopedia: Concrete Biosand Filter, available at: http://www.akvopedia.org/wiki/Concrete_Biosand_Filter, last access: 21 July 2014c.

Basic Water Needs: Tulip Waterfilters, available at: www.basicwaterneeds.com/, last access: 1 December 2014.

Baumgartner, J., Murcott, S., and Ezzati, M.: Reconsidering "appropriate technology": the effects of operating conditions on the bacterial removal performance of two household drinking-water filter systems, Environ. Res. Lett., 2, 024003, 2007.

Bielefeldt, A. R., Kowalski, K., Schilling, C., Schreier, S., Kohler, A., and Scott Summers, R.: Removal of virus to protozoan sized particles in point-of-use ceramic water filters, Water Res., 44, 1482–1488, 2010.

Bloem, S. C., van Halem, D., Sampson, M. L., Huoy, L.-S., and Heijman, B.: Silver impregnated ceramic pot filter: flow rate versus the removal efficiency of pathogens, Disinfection, 2009.

Bloomberg: Markets, available at: http://www.bloomberg.com/markets/, last access: 6 December 2014.

Brown, Proum, S., and Sobsey, M.: Sustained use of a household-scale water filtration device in rural Cambodia, J. Water Health, 7, 404–412, 2009.

Brown Sobsey, M. D. and Loomis, D.: Local drinking water filters reduce diarrheal disease in Cambodia: a randomized, controlled trial of the ceramic water purifier, Am. J. Trop. Med. Hyg., 79, 394–400, 2008.

Brown, J.: Effectiveness of ceramic filtration for drinking water treatment in Cambodia, ProQuest, 2007.

Brown, J. and Sobsey, M.: Microbiological effectiveness of locally produced ceramic filters for drinking water treatment in Cambodia, Journal Water Health, 8, 1–10, 2010.

CAWST: Household Water Treatment and Safe Storage Fact Sheet, available at: www.resources.cawst.org/package/household-water-treatment-and-safe-storage-fact-sheets-detailed_en (last access: 1 December 2014), 2012.

Chilvers, K., Reed, R., and Perry, J.: Phototoxicity of Rose Bengal in mycological media–implications for laboratory practice, Lett. Appl. Microbiol., 28, 103–107, 1999.

Dejung, S., Fuentes, I., Almanza, G., Jarro, R., Navarro, L., Arias, G., and Iriarte, M.: Effect of solar water disinfection (SODIS) on model microorganisms under improved and field SODIS conditions, J. Water Supply Res. T., 56, 245–256, 2007.

Duffy, E., Al Touati, F., Kehoe, S., McLoughlin, O., Gill, L., Gernjak, W., and Cassidy, J.: A novel TiO$_2$-assisted solar photocatalytic batch-process disinfection reactor for the treatment of biological and chemical contaminants in domestic drinking water in developing countries, Solar Energ., 77, 649–655, 2004.

Duke, W., Nordin, R., Baker, D., and Mazumder, A.: The use and performance of BioSand filters in the Artibonite Valley of Haiti: a field study of 107 households, Rural Remote Health, 6, 570, 2006.

Elliott, M., DiGiano, F., and Sobsey, M.: Virus attenuation by microbial mechanisms during the idle time of a household slow sand filter, Water Res., 45, 4092–4102, 2011.

Elliott, M., Stauber, C., Koksal, F., DiGiano, F., and Sobsey, M.: Reductions of E. coliechovirus type 12 and bacteriophages in an intermittently operated household-scale slow sand filter, Water Res., 42, 2662–2670, 2008.

Fujioka, R. and Yoneyama, B.: Sunlight inactivation of human enteric viruses and fecal bacteria, Water Sci. Technol., 46, 291–295, 2002.

Heaselgrave, W., Patel, N., Kilvington, S., Kehoe, S. C., and McGuigan, K. G.: Solar disinfection of poliovirus and Acanthamoeba polyphaga cysts in water – a laboratory study using simulated sunlight, Lett. Appl. Microbiol., 43, 125–130, doi:10.1111/j.1472-765X.2006.01940.x, 2006.

Heinsbroek, A. and Peters, T.:Solar Water DIsinfection in Developing Countries, CIE5421 Water & Health Technology Review Essay, 2014.

Huisman, L., Wood, W., and Organization, W. H.: Slow sand filtration, World Health Organization Geneva, Vol. 16, 1974.

Hunter, P. R.: Household water treatment in developing countries: comparing different intervention types using meta-regression, Environ. Sci. Technol., 43, 8991–8997, 2009.

Hwang, H. G., Kim, M. S., Shin, S. M., and Hwang, C. W.: Risk Assessment of the Schmutzdecke of Biosand Filters: Identification of an Opportunistic Pathogen in Schmutzdecke Developed by an Unsafe Water Source, Int. J. Environ. Res. public health, 11, 2033–2048, 2014.

Joyce, T., McGuigan, K., Elmore-Meegan, M., and Conroy, R.: Inactivation of fecal bacteria in drinking water by solar heating, Appl. Environ. Microbiol., 62, 399-402, 1996.

Lantagne, D.: Investigation of the Potters for Peace colloidal silver impregnated ceramic filter: Report 1: Intrinsic effectiveness, lethia Environmental, Boston, MA:, USA, 2001.

Lonnen, J. K. S., Kehoe, S. C., Al-Touati, F., and McGuigan, K. G.: Solar and photocatalytic disinfection of protozoan, fungal and bacterial microbes in drinking water, Water Resour. Res., 39, 877–883, 2005.

Martín-Domínguez, A., Alarcón-Herrera, M. T., Martín-Domínguez, I. R., and González-Herrera, A.: Efficiency in the disinfection of water for human consumption in rural communities using solar radiation, Solar Energ., 78, 31–40, doi:10.1016/j.solener.2004.07.005, 2005.

McGuigan, K. G., Conroy, R. M., Mosler, H.-J., Preez, M. d., Ubomba-Jaswa, E., and Fernandez-Ibañez, P.: Solar water disinfection (SODIS): A review from bench-top to roof-top, J. Hazard. Materials, 235, 29–46, 2012.

Meierhofer, R. and Landolt, G.: Factors supporting the sustained use of solar water disinfection – Experiences from a global promotion and dissemination programme, Desalination, 248, 144–151, 2009.

Meierhofer, R.: Establishing solar water disinfection as a water treatment method at household level, Madagascar Conservation & Development, 1, 25–30, 2006.

Meyer, V. and Reed, R.: SOLAIR disinfection of coliform bacteria in hand-drawn drinking water, Water SA, 27, 49–52, 2004.

Murphy, H., McBean, E., and Farahbakhsh, K.: A critical evaluation of two point-of-use water treatment technologies: can they provide water that meets WHO drinking water guidelines?, J. Water Health, 8, 611–630, 2010a.

Murphy, H., McBean, E., and Farahbakhsh, K.: Nitrification, denitrification and ammonification in point-of-use biosand filters in rural Cambodia, J. Water Health, 8, 803–817, 2010b.

NWP: Netherlands Water Partnership Smart Disinfection Solutions Smart Solutions, 2010.

Oyanedel-Craver, V. A. and Smith, J. A.: Sustainable colloidal-silver-impregnated ceramic filter for point-of-use water treatment, Environ. Sci. Technol., 42, 927–933, 2007.

Palmateer, G.: Toxicant and parasite challenge of Manz intermittent slow sand filter, Environ. Toxicol., 14, 217–225, 1999.

Potters for Peace: Ceramic Water Filter Project, available at: http://pottersforpeace.com/?page_id=8 (last access: 5 April 2016), 2014.

Rayner, J., Skinner, B., and Lantagne, D.: Current practices in manufacturing locally-made ceramic pot filters for water treatment in developing countries, Journal of Water, Sanitation and Hygiene for Development, 3, 252–261, 2013.

Reed, R. H.: The inactivation of microbes by sunlight: solar disinfection as a water treatment process, Adv. Appl. Microbiol., 54, 333–366, 2004.

Ren, D. and Smith, J. A.: Retention and transport of silver nanoparticles in a ceramic porous medium used for point-of-use water treatment, Environ. Sci. Technol., 47, 3825–3832, 2013.

Roberts, M.: Ceramic Water Purifier Cambodia Field Tests, Phnom Penh: International Development Enterprises Working Paper, 2013.

Russell, A. D. and Hugo, W. B.: 7 Antimicrobial Activity and Action of, Progress in Medicinal Chemistry, 31, 351–366, 1994.

Simonis, J. J. and Basson, A. K.: Evaluation of a low-cost ceramic micro-porous filter for elimination of common disease microorganisms, Phys. Chem. Earth, 36, 1129–1134, 2011.

Sobsey, M. D., Stauber, C. E., Casanova, L. M., Brown, J. M., and Elliott, M. A.: Point of use household drinking water filtration: a practical, effective solution for providing sustained access to safe drinking water in the developing world, Environ. Sci. Technol., 42, 4261–4267, 2008.

Sodis.ch.: Microbiology, available at: http://www.sodis.ch/methode/forschung/mikrobio/index_EN (last access: 1 March 2015), 2011.

Sswm.info.: Biosand Filter, available at: www.sswm.info/category/implementation-tools/water-purification/hardware/point-use-water-treatment/bio-sand-filtrat, last access: 1 December 2014.

Stauber, C. E., Ortiz, G. M., Loomis, D. P., and Sobsey, M. D.: A randomized controlled trial of the concrete biosand filter and its impact on diarrheal disease in Bonao, Dominican Republic, Am. J. Trop. Med. Hyg., 80, 286–293, 2009.

Stuurman, D. J., Tielemans, M. W. M., and Nederstigt, J. M. J.: Marketing first: Getting ceramic pot filters to the target group, edited by: Initiative of Aqua for All research group of Cermaic Pot Filters, 1–22, available at: http://practica.org/wp-content/uploads/2014/08/Marketing-first-Getting-Ceramic-Pot-Filters-to-the-Target-Group.pdf (last access: 5 April 2016), 2010.

Tulipfilter: Product Details, Wat je weten wilt over een Tulip water filter, available at: http://www.tulipwaterfilter.nl/?s_page_id=575 (last access: 26 November 2014), 2013.

Unicef and WHO: World Health Organization: Progress on drinking water and sanitation: update, Nueva York: UNICEF, available at: http://apps.who.int/iris/bitstream/10665/81245/1/9789241505390_eng.pdf (last access: 5 April 2016), 2012.

van der Laan, H., van Halem, D., Smeets, P., Soppe, A., Kroesbergen, J., Wubbels, G., and Heijman, S.: Bacteria and virus removal effectiveness of ceramic pot filters with different silver applications in a long term experiment, Water Res., 51, 47–54, 2014.

Van Halem, D., Heijman, S., Soppe, A., van Dijk, J., and Amy, G.: Ceramic silver-impregnated pot filters for household drinking water treatment in developing countries: material characterization and performance study, Water Sci. Technol., 7, 9–17, 2007.

Vanderzwaag, J. C., Atwater, J. W., Bartlett, K. H., and Baker, D.: Field evaluation of long-term performance and use of biosand filters in Posoltega, Nicaragua, Water Quality Research Journal of Canada, 44, 111–121, 2009.

Wang, H., Narihiro, T., Straub, A. P., Pugh, C. R., Tamaki, H., Moor, J. F., and Nguyen, T. H.: MS2 Bacteriophage Reduction and Microbial Communities in Biosand Filters, Environ. Sci. Technol., 48, 6702–6709, 2014.

Weber-Shirk, M. L. and Dick, R. I.: Physical-chemical mechanisms in slow sand filters, Journal-American Water Works Association, 89, 87–100, 1997.

Wegelin, M., Canonica, S., Alder, A., Marazuela, D., Suter, M., Bucheli, T. D., and Kelly, M.: Does sunlight change the material and content of polyethylene terephthalate (PET) bottles?, Aqua-Journal of Water Supply, 50, 125–134, 2001.

Wegelin, M., Canonica, S., Mechsner, K., Fleischmann, T., Pesaro, F., and Metzler, A.: Solar water disinfection: scope of the process and analysis of radiation experiments, Aqua, 43, 154–169, 1994.

Wegelin, M. and De Stoop, C.: Potable water for all: promotion of solar water disinfection, Paper presented at the WEDC Conference, 1999.

Wegelin, M. and Sommer, B.: Solar Water Disinfection (SODIS) – destined for worldwide use?, Waterlines, 16, 30–32, 1998.

WHO: Database on Water, Sanitation and Hygiene, Database WHO, availablte at: http://www.who.int/water_sanitation_health/publications/2011/evaluating_water_treatment.pdf (last access: 5 April 2016), 2004.

Young-Rojanschi, C. and Madramootoo, C: Intermittent versus continuous operation of biosand filters, Water Res., 49, 1–10, 2014.

Permissions

List of Contributors

T. Grischek, D. Schoenheinz, C. Syhre and K. Saupe
University of Applied Sciences, Faculty of Civil Engineering/Architecture, Division of Water Sciences, Friedrich-List-Platz 1, 01069 Dresden, Germany

R. Jacinto, M. J. Cruz and F. D. Santos
CCIAM, SIM, Faculty of Sciences, University of Lisbon, C1, Sala 1.4.39, Campo Grande, 1749-016 Lisbon, Portugal

M. F. Fadal and J. Haarhoff
Department of Civil Engineering Science, University of Johannesburg, South Africa

S. Marais
Process Technology Department, Rand Water, South Africa

S. Nazarovs, S. Dejus and T. Juhna
Riga Technical University, Riga, Latvia

S. Meniconi, B. Brunone, M. Ferrante and C. Massari
Dipartimento di Ingegneria Civile ed Ambientale, Universit`a degli Studi di Perugia, Perugia, Italy

L. T. J. van der Aa
Waternet, P.O. Box 94370, 1090 GJ Amsterdam, The Netherlands
Delft University of Technology, P.O. Box 5048, 2600 GA Delft, The Netherlands

L. C. Rietveld and J. C. van Dijk
Delft University of Technology, P.O. Box 5048, 2600 GA Delft, The Netherlands

P. K. Mutiyar and A. K. Mittal
Department of Civil Engineering, Indian Institute of Technology Delhi, 110016 New Delhi, India

F. Nejjari, R. Pérez, V. Puig, J. Quevedo, R. Sarrate, M. A. Cugueró and G. Sanz
Technical University of Catalonia (UPC), 10 Rambla Sant Nebridi, 08222 Terrassa, Spain

J. M. Mirats
CETaqua Water Technology Centre, Crta. d'Esplugues n.75, 08940 Cornell`a de Llobregat, Spain

T. L. A. Driessen
Royal HaskoningDHV, Rivers, Deltas & Coasts, Nijmegen, the Netherlands

M. van Ledden
Royal HaskoningDHV, Rivers, Deltas & Coasts, Rotterdam, the Netherlands

A. Marchi and A. R. Simpson
School of Civil, Environmental and Mining Engineering, The University of Adelaide, SA 5005, Australia

N. Ertugrul
School of Electrical and Electronic Engineering, The University of Adelaide, SA 5005, Australia

K. Dutta Roy
Kolkata Metropolitan Water and Sanitation Authority, Kolkata, India

B. Thakur
Meghnad Saha Institute of Technology, Kolkata, India

T. S. Konar
Civil Engineering Dept., Jadavpur University, Kolkata, India

S. N. Chakrabarty
Civil Engineering Dept., Jadavpur University, Kolkata, India

E. Lee
Department of Civil, Environmental and Architectural Engineering, University of Colorado Boulder, Boulder, CO 80309, USA

S. Lee
Woongjin Chemical Co., Ltd, KANC 906-10, lui-dong, Yeongtong-gu, Suwon-si, Gyeonggi-do 443-270, Korea

J. Park and J. Cho
School of Civil and Environmental Engineering, Yonsei University, Yonsei-ro 50, Seodaemun-gu, Seoul 120-749, Korea

Y. Kim
Department of Environmental Science and Engineering, Gwangju Institute of Science and Technology (GIST), 1 Oryong-dong, Buk-gu, Gwangju, 500-712, Korea

K. Tihomirova, A. Briedis, J. Rubulis and T. Juhna
Department of Water Engineering and Technology, Riga Technical University, Azenes street 16/20-263, LV-1048, Riga, Latvia

P. van Thienen, R. Floris and S. Meijering
KWR Watercycle Research Institute, P.O. Box 1072, 3430 BB Nieuwegein, The Netherlands

R. Floris, P. van Thienen and H. Beverloo
KWR Watercycle Research Institute, P.O. Box 1072, 3430 BB Nieuwegein, the Netherlands

M. A. Tahir and H. Rasheed
National Water Quality Laboratory, Pakistan Council of Research in Water Resources, Kheyaban-e-Johar, H-8/1, Islamabad, Pakistan

A. Malana
Department of Chemistry, Bahawuddin Zikriya University, Multan, Pakistan

P. J. de Moel
Omnisys, Eiberlaan 23, 3871 TG, Hoevelaken, the Netherlands
Delft University of Technology, Faculty of Civil Engineering and Geosciences, Department of Water Management, P.O. Box 5048, 2600 GA, Delft, the Netherlands

A. W. C. van der Helm
Delft University of Technology, Faculty of Civil Engineering and Geosciences, Department of Water Management, P.O. Box 5048, 2600 GA, Delft, the Netherlands
Waternet, P.O. Box 94370, 1090 GJ, Amsterdam, the Netherlands

M. van Rijn
Vitens, P.O. Box 1205, 8001 BE Zwolle, the Netherlands

J. C. van Dijk
Delft University of Technology, Faculty of Civil Engineering and Geosciences, Department of Water Management, P.O. Box 5048, 2600 GA, Delft, the Netherlands
VanDijkConsulting, Rossenberglaan 9, 3833 BN, Leusden, the Netherlands

W. G. J. van der Meer
Delft University of Technology, Faculty of Civil Engineering and Geosciences, Department of Water Management, P.O. Box 5048, 2600 GA, Delft, the Netherlands
Oasen, P.O. Box 122, 2800 AC, Gouda, the Netherlands

A. H. Knol and K. Lekkerkerker-Teunissen
Dunea N.V., P.O. Box 756, 2700 AT Zoetermeer, the Netherlands

C. J. Houtman
Het Waterlaboratorium, P.O. Box 734, 2003 RS Haarlem, the Netherlands

J. Scheideler and A. Ried
Xylem, Boschstrasse 4–14, 32051 Herford, Germany

J. C. van Dijk
TU Delft, Stevinweg 1, 2628 CN Delft, the Netherlands

J. Machell, S. R. Mounce, B. Farley and J. B. Boxall
Pennine Water Group, University of Sheffield, Civil and Structural Engineering, Mappin Street, Sheffield S1 3JD, UK

L. Bonzanigo and G. Sinnona
Sustainable Development Programme, Fondazione Eni Enrico Mattei (FEEM), Isola di San Giorgio Maggiore, 30124 Venezia, Italia

P. Skworcow, D. Paluszczyszyn and B. Ulanicki
Water Software Systems, De Montfort University, The Gateway, Leicester LE1 9BH, UK

C. Oliveira, A. Alves and L. M. Madeira
LEPABE – Laboratory for Process Engineering, Environment, Biotechnology and Energy, Department of Chemical Engineering, Faculty of Engineering, University of Porto, R. Dr. Roberto Frias, s/n, 4200-465 Porto, Portugal

K. Gruskevica, T. Juhna and K. Tihomirova
Department of Water Engineering and Technology, Riga Technical University, 16, Azenes street, Riga, Latvia

K. Chon
Department of Civil and Environmental Engineering, Yonsei University, Yonsei-ro 50, Seoul 120-749, Korea
Chemical Safety Division, National Institute of Agricultural Science, 166, Nongsaengmyeong-ro, Iseo-myeon, Wanju-gun, Jeollabuk-do, 565-851, Korea

Y. Kim
School of Environmental Science and Engineering, Gwangju Institute of Science and Technology (GIST), 123 Cheomdangwagi-ro, Buk-gu, Gwangju 500-712, Korea

D. H. Bae
Department of Civil and Environmental Engineering, Sejong University, Neungdong-ro 209, Gwangjin-gu, Seoul 143-747, Korea

J. Cho
5School of Urban and Environmental Engineering, Ulsan Institute of Science and Technology (UNIST), UNIST-gil 50, Ulsan 689-798, Korea

A. S. El-Kalliny
Product and Process Engineering, ChemE, Delft University of Technology, Julianalaan 136, 2628 BL Delft, the Netherlands
Sanitary Engineering, Faculty of Civil Engineering and Geosciences, Delft University of Technology, Stevinweg 1, 2628 CN Delft, the Netherlands
Department of Water Pollution Research, National Research Centre, Dokki, 12622 Giza, Egypt

S. F. Ahmed and P. W. Appel
Product and Process Engineering, ChemE, Delft University of Technology, Julianalaan 136, 2628 BL Delft, the Netherlands

L. C. Rietveld
Sanitary Engineering, Faculty of Civil Engineering and Geosciences, Delft University of Technology, Stevinweg 1, 2628 CN Delft, the Netherlands

Stefanie M. L. Stubbé and Doris van Halem
TU Delft, Civil Engineering, Sanitary Department Stevinweg 1 (Building 23), 2628 CN Delft, the Netherlands

Alida Pelgrim-Adams and Gabor L. Szántó
PRACTICA Foundation, Geulwg 16, 3356 LB, Papendrecht, the Netherlands

Index